ARITHMETIC: A Problem Solving Approach

ARITHMETIC: A Problem Solving Approach

Joseph Cleary

MASSASOIT COMMUNITY COLLEGE

Walter Gleason

BRIDGEWATER STATE COLLEGE

WEST PUBLISHING COMPANY
St. Paul New York Los Angeles San Francisco

Cover Photograph: Co Rentmeester/FPG International

Production Coordination: COBB/DUNLOP Publisher Services Incorporated

50 West Kellogg Boulevard
P.O. Box 43526
St. Paul, Minnesota 55164

Library of Congress Cataloging in Publication Data

Cleary, Joseph.
Arithmetic: a problem solving approach.

Includes index.
1. Arithmetic—1961– I. Gleason, Walter
II. Title.
QA107.C64 1985 513 84-17212
ISBN 0-314-78011-4

We would like to dedicate this book to our families.

My parents Frances and Joseph Cleary
My wife Helen
My children Allison
Matthew
Meglian

My parents Mary and John Gleason
My wife Nancy
My children Stephen
Brian
Gary

Contents

Preface

We have written a text that provides all the basic concepts of arithmetic that are necessary for the further study of mathematics and enables the student to solve everyday problems.

To assist students in accomplishing their goals, we have written the text in a two-part workbook format that is *readable.* In Part I, *Arithmetic Revisited,* arithmetic concepts are developed with a clear concise explanation of the subject matter followed by a number of illustrative examples. Part II consists of three chapters dealing with real life situation applications. These chapters may be studied in any order; they are independent of one another.

To assist instructors in helping students reach their goals, the text can be used in the *traditional* classroom approach or in an individualized mathematics learning laboratory. Each chapter in Part I has a pretest where student weaknesses can be pinpointed. Each section contains an explicit explanation of the material with a large number of graded and illustrated examples followed by exercises that are correlated to the examples. Answers to all section exercises follow the exercise set. At the end of each chapter, there are two tests: a *Warmup Test* and a *Challenge Test.* Part I concludes with a set of cumulative tests.

PURPOSE OF PRETESTS

If a student is enrolled in a self-paced or nonlecture class and if the student completes the pretest without any errors he or she may proceed to the chapter review and complete those problems before continuing on to the next chapter. If the student cannot complete the pretest without error, he or she will be referred to the appropriate section in the chapter so that he or she may find the necessary help. *However, before moving on, the student should remember to take the time to thoroughly understand any errors.* Those students who want to be certain that they learn all the material in the chapter may proceed directly to the first section of the text.

PEDAGOGICAL FEATURES

Pretests Each chapter of Part I has a pretest with all questions referenced to the appropriate section in the chapter and a set of succinct learning objectives correlated to subject topics.

Illustrative Examples and Exercises The text contains many examples with illustrative solutions and hundreds of exercises with answers at the end of each section including numerous word problems. Exercises are arranged in order of difficulty with a real effort made to achieve a balance between odd and even numbered exercises. Word problems are identified by a picture key indicating applications from the following fields.

Environmental Studies

Engineering, Technical

Business/Economics

Political Science

Agriculture

Psychology

Education

Sociology

Chemistry

Medicine

Biology

Sports

Physical Science

Physics

Law

Focus on Problem Solving A detailed illustrated and solved application of arithmetic at the end of each section.

Building Your Math Vocabulary A glossary type of section summarizing all of the important terms, with an illustrative example, at the end of each chapter.

Chapter Review A set of review exercises, at the end of each chapter, that are based on the key concepts of the chapters. Answers to all exercises are included.

Chapter Tests At the end of each chapter, there is a *Warmup Test* with answers included (at the back of the text), and a *Challenge Test* for which answers are not included.

Cumulative Tests A set of three cumulative review tests covering the material of Part I. Answers to these cumulative tests are located in the back of the text.

Problem Sets Appendix An appendix consisting of three collections (A, B, and C) of 1,800 arithmetic drill exercises with answers. This feature is included for students who need work in the basic arithmetic operations.

Arithmetic in A Nutshell An appendix reviewing all of the key concepts of the text with an illustrative example of each.

ACKNOWLEDGEMENTS

There are many people who contributed to the publication of this text. We wish to thank Professor John Tobey of North Shore Community College for his exhaustive reviews and many helpful suggestions. To Professors Linda Kelley and Theresa Macanka, thank you for helping us class test the text. A special thank you to Gary Gleason for his many hours spent solving text exercises and to Helen Cleary for her hard work typing the manuscript. The people at West Publishing have been supportive of our efforts especially our editor, Pat Fitzgerald, whose cooperation and encouragement we sincerely appreciate.

We would also like to thank our reviewers

Jeff Mock—Diablo Valley College
John Rose—Miami Dade Community College
John Tobey—North Shore Community College
Kay Hudspeth—Penn State
Eugenie Trow—Antelope Valley College
Clifton Gary—Oscar Rose Junior College
Raymond Boersema—Community College of Denver, No. Campus
Susan Armfield, Austin Community College
Susan Wagner—North Virginia Community College
Jim Marcotte—Cincinnati Technical College
Ed Zinella—Community College of Rhode Island
Sandy Wager—Penn State
Barbara Treadwell—W. Michigan University
Martha Clutter—Piedmont Community College
Reginald Luke—Middlesex Community College
Ned Shillow—Lehigh County Community College

Part

ARITHMETIC REVISITED

Whole Numbers

CHAPTER OUTLINE

This chapter deals with adding, subtracting, multiplying, and dividing whole numbers (0, 1, 2, 3, . . .). When you deal with these numbers, both speed and accuracy of computation are important. At this point we will analyze your present level of proficiency and, if necessary, provide you with an opportunity to sharpen your skills. This goal may be achieved by completing the following test. Before you take the test remember (1) to strive for both speed and accuracy and (2) to review your solution to each problem.

1–0 PRETEST

Do all work on these pages. Do not use a calculator.

Problem	Section Reference	Learning Objective
1. To express the fact that 20 is greater than 15 we write $20 > 15$. How would you express the fact that 18 is less than 22? 18 < 22	1.1	Understanding of "greater than" and "less than"
In problems 2 and 3 what does the digit 3 mean in each number?		
2. 63,521 Handreths	1.2	Understanding the place value of a digit in a number
3. 20,035 hundreths	1.2	Understanding the place value of a digit in a number
4. Express the number 4,062 in written form. four thousand sixty-two	1.2	Writing the name of a number
5. Express the number forty thousand, two hundred twenty in numerical form. 40,220	1.2	Writing a number in numerical form
6. Round off 3,765 to the nearest hundred. 3,800	1.2	Rounding off whole numbers

In problems 7–20 perform the indicated operation if possible. If the operation is not possible, state "no solution."

Problem		Section Reference	Learning Objective
7. $\begin{array}{r} 6 \\ 2 \\ 9 \\ 8 \\ 6 \\ 5 \\ +\ 2 \\ \hline \end{array}$ 38	**8.** $\begin{array}{r} 53{,}634 \\ +\ 17{,}009 \\ \hline \end{array}$ 70643	1.3	Adding whole numbers

Problem		Section Reference	Learning Objective
9. $\begin{array}{r} 742 \\ 10 \\ 46{,}733 \\ 32 \\ +\ \ 789 \\ \hline \end{array}$		1.3	Adding whole numbers
10. $\begin{array}{r} 898 \\ -\ \ 62 \\ \hline \end{array}$	**11.** $\begin{array}{r} 96 \\ -\ 78 \\ \hline \end{array}$	1.3	Subtracting whole numbers
12. $\begin{array}{r} 5{,}000 \\ -\ 3{,}241 \\ \hline \end{array}$	**13.** $\begin{array}{r} 47{,}346 \\ -\ 28{,}137 \\ \hline \end{array}$	1.3	Subtracting whole numbers
14. $\begin{array}{r} 890 \\ \times\ \ 38 \\ \hline \end{array}$	**15.** $\begin{array}{r} 3{,}762 \\ \times\ \ \ 200 \\ \hline \end{array}$	1.5	Multiplying whole numbers
16. $\begin{array}{r} 6{,}743 \\ \times\ \ 39 \\ \hline \end{array}$		1.5	Multiplying whole numbers
17. $9\overline{)8{,}672}$	**18.** $4\overline{)0}$	1.6	Dividing whole numbers
19. $0\overline{)4}$	**20.** $67\overline{)7{,}372}$	1.6	Dividing whole numbers
21. Write the following in exponential form: $2 \times 2 \times 2$		1.7	Meaning of exponents
22. Evaluate: 5^3 ______		1.7	Evaluating exponential numbers
23. Evaluate: $15 \div 3 + 4 - 3^2$ ______		1.7	Order of operations
24. Evaluate: $2^3 + 3 \times 2 - 12$ ______		1.7	Order of operations

1.1 BASIC SYMBOLS

Historically, humans first used numbers to count objects. A person had five sheep, or must cross three hills to reach a destination, or owned eight arrows. Our first numbers arose "naturally" out of a desire to count. Today, we have the following collection of **natural numbers,** denoted *N:*

$$N = 1, 2, 3, 4, 5, \ldots$$

The three dots ". . ." signify that the pattern of numbers is to be continued without end. This is important because the smallest natural number is 1 and there is no largest natural number. On the other hand, 1, 2, 3, . . ., 100 means to continue the preceding pattern of numbers from 3 to 100 and to stop at 100.

It is helpful to visualize each natural number as corresponding to a unique point on a line. When all of our natural numbers are represented by equally spaced points on a line, we have a **number line** of natural numbers. This is illustrated in Figure 1.1, where the line begins at one and the arrow points in the direction in which the numbers are getting larger.

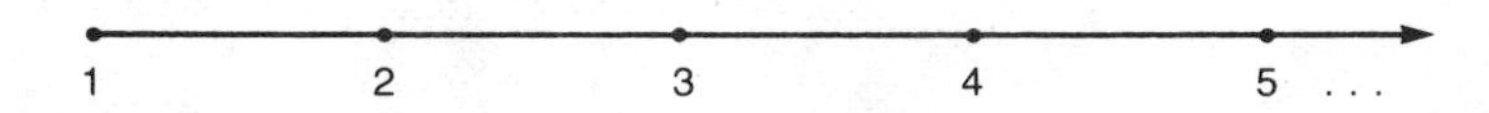

Figure 1.1

A baseball team fails to score in a particular inning, there aren't any days left before your final exam, or your automobile gas tank is empty. Zero, 0, is a number that has been introduced as a convenient symbol to denote the absence of an amount. If you have zero dollars in your pocket, your pocket is empty. When 0 is included with the natural numbers we have the following collection of **whole numbers,** denoted *W:*

$$W = 0, 1, 2, 3, 4, 5, \ldots$$

This chapter is all about whole numbers. Is it clear that all natural numbers are whole numbers? Figure 1.2 depicts the number line of whole numbers.

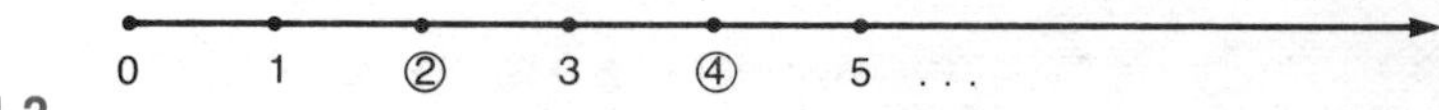

Figure 1.2

Given two whole numbers, the number associated with the point farthest to the right, on the number line, is the largest. Four is to the right of two on the whole number line, so 4 is greater than 2. This may be expressed by writing either $4 > 2$ or $2 < 4$. To express the fact that \$9 is greater than \$5, we write \$9 > \$5. For the collection of whole numbers, *W*, it is true that $0 < 1 < 2 < 3. \ldots$ The symbol "<" (read "is less than") and the symbol ">" (read "is greater than") are called **inequality symbols.** An easy way to remember the meaning of the inequality symbol is to notice that the wide part of the symbol is next to the larger number. Hence, $15 > 12$ indicates that 15 is greater than 12 or 12 is less than 15.

Let the letters a and b represent any two whole numbers.
Less than: $a < b$ indicates that a is to the left of b on the number line. For example, $3 < 10$.
Greater than: $a > b$ indicates that a is to the right of b on the number line. For example, $11 > 5$.

Example 1 Given two numbers 3 and 6, 6 is the larger number because it lies on the number line to the right of 3. Thus, you can write either $6 > 3$ or $3 < 6$.

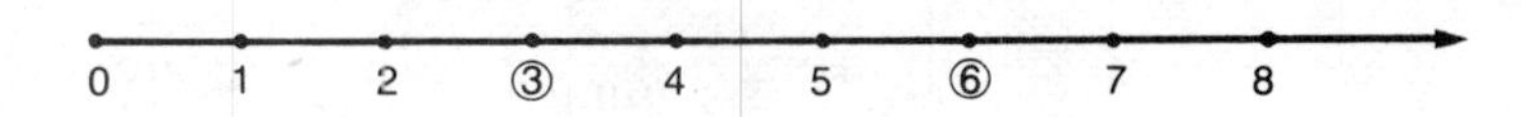

Example 2 $9 < 14$ means "9 is less than 14."

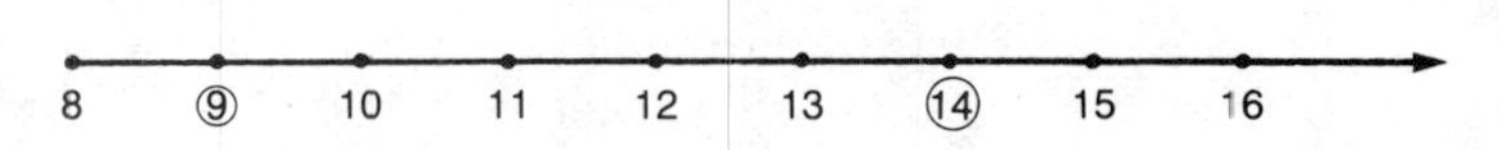

If you wish to say that five is less than or equal to some whole number, x, you can write $5 \leq x$. Here the symbol "$\leq$" is an abbreviation of the two symbols $<$ and $=$. When you write $x \geq 10$, this means that x can be any whole number greater than or equal to 10. That is, x can be 10, 11, 12, and so on.

Example 3 **(a)** $y \leq 5$ means y is less than or equal to 5. Hence $y = 0, 1, 2, 3, 4$, or 5.

(b) To express the fact that the time t, is greater than or equal to 60 minutes, we can write $t \geq 60$ minutes.

Our final symbol in this section is the "not equal to" symbol, $\neq$. For example, the area A of a circle is not equal to 0. This may be expressed $A \neq 0$. The present social security deduction rate r, is not equal to 5%. That is, $r \neq 5\%$. Table 1.1 is a review of the important symbols of this section. Remember we are using the letters a and b to represent any two whole numbers.

Table 1.1

Symbol	Meaning	Example
N	natural numbers	1, 2, 3, . . .
W	whole numbers	0, 1, 2, 3, . . .
$a < b$	a is less than b	$3 < 6$
$a \leq b$	a is less than or equal to b	$a \leq 4$ would mean that a is 0, 1, 2, 3, or 4
$a > b$	a is greater than b	$8 > 2$
$a \geq b$	a is greater than or equal to b	$a \geq 7$ would mean that a is 7, 8, 9, or larger
$a \neq b$	a is not equal to b	$7 \neq 2$

1.1 EXERCISES

1. Explain why 9 is greater than 7. ______________________________

2. Explain why 4 is less than 5. ______________________________

3. (true or false) If $2 < 7$, then 2 must be to the left of 7 on the number line.

4. (true or false) If $x \neq 7$, then x must be to the left of 7 on the number line.

5. What is the smallest whole number? ______________________________

6. What is the fifth smallest whole number? ______________________________

7. How many whole numbers are in the collection $\{0, 1, 2, 3, 4, 5, 6, 7, 8\}$?

____

How many natural numbers are in this collection? ______________________________

8. Write all the whole numbers less than 3. ______________________________

9. Write all the whole numbers less than or equal to 3. ______________________________

10. Write all the whole numbers greater than 20. ______________________________

11. Write all the whole numbers greater than or equal to 20. ______________________________

12. Rewrite the following statements using inequality symbols.

(a) 33 is greater than 19. ______________________________

(b) 17 exceeds 11. ______________________________

13. Rewrite the following statements using inequality symbols.

(a) 8 is less than 48. ______________________________

(b) 27 is more than 25. ______________________________

14. All natural numbers are whole numbers. Explain why all whole numbers are not natural numbers.

____

In problems 15 to 22 determine which of the two symbols $>$ or $<$ should be used to make each statement true.

15. 5 ______ 3 **16.** 4 ______ 11 **17.** 0 ______ 1

18. 100 ______ 98 **19.** 17 ______ 27 **20.** 1 ______ 0

21. The volume of a basketball is ______ the volume of a baseball.

22. The distance from Boston to New York City is ______ the distance from Boston to Atlanta.

23. Use symbols to express the fact that "time, t, is greater than or equal to 5 hours."

____

24. Use symbols to express the fact that "weight, w, is less than or equal to 150 pounds."

____

ANSWERS

1. 9 is to the right of 7 on the number line **2.** 4 is to the left of 5 on the number line **3.** true **4.** false, x could be to the right of 7 on the number line **5.** 0 **6.** 4 **7.** 9; 8 **8.** 2, 1, 0 **9.** 3, 2, 1, 0 **10.** 21, 22, 23, . . . **11.** 20, 21, 22, . . . **12.** (a) $33 > 19$ (b) $17 > 11$ **13.** (a) $8 < 48$ (b) $27 > 25$ **14.** 0 is a whole number but not a natural number **15.** $>$ **16.** $<$ **17.** $<$ **18.** $>$ **19.** $<$ **20.** $>$ **21.** $>$ **22.** $<$ **23.** $t \geq 5$ **24.** $w \leq 150$

1.2 PLACE VALUE, WRITING, AND ROUNDING OFF

Place Value

Our number system is based on ten **digits:** 0, 1, 2, 3, 4, 5, 6, 7, 8, and 9. To express ten we write the digit 1 followed by the digit 0, or "10." There is no single symbol for ten. Our number system, called a **decimal system,** is based on expressing numbers in "groups" or "packages" of 1, $1 \times 10 = 10$, $10 \times 10 = 100$, $10 \times 10 \times 10 = 1{,}000$ and so on placed in locations from right to left of a starting point called a **decimal point,** denoted ".". See Table 1.2.

Table 1.2

. . .	Hundred thousands	Ten thousands	Thousands	Hundreds	Tens	Units (or Ones)	Decimal Point
	100,000	10,000	1,000	100	10	1	.

As a writing convenience, we generally omit the decimal point in whole numbers. Thus we would write 4,763 and not 4,763., or 705 and not 705., for example. The decimal system is one of the great inventions in mathematics for it allows us to express large numbers in a compact form. A comma is used every third place starting at the decimal point and moving left. This is done to help us read the number. Below is a table examining the place value of the number 5,678,944,236.

Billions	Hundred millions	Ten millions	Millions	Hundred thousands	Ten thousands	Thousands	Hundreds	Tens	Units or ones	Decimal point
5,	6	7	8,	9	4	4,	2	3	6	.

Observe that the place value assigned each location is 10 times the value of the location to its right. This is illustrated in the next examples.

Example 1

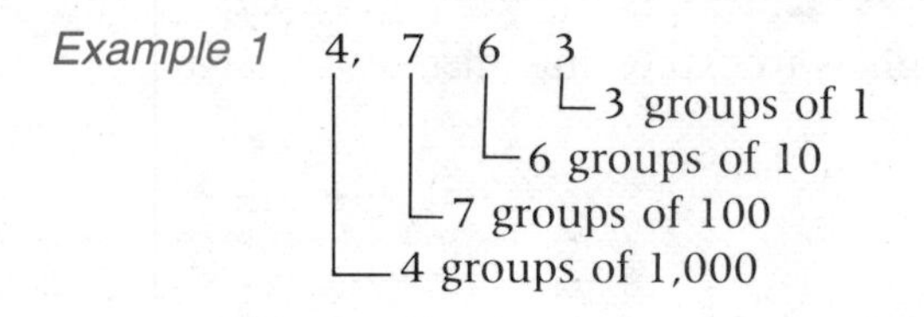

Example 2

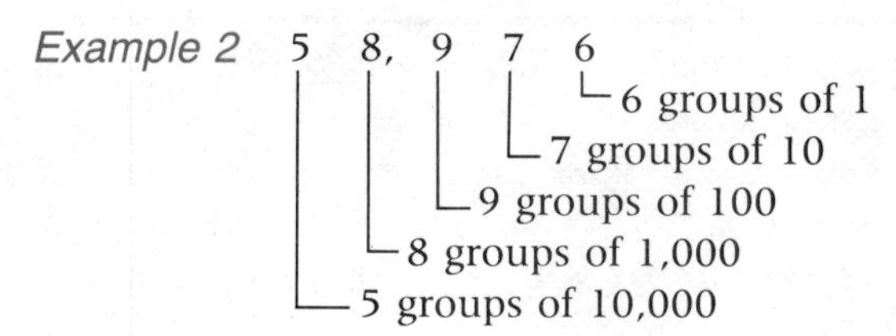

Writing Whole Numbers

How do we read the following number: 5,678,944,236? Starting at the right, group the digits in sets of three. The last set formed may contain 1, 2, or 3 digits. Each group of three digits has a name as shown below:

5	6 7 8	9 4 4	2 3 6
. . . billions	millions	thousands	units

The number 5,678,944,236 is written, "five billion, six hundred seventy-eight million, nine hundred forty-four thousand, two hundred thirty-six." This is the **written form** of the number. Have you noticed that:

1. We used commas to separate groups of digits.
2. The numbers 21–99 are written with hyphens.
3. The word "and" is not used.

Example 3 Express the following numbers in written form.
(a) 82 (b) 25,073 (c) 1,456,298

Solution
(a) eighty-two
(b) twenty-five thousand, seventy-three
(c) one million, four hundred fifty-six thousand, two hundred ninety-eight

Example 4 Express the following amounts in written form.
(a) $906 (b) $7,056 (c) $438,761

Solution
(a) nine hundred six dollars
(b) seven thousand, fifty-six dollars
(c) four hundred thirty-eight thousand, seven hundred sixty-one dollars

Rounding Off

A container holds 2 quarts of liquid, your weight is approximately 110 pounds, and your home is nearly 400 miles from the college campus. The first value was rounded off to the nearest unit, the second value was rounded off to the nearest ten, and the last value was rounded off to the nearest hundred. For most purposes, such values are sufficiently accurate to convey a numerical message. Banks frequently record assests in thousands of dollars and the national debt is given in billions of dollars.

To Round Off a Whole Number

1. Find the digit in the place to which you are rounding.
2. Look at the next digit to the right.
3. (a) *If it is less than 5,* round down by replacing it and any other digit to the right with zeros.
   (b) *If it is 5 or more,* round up by adding 1 to the digit in the place to which you are rounding, then replace all digits to the right of it with zeros.

Example 5 Round off 3,457 to the nearest *ten*.

Solution
1. Find the tens column 3,4,[5]7
2. Find the number to its right in the units column 3,4 5[7]
3. 7 > 5, so we round off to 3,4[6 0]

Example 6 Round off 78,623 to the nearest *hundred*.

Solution
1. Find the hundreds column 7 8,[6]2 3
2. Find the number to its right in the tens column 7 8,6[2]3
3. Find 2 < 5, so we round off to 7 8,[6 0 0]

Example 7 Round off 47,953 to the nearest *thousand*.

Solution
1. Find the thousands column 4 [7],9 5 3
2. Find the number to its right in the hundreds column 4 7,[9]5 3
3. 9 > 5, so we round off to 4[8,0 0 0]

1.2 EXERCISES

In problems 1–8 express the number in written form.

1. 96 ninety six
2. 2,498 two thousand, four Hundred ninety-eight
3. 45,607 ______
4. 807,576 ______
5. 6,385,314 ______
6. 493 ______
7. 8,328,715 ______
8. 29,781 ______

In problems 9–16 use digits to express the number in numerical form.

9. fifty-two thousand, six hundred forty-seven ______
10. four hundred ninety-five ______

11. seven hundred thirty-six thousand, two hundred one ____________

12. five million, eight hundred forty-four thousand, one hundred seventy

____

13. sixteen thousand, nine hundred twenty-two ____________

14. two million, five hundred eighty-eight thousand, four hundred sixty-three

____

15. twenty-nine ____________

16. one hundred sixty-two thousand, five hundred seventy-seven

____

In problems 17–24 you are to round off each of the given numbers to the indicated place value.

	nearest ten	nearest hundred	nearest thousand
17. 21,485	______	______	______
18. 7,763	______	______	______
19. 16,455	______	______	______
20. 982	______	______	______
21. 3,016	______	______	______
22. 770	______	______	______
23. 47,658	______	______	______
24. 8,072	______	______	______

25. Express $65,723 in written form.

____

26. Express $4,081 in written form.

____

27. Round off $8,456,098 to the nearest million. ____________

To the nearest thousand. ____________

To the nearest hundred. ____________

28. A woman has saved seventeen thousand, four hundred ninety-four dollars.

Express this sum in numerical form. ____________

ANSWERS

1. ninety-six **2.** two thousand, four hundred ninety-eight **3.** forty-five thousand, six hundred seven **4.** eight hundred seven thousand, five hundred seventy-six **5.** six million, three hundred eighty-five thousand, three hundred fourteen **6.** four hundred ninety-three **7.** eight million, three hundred twenty-eight thousand, seven hundred fifteen **8.** twenty-nine thousand, seven hundred eighty-one **9.** 52,647 **10.** 495 **11.** 736,201 **12.** 5,844,170 **13.** 16,922 **14.** 2,588,463 **15.** 29 **16.** 162,577 **17.** 21,490; 21,500; 21,000 **18.** 7,760; 7,800; 8,000 **19.** 16,460; 16,500; 16,000 **20.** 980; 1,000; 1,000 **21.** 3,020; 3,000; 3,000 **22.** 770; 800; 1,000 **23.** 47,660; 47,700; 48,000 **24.** 8,070; 8,100; 8,000 **25.** sixty-five thousand, seven hundred twenty-three dollars **26.** four thousand, eighty-one dollars **27.** $8,000,000; $8,456,000; $8,456,100 **28.** $17,494

1.3 ADDITION (+)

Mercy Hospital purchases aspirin tablets in bottles that contain 750 tablets. Currently there are 4 bottles, containing 53 tablets, 246 tablets, 87 tablets, and 192 tablets, respectively. If these bottles are combined, how many tablets does the hospital have in all? Can all of these tablets be placed in one 750-unit bottle? These questions will be answered in this section.

Addition involves combining like amounts. To add numbers, write them in a column with units below units, tens below tens, hundreds below hundreds and so on. The numbers being added are called **addends.** Add by moving down each column beginning with the column on the right. Your answer is called the **sum** or **total.**

Example 1 Add

$$\begin{array}{r} 34 \\ +\ 52 \\ \hline \end{array}$$

Solution

3 4	=	3 tens +	4 units	(or ones)
+ 5 2	=	5 tens +	2 units	(or ones)
6			6 units	(or ones)

3 4	=	3 tens	+ 4 units
+ 5 2	=	5 tens	+ 2 units
8 6	=	8 tens	+ 6 units

Example 2 Add

$$\begin{array}{r} 46 \\ +\ 37 \\ \hline \end{array}$$

Solution

4 6	=	4 tens +	6 units	Remember *we can only add like amounts*
+ 3 7	=	3 tens +	7 units	
3			13 units	13 units = 1 ten + 3 units

1		*1 ten*		
4 6	=	4 tens	+ 6 units	Leave 3 units and place one package of 10 with the other packages of ten and proceed to add. This is called *"carrying"* a package of ten.
+ 3 7	=	3 tens	+ 7 units	
8 3	=	8 tens	+ 3 units	

Example 3 Add

$$\begin{array}{r} 76 \\ 95 \\ +\ 788 \\ \hline \end{array}$$

Solution

$$\begin{array}{rrrcrrr} & 7 & 6 & = & & 7 \text{ tens} + & 6 \text{ units} \\ & 9 & 5 & = & & 9 \text{ tens} + & 5 \text{ units} \\ +\ 7 & 8 & 8 & = & 7 \text{ hundreds} + & 8 \text{ tens} + & 8 \text{ units} \\ \hline & & 9 & & & & 19 \text{ units} \end{array}$$

19 units = 1 ten + 9 units. Leave 9 units and carry one package of 10.

$$\begin{array}{rrrcrrr} & \mathit{1} & & & & \mathit{1\ ten} & \\ & 7 & 6 & = & & 7 \text{ tens} & + 6 \text{ units} \\ & 9 & 5 & = & & 9 \text{ tens} & + 5 \text{ units} \\ +\ 7 & 8 & 8 & = & 7 \text{ hundreds} + & 8 \text{ tens} & + 8 \text{ units} \\ \hline & 3 & 9 & & & 25 \text{ tens} & + 9 \text{ units} \end{array}$$

25 tens = 2 hundreds + 5 tens. Leave 5 tens and carry two packages of a hundred.

$$\begin{array}{rrrcrrr} & 7 & 6 & = & & 7 \text{ tens} & + 6 \text{ units} \\ \mathit{2} & 9 & 5 & = & \mathit{2\ hundreds} + & 9 \text{ tens} & + 5 \text{ units} \\ +\ 7 & 8 & 8 & = & 7 \text{ hundreds} + & 8 \text{ tens} & + 8 \text{ units} \\ \hline 9 & 3 & 9 & & 9 \text{ hundreds} + & 5 \text{ tens} & + 9 \text{ units} \end{array}$$

In actual practice we abbreviate these steps as follows when we perform addition.

Example 4 Add

$$\begin{array}{r} 567 \\ 418 \\ 824 \\ +\ 117 \\ \hline \end{array}$$

Solution

$$\begin{array}{rl} \mathit{12} & \\ 567 & \text{addend} \\ 418 & \text{addend} \\ 924 & \text{addend} \\ +\ 117 & \text{addend} \\ \hline 2026 & \text{sum or total} \end{array}$$

The column 1 total is 26; write 6 and carry 2.
The column 2 total is 12; write 2 and carry 1.
The column 3 total is 20; write 0 and carry 2.

To check an addition, repeat the addition by moving upward along each column beginning with the column on the right.

Example 5 Add 16 + 3 + 127 + 1,008

Solution Remember to place numbers in a column, with units below units, tens below tens, etc.

$$\begin{array}{r} 16 \\ 3 \\ 127 \\ +\ 1{,}008 \\ \hline 1{,}154 \end{array}$$

Example 6 Mercy Hospital purchases aspirin in bottles that contain 750 tablets. Currently there are 4 bottles containing 53 tablets, 246 tablets, 87 tablets, and 192 tablets, respectively. If these bottles are combined, how many tablets does the hospital have in all? Can all of these tablets be placed in one 750-unit bottle?

Solution

$$\begin{array}{r} \scriptstyle 21 \\ 53 \\ 246 \\ 87 \\ +\ 192 \\ \hline 578 \end{array}$$

Yes, all the tablets may be placed in one bottle, since 578 < 750.

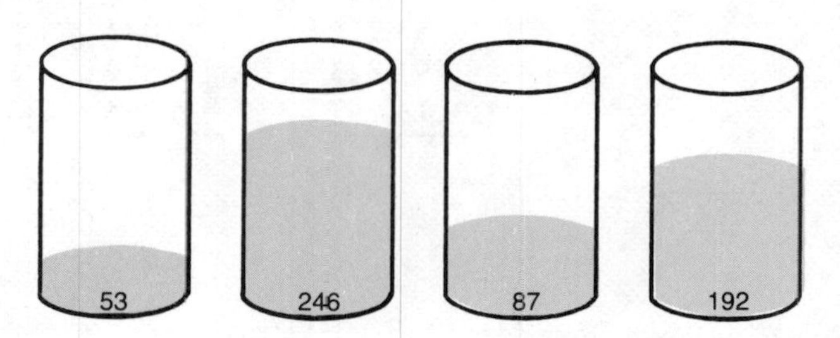

1.3 EXERCISES

Perform the indicated additions in problems 1–15.

1. $\begin{array}{r} 12 \\ +\ 5 \\ \hline \end{array}$ **2.** $\begin{array}{r} 15 \\ +\ 9 \\ \hline \end{array}$ **3.** $\begin{array}{r} 18 \\ +\ 12 \\ \hline \end{array}$ **4.** $\begin{array}{r} 48 \\ +\ 56 \\ \hline \end{array}$ **5.** $\begin{array}{r} 608 \\ +\ 847 \\ \hline \end{array}$

6. $\begin{array}{r} 18 \\ +\ 7 \\ \hline \end{array}$ **7.** $\begin{array}{r} 37 \\ +\ 45 \\ \hline \end{array}$ **8.** $\begin{array}{r} 63 \\ +\ 75 \\ \hline \end{array}$ **9.** $\begin{array}{r} 49 \\ +\ 51 \\ \hline \end{array}$ **10.** $\begin{array}{r} 482 \\ +\ 516 \\ \hline \end{array}$

11. $\begin{array}{r} 508 \\ +\ 371 \\ \hline \end{array}$ **12.** $\begin{array}{r} 725 \\ +\ 561 \\ \hline \end{array}$ **13.** $\begin{array}{r} 1{,}067 \\ +\ 985 \\ \hline \end{array}$ **14.** $\begin{array}{r} 9{,}432 \\ +\ 8{,}998 \\ \hline \end{array}$ **15.** $\begin{array}{r} 15{,}842 \\ +\ 6{,}976 \\ \hline \end{array}$

Check your answers to problems 1–15. If you did not get at least 12 problems correct, turn to Practice Problem Set 1A in the Appendix.

Perform the indicated additions in problems 16–35.

16. $\begin{array}{r} 72{,}999 \\ +\ 57{,}321 \\ \hline \end{array}$ **17.** $\begin{array}{r} 9 \\ 7 \\ 5 \\ 6 \\ +\ 8 \\ \hline \end{array}$ **18.** $\begin{array}{r} 15 \\ 84 \\ 28 \\ 35 \\ +\ 72 \\ \hline \end{array}$ **19.** $\begin{array}{r} 268 \\ 354 \\ 107 \\ 362 \\ +\ 758 \\ \hline \end{array}$ **20.** $\begin{array}{r} 111 \\ 794 \\ 85 \\ +\ 6379 \\ \hline \end{array}$

21. $\begin{array}{r} 7{,}224 \\ 391 \\ 26 \\ +\ 5{,}869 \\ \hline \end{array}$ **22.** $\begin{array}{r} 457 \\ 99 \\ 8{,}342 \\ +\ \ \ 413 \\ \hline \end{array}$ **23.** $\begin{array}{r} 7 \\ 16 \\ 13 \\ +\ \ 4 \\ \hline \end{array}$ **24.** $\begin{array}{r} 324 \\ 516 \\ 789 \\ 775 \\ +\ 908 \\ \hline \end{array}$ **25.** $\begin{array}{r} 967 \\ 91 \\ 29 \\ +\ 337 \\ \hline \end{array}$

26. $\begin{array}{r} 45{,}609 \\ 53{,}772 \\ 91{,}843 \\ +\ 16{,}067 \\ \hline \end{array}$ **27.** $\begin{array}{r} 4{,}576 \\ 8{,}093 \\ 559 \\ +\ 6{,}704 \\ \hline \end{array}$ **28.** $\begin{array}{r} 623{,}750 \\ 12{,}894 \\ 835{,}926 \\ +\ 359{,}432 \\ \hline \end{array}$ **29.** $\begin{array}{r} 90{,}003 \\ 425{,}016 \\ 398{,}058 \\ +\ \ 86{,}923 \\ \hline \end{array}$ **30.** $\begin{array}{r} 341{,}876 \\ 612{,}915 \\ 98{,}473 \\ +\ 700{,}389 \\ \hline \end{array}$

31. $34 + 48 + 72 =$ ____________________

32. $28 + 112 + 315 + 46 =$ ____________________

33. $116 + 380 + 792 + 94 =$ ____________________

34. $72 + 89 + 604 + 10 + 147 =$ ____________________

35. $453 + 22{,}312 + 406{,}523 + 78{,}512 =$ ____________________

36. The number of people requiring treatment in the emergency room of a hospital over a 5-day period was: 17 people, 23 people, 18 people, 34 people, and 29 people. What was the total number of people treated?

____

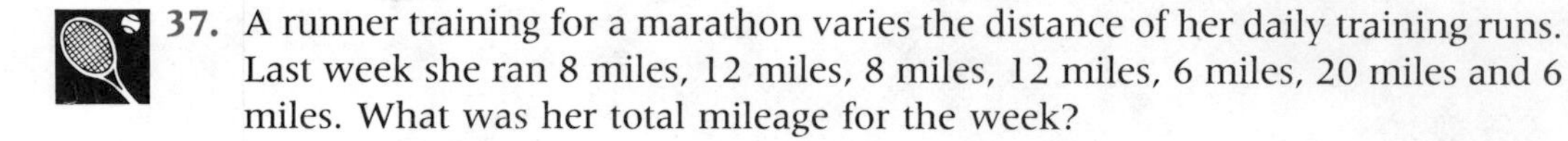

37. A runner training for a marathon varies the distance of her daily training runs. Last week she ran 8 miles, 12 miles, 8 miles, 12 miles, 6 miles, 20 miles and 6 miles. What was her total mileage for the week?

____

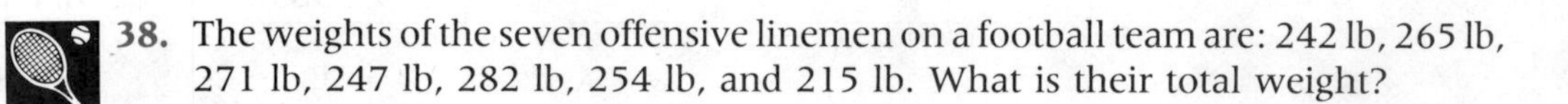

38. The weights of the seven offensive linemen on a football team are: 242 lb, 265 lb, 271 lb, 247 lb, 282 lb, 254 lb, and 215 lb. What is their total weight?

____

39. The following attendance figures were announced for a 4-game Red Sox–Yankees series: 35,043; 33,978; 34,685; 35,212. What was the total attendance for the four games?

____

40. The number of Canadian $100 gold coins in proof condition minted from 1977–1980 is as follows: 1977: 180,396; 1978: 180,009; 1979: 273,920; and 1980: 300,000. What is the total mintage for the years 1977–1980?

____

ANSWERS

1. 17 **2.** 24 **3.** 30 **4.** 104 **5.** 1,455 **6.** 25 **7.** 82 **8.** 138 **9.** 100 **10.** 998 **11.** 879 **12.** 1,286 **13.** 2,052 **14.** 18,430 **15.** 22,818 **16.** 130,320 **17.** 35 **18.** 234 **19.** 1,849 **20.** 7,369 **21.** 13,510 **22.** 9,311 **23.** 40 **24.** 3,312 **25.** 1,424 **26.** 207,291 **27.** 19,932 **28.** 1,832,002 **29.** 1,000,000 **30.** 1,753,653 **31.** 154 **32.** 501 **33.** 1,382 **34.** 922 **35.** 507,800 **36.** 121 people **37.** 72 miles **38.** 1,776 lb. **39.** 138,918 people **40.** 934,325

1.4 SUBTRACTION (−)

During a certain time interval a bank teller accepts deposits of $183, $52, $400, and $95 while, at the same time, paying out withdrawals of $75 and $150. What is the net amount of money that she has taken into the bank? This question will be answered in this section.

Subtraction is the process of finding the difference between two numbers. Place the numbers to be subtracted in a column with units below units, tens below tens, hundreds below hundreds and so on. Begin subtracting in the column, at the right. Take the **subtrahend** (bottom number) from the **minuend** (top number). Your answer is called the **difference.**

Example 1 Subtract

$$\begin{array}{r} 58 \\ -\ 32 \\ \hline \end{array}$$

Solution

58	= 5 tens +	8 units (or ones)	← minuend
− 32	= 3 tens +	2 units (or ones)	← subtrahend
6		6 units (or ones)	

58	= 5 tens	+ 8 units	
− 32	= 3 tens	+ 2 units	
26	2 tens	+ 6 units	← difference

Example 2 Subtract

$$\begin{array}{r} 61 \\ -\ 43 \\ \hline \end{array}$$

Solution

5	= *5 tens* +	*11 units*
~~6~~ ^{1}1	= ~~6~~ tens +	~~1~~ unit
− 4 3	= 4 tens +	3 units
8		8 units

We cannot take 3 units from 1 unit. "Borrow" one of the 6 packages of ten (leaving 5), which equals 10 units.
(10 units + 1 unit) = 11 units
11 units − 3 units = 8 units
Finish subtracting as indicated.

$$\begin{array}{rrcll} & 5 & & \textit{5 tens} & \textit{11 units} \\ & \cancel{6} & {}^{1}1 \;= & \cancel{6}\text{ tens} & +\ \cancel{1}\text{ unit} \\ - & 4 & 3 \;= & 4\text{ tens} & +\ 3\text{ units} \\ \hline & 1 & 8 \;= & 1\text{ ten} & +\ 8\text{ units} \end{array}$$

Example 3 Subtract

$$\begin{array}{r} 900 \\ -\ 357 \\ \hline \end{array}$$

Solution

$$\begin{array}{rrrrcllll} & 8 & 9 & 1 & & \textit{8 hundreds} & \textit{9 tens} & & \textit{10 units} \\ & & & & & & \cancel{\textit{10}}\textit{ tens} & & \\ & \cancel{9} & {}^{1}\cancel{0} & 0 & = & \cancel{9}\text{ hundreds} & +\ \cancel{0}\text{ tens} & + & \cancel{0}\text{ units} \\ - & 3 & 5 & 7 & = & 3\text{ hundreds} & +\ 5\text{ tens} & + & 7\text{ units} \\ \hline & & & 3 & = & & & & 3\text{ units} \end{array}$$

We cannot take 7 units from zero units. We cannot borrow a package of ten (there are zero tens). We borrow one of the 9 packages of a hundred (leaving 8) which equals 10 tens; that is, 10 tens + 0 tens = 10 tens. Now we can borrow one of the 10 packages of ten (leaving 9) giving us 10 units: 10 units + 0 units = 10 units, and subtract (10 units − 7 units = 3 units). Subtraction is finished as indicated.

$$\begin{array}{rrrrcllll} & & & & & & \textit{9 tens} & & \\ & 8 & 9 & & & \textit{8 hundreds} & \cancel{\textit{10}}\textit{ tens} & & \textit{10 units} \\ & \cancel{9} & {}^{1}\cancel{0} & {}^{1}0 & = & \cancel{9}\text{ hundreds} & +\ \cancel{0}\text{ tens} & + & \cancel{0}\text{ units} \\ - & 3 & 5 & 7 & = & 3\text{ hundreds} & +\ 5\text{ tens} & + & 7\text{ units} \\ \hline & & 4 & 3 & = & & +\ 4\text{ tens} & + & 3\text{ units} \end{array}$$

$$\begin{array}{rrrrcllll} & & & & & & \textit{9 tens} & & \\ & 8 & 9 & & & \textit{8 hundreds} & \cancel{\textit{10}}\textit{ tens} & & \textit{10 units} \\ & \cancel{9} & {}^{1}\cancel{0} & {}^{1}0 & = & \cancel{9}\text{ hundreds} & +\ \cancel{0}\text{ tens} & + & \cancel{0}\text{ units} \\ - & 3 & 5 & 7 & = & 3\text{ hundreds} & +\ 5\text{ tens} & + & 7\text{ units} \\ \hline & 5 & 4 & 3 & = & 5\text{ hundreds} & +\ 4\text{ tens} & + & 3\text{ units} \end{array}$$

Example 4 Subtract

$$\begin{array}{r} 3{,}754 \\ -\ \ \ 298 \\ \hline \end{array}$$

Solution

$$\begin{array}{rrrrr} & & & \textit{14} & \\ & & \textit{6} & \cancel{\textit{4}} & \textit{14} \\ \$ & 3, & \cancel{7} & \cancel{5} & \cancel{4} \\ -\ \$ & & 2 & 9 & 8 \\ \hline \$ & 3, & 4 & 5 & 6 \end{array}$$

Because we cannot take 8 from 4, we must "borrow" a package of ten from the 5 packages of ten in the next column to the left. Thus, (4 + 10) − 8 = 6 and so on.

Example 5a Subtract

$$\begin{array}{r} 6{,}582 \\ -\ 4{,}937 \\ \hline \end{array}$$

Solution

$$\begin{array}{rrrrrl} & \textit{5} & \textit{15} & \textit{7} & \textit{12} & \\ & \cancel{6}, & \cancel{5} & \cancel{8} & \cancel{2} & \text{minuend} \\ - & 4, & 9 & 3 & 7 & \text{subtrahend} \\ \hline & 1, & 6 & 4 & 5 & \text{difference} \end{array}$$

To check a subtraction add the difference to the subtrahend; the sum should be the minuend.

Difference + Subtrahend = Minuend

Example 5b Check the work of Example 5a.

```
  4, 9 3 7   subtrahend
+ 1, 6 4 5   difference
  6, 5 8 2   minuend
```

Example 6 During a certain time interval a bank teller accepts deposits of $183, $52, $400, and $95 while at the same time paying out withdrawals of $75 and $150. What is the net amount of money that she has taken into the bank?

Solution She has taken in $183 + $52 + $400 + $95 = $730.
She has paid out $75 + $150 = $225.
Therefore, we subtract to find the net amount.

```
  $ 730
  - 225
  $ 505
```

Example 7 Subtract

```
  8,004
- 7,685
```

Solution

```
      9  9
  7, 10 10 14
  8,  0  0  4
- 7,  6  8  5
      3  1  9
```

1.4 EXERCISES

Perform the indicated subtractions in problems 1–15.

1. 95 − 34	**2.** 89 − 62	**3.** 967 − 852	**4.** 423 − 201	**5.** 487 − 361
6. 63 − 37	**7.** 451 − 260	**8.** 800 − 374	**9.** 1790 − 358	**10.** 446 − 375
11. 372 − 48	**12.** 204 − 67	**13.** 826 − 291	**14.** 635 − 487	**15.** 8,000 − 2,764

Check your answers to problems 1–15. If you did not get at least 12 problems correct, turn to Practice Problem Set 2A in the Appendix.

Perform the indicated subtractions in problems 16–35.

16. 15,946	**17.** 36,589	**18.** 412,959	**19.** 345,625	**20.** 72,063
− 7,783	− 18,491	− 351,744	− 64,335	− 51,982
21. 235,604	**22.** 7,801	**23.** 19,463	**24.** 786,304	**25.** 89,911
− 78,921	− 2,999	− 6,481	− 207,634	− 76,344
26. 83,695	**27.** 10,048	**28.** 32,356	**29.** 6,005	**30.** 100,700
− 64,596	− 6,539	− 29,999	− 3,247	− 80,395
31. 10,000	**32.** 65,196	**33.** 47,009	**34.** 71,092	**35.** 419,388
− 9,345	− 42,880	− 28,557	− 53,226	− 274,959

36. Robert, Francine, and Carl ran for president of their union local. Robert received 243 votes, Francine received 388 votes, and Carl received 289 votes. What was the total number of votes cast? How many more votes did Francine receive than Carl?

37. Raymond drove from Detroit to St. Louis for a sales meeting. On the way he stopped to see friends in Chicago. The distance from Detroit to Chicago is 435 km and from Chicago to St. Louis is 470 km. On the return trip he drove directly home and covered 819 km. How many kilometers did he drive round-trip? How much shorter was his return trip?

38. An empty truck weighs 3,750 kilograms. When fully loaded this same truck weighs 9,955 kilograms. What is the weight of the load that this truck is carrying?

39. The sticker price of a new car is $8,750. The dealer will give Bob $3,075. on a trade-in. What will the new car cost Bob?

40. Last week the balance in Tom's checkbook was $634. He wrote checks for $75, $212, $30, and $56. He made deposits of $105 and $62. What is Tom's balance at the end of the week?

41. Mr. Jones bought a new car. The list price was $8,025. In addition, he purchased the following options: $125 for an AM-FM radio, $165 for power steering and $115 for rustproofing. The dealer offered a $300 rebate for the purchase of a new car. What did Mr. Jones pay for the car?

ANSWERS

1. 61 **2.** 27 **3.** 115 **4.** 222 **5.** 126 **6.** 26 **7.** 191 **8.** 426 **9.** 1,432
10. 71 **11.** 324 **12.** 137 **13.** 535 **14.** 148 **15.** 5,236 **16.** 8,163
17. 18,098 **18.** 61,215 **19.** 281,290 **20.** 20,081 **21.** 156,683
22. 4,802 **23.** 12,982 **24.** 578,670 **25.** 13,567 **26.** 19,099 **27.** 3,509
28. 2,357 **29.** 2,758 **30.** 20,305 **31.** 655 **32.** 22,316 **33.** 18,452
34. 17,866 **35.** 144,429 **36.** total: 920; 99 more **37.** 1,724 total; 86 km shorter
38. 6,205 kilograms **39.** $5,675 **40.** $428 **41.** $8,130

1.5 MULTIPLICATION (×)

A glass commonly used in a hospital for the intake of fluids is calculated to hold 205 milliliters of fluid. If a patient drank 7 glasses of fluid, what would be his fluid intake? This question will be answered in this section.

Example 1 Four nickels, each containing 5 cents; total, 4 × 5 cents = 20 cents.

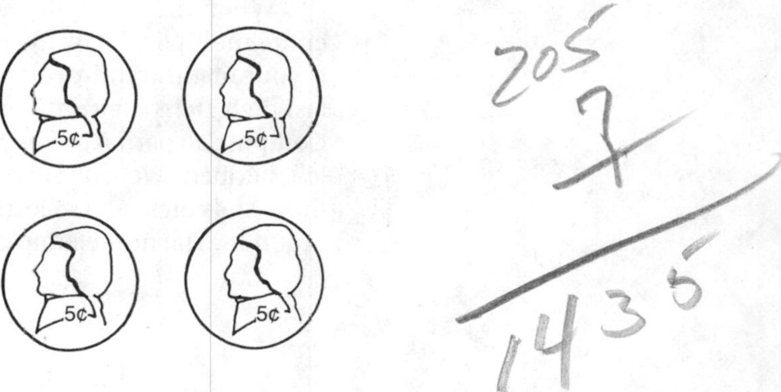

Multiplication may be thought of as a shorthand method of adding equal addends. For example, the sum 4 + 4 + 4 + 4 + 4 + 4 consists of six equal addends. The sum may be determined by computing 6 × 4 = 24. Also, 7 × 205 = 205 + 205 + 205 + 205 + 205 + 205 + 205 = 1,435.

When you multiply two numbers, with one located above the other, the top number is called the **multiplicand** and the bottom number is called the **multiplier.** The answer is called the **product.** (You will find it beneficial to memorize the multiplication facts given in the table in the Appendix.)

$$\begin{array}{r l} 8 & \leftarrow \text{multiplicand} \\ \times \quad 6 & \leftarrow \text{multiplier} \\ \hline 48 & \leftarrow \text{product} \end{array}$$

Example 2 A glass commonly used in a hospital for the intake of fluids is calculated to hold 205 milliliters of fluid. If a patient drank 7 glasses of fluid, what would be his fluid intake?

Solution This is a multiplication problem involving a one digit multiplier.

$$\begin{array}{rcr} 2\ 0\ 5 & = & 2 \text{ hundreds} + 0 \text{ tens} + \quad 5 \text{ units} \\ \times \quad 7 & = & 7 \text{ units} \\ \hline 5 & & 35 \text{ units} \end{array}$$

$$\left\{ \begin{array}{l} 7 \text{ units} \times 5 \text{ units} = 35 \text{ units} \\ 35 \text{ units} = 3 \text{ tens} + 5 \text{ units} \\ \text{Leave 5 units and carry the 3 packages of ten} \end{array} \right.$$

$$
\begin{array}{rcr}
\overset{3}{2\ 0\ 5} & = & \overset{3\ tens}{2 \text{ hundreds} + 0 \text{ tens} + 5 \text{ units}} \\
\times \quad 7 & = & 7 \text{ units} \\
\hline
3\ 5 & & 3 \text{ tens} + 5 \text{ units}
\end{array}
$$

$$
\begin{cases}
7 \text{ units} \times 0 \text{ tens} = 0 \text{ tens} \\
0 \text{ tens} + 3 \text{ tens} = 3 \text{ tens}
\end{cases}
$$

$$
\begin{array}{rcr}
\overset{3}{2\ 0\ 5} & = & 2 \text{ hundreds} + 0 \text{ tens} + 5 \text{ units} \\
\times \quad 7 & = & 7 \text{ units} \\
\hline
1,\ 4\ 3\ 5 & & 14 \text{ hundreds} + 3 \text{ tens} + 5 \text{ units}
\end{array}
$$

$$
\begin{cases}
7 \text{ units} \times 2 \text{ hundreds} = 14 \text{ hundreds} \\
14 \text{ hundreds} = 1 \text{ thousand} + 4 \text{ hundreds} \\
\text{Leave 4 hundreds and carry the one package of a thousand.}
\end{cases}
$$

Answer 1 thousand + 4 hundreds + 3 tens + 5 units or 1,435

When the multiplier consists of two or more digits, begin by multiplying the entire multiplicand by the units digit (right most digit) of the multiplier; this product is your first **partial product.** Next multiply the entire multiplicand by the second-last digit (tens digit) of the multiplier; this product is your second partial product. Compute all partial products. Each partial product must be indented one space to the left because we are successively multiplying by the tens digit, hundreds digit, thousands digit, and so forth. Your answer, the product, is found by adding all partial products. The next examples will further illustrate multiplication of whole numbers.

Example 3a Multiply

$$
\begin{array}{r}
56 \\
\times\ 72 \\
\hline
\end{array}
$$

Solution

$$
\begin{array}{rl}
5\ 6 & \text{multiplicand} \\
\times\ \textcircled{7}\textcircled{2} & \text{multiplier} \\
\hline
1\ 1\ \textcircled{2} & \text{first partial product} \leftarrow 56 \times 2 \\
3\ 9\ \textcircled{2}\ \ & \text{second partial product} \leftarrow 56 \times 70 \\
\hline
4,\ 0\ 3\ 2 & \text{product}
\end{array}
$$

To check a multiplication, reverse the order of multiplicand and multiplier and compute the product again.

Example 3b Check the work of Example 3a.

$$
\begin{array}{rl}
7\ 2 & \text{multiplicand} \\
\times\ \textcircled{5}\textcircled{6} & \text{multiplier} \\
\hline
4\ 3\ \textcircled{2} & \\
3\ 6\ \textcircled{0}\ \ & \\
\hline
4,\ 0\ 3\ 2 & \text{product}
\end{array}
$$

Memory Device: The right most digit in each partial product lies directly below the digit being used in the multiplier. These digits are circled here for emphasis.

Example 4 Multiply

$$\begin{array}{r} 454 \\ \times\ 389 \\ \hline \end{array}$$

Solution

$$\begin{array}{rl} 4\ 5\ 4 & \text{multiplicand} \\ \times\ ③⑧⑨ & \text{multiplier} \\ \hline 4\ 0\ 8⑥ & \text{first partial product} \\ 3\ 6\ 3② & \text{second partial product} \\ 1\ 3\ 6\ ② & \text{third partial product} \\ \hline 1\ 7\ 6,\ 6\ 0\ 6 & \text{product} \end{array}$$

Example 5 Multiply

$$\begin{array}{r} 639 \\ \times\ 402 \\ \hline \end{array}$$

Solution

$$\begin{array}{r} 6\ 3\ 9 \\ \times\ ④⓪② \\ \hline 1\ 2\ 7⑧ \\ 2\ 5\ 5\ ⑥⓪ \\ \hline 2\ 5\ 6,\ 8\ 7\ 8 \end{array}$$

When zero is a digit of the multiplier, we can place a zero directly below the 0 in the multiplier and multiply by the next digit in the multiplier, writing the partial product on the same line as the zero.

Observe what happens when a number, such as 73, is multiplied by 10, 100, 1,000, 10,000. and so on.

73 × 10 = 730	73 followed by *one zero*
73 × 100 = 7,300	73 followed by *two zeros*
73 × 1,000 = 73,000	73 followed by *three zeros*
73 × 10,000 = 730,000	73 followed by *four zeros*

In this special case, you may quickly determine the answer to your product by writing the original number followed by an appropriate number of zeros.

Example 6 Multiply 426 by
(a) 10 (b) 100 (c) 1,000 (d) 1,000,000

Solution (a) 4,260 (b) 42,600 (c) 426,000 (d) 426,000,000

Example 7 Lisa is paid $8 per hour for the first 40 hours that she works in any given week and $12 (time-and-a-half) for each hour over 40 worked in a week. For a week in which she works 44 hours, what is her gross pay?

Solution gross pay = 40 × $8 + 4 × $12

= $320 + $48

= $368

Multiplication Symbols

If a and b are any two numbers, the product of a and b can be expressed:

1. $a \times b$ or
2. $a \cdot b$ using a "raised dot" or
3. $(a)(b)$ using parentheses

Thus $6 \times 7 = 6 \cdot 7 = (6)(7)$, $5 \cdot 9 = 5 \times 9 = (5)(9)$, and $(17)(19) = 17 \cdot 19 = 17 \times 19$. The product of 5, 13, and 44 may be denoted $(5)(13)(44)$ or $5 \cdot 13 \cdot 44$. *These ways of denoting a product may be used whenever it is convenient throughout the remainder of this text.*

Example 8 (a) 4 times 8 may be expressed as $4 \cdot 8$ or as $(4)(8)$.
(b) 12 times 12 may be expressed as $12 \cdot 12$ or as $(12)(12)$.

Example 9 The product $22 \times 5 \times 111$ may be expressed either as $22 \cdot 5 \cdot 111$ or as $(22)(5)(111)$. To find this product we can first find $(22)(5) = 110$ and then multiply $(110)(111) = 12{,}210$.

1.5 EXERCISES

Perform the indicated multiplications in problems 1–15.

1. 34×2 **2.** 42×3 **3.** 35×6 **4.** 86×8

5. 16×7 **6.** 39×5 **7.** 52×9 **8.** 77×8

9. 643×7 **10.** 812×9 **11.** 921×31 **12.** 504×48

13. $8{,}216 \times 45$ **14.** $7{,}891 \times 55$ **15.** $9{,}207 \times 74$

Check your answers to problems 1–15. If you did not get at least 12 problems correct, turn to Practice Problem Set 3A in the Appendix.

Perform the indicated multiplications in problems 16–32.

16. $9{,}359 \times 47$ **17.** 398×40 **18.** $6{,}237 \times 318$ **19.** $12{,}682 \times 563$ **20.** $5{,}004 \times 608$

21. $\begin{array}{r} 612 \\ \times\ 300 \\ \hline \end{array}$ **22.** $\begin{array}{r} 8{,}478 \\ \times\ 165 \\ \hline \end{array}$ **23.** $\begin{array}{r} 675 \\ \times\ 204 \\ \hline \end{array}$ **24.** $\begin{array}{r} 83{,}476 \\ \times\ 928 \\ \hline \end{array}$ **25.** $\begin{array}{r} 4{,}768 \\ \times\ 359 \\ \hline \end{array}$

26. $\begin{array}{r} 87{,}253 \\ \times\ 1{,}005 \\ \hline \end{array}$ **27.** $\begin{array}{r} 6{,}631 \\ \times\ 2{,}000 \\ \hline \end{array}$ **28.** $\begin{array}{r} 3{,}980 \\ \times\ 420 \\ \hline \end{array}$ **29.** $\begin{array}{r} 74{,}361 \\ \times\ 228 \\ \hline \end{array}$ **30.** $\begin{array}{r} 35{,}000 \\ \times\ 2{,}200 \\ \hline \end{array}$

31. $\begin{array}{r} 98{,}000 \\ \times\ 360 \\ \hline \end{array}$ **32.** $\begin{array}{r} 72{,}404 \\ \times\ 562 \\ \hline \end{array}$

In problems 33–36 express the given product by using (a) a "dot" and (b) parentheses.

33. 19×5 ______ **34.** 17×330 ______

35. $10 \times 10 \times 12$ ______ **36.** $5 \times 18 \times 25$ ______

In problems 37–44 find the indicated product.

37. $(9)(12)$ ______ **38.** $(36)(20)$ ______

39. $15 \cdot 64$ ______ **40.** $42 \cdot 185$ ______

41. $(3)(4)(9)$ ______ **42.** $(6)(5)(10)$ ______

43. $3 \cdot 8 \cdot 54$ ______ **44.** $96 \cdot 7 \cdot 83$ ______

45. Multiply 25 by: (a) 10 ____ (b) 100 ____ (c) 1,000 ____

46. Multiply 139 by: (a) 10 ____ (b) 10,000 ____

47. ($57 + $19) × 100 = (*Hint:* Work inside the parentheses first.)

______

48. A student pays $98 for each of 9 months for his dorm room. How much is paid in all?

______

49. A particular car with a four-cylinder engine gets 10 miles per liter of gasoline. How far can the car travel on 65 liters of gasoline?

______

50. The distance from the planet Neptune to the sun is approximately 30 times that of the earth. If the distance from the earth to the sun is 93,000,000 miles, approximately how far is Neptune from the sun?

____

51. In a particular year a company pays its employees an average salary of $23,516. If the company has 463 employees, what is the total salary paid?

____

52. Each day an industrial plant pollutes 153,267 gallons of water. Assuming that a year has 365 days, how many gallons are polluted in a year?

____

ANSWERS

1. 68 **2.** 126 **3.** 210 **4.** 688 **5.** 112 **6.** 195 **7.** 468 **8.** 616 **9.** 4,501 **10.** 7,308 **11.** 28,551 **12.** 24,192 **13.** 369,720 **14.** 434,005 **15.** 681,318 **16.** 439,873 **17.** 15,920 **18.** 1,983,366 **19.** 7,139,966 **20.** 3,042,432 **21.** 183,600 **22.** 1,398,870 **23.** 137,700 **24.** 77,465,728 **25.** 1,711,712 **26.** 87,689,265 **27.** 13,262,000 **28.** 1,671,600 **29.** 16,954,308 **30.** 77,000,000 **31.** 35,280,000 **32.** 40,691,048 **33. (a)** $19 \cdot 5$ **(b)** (19)(5) **34. (a)** $17 \cdot 330$ **(b)** (17)(330) **35. (a)** $10 \cdot 10 \cdot 12$ **(b)** (10)(10)(12) **36. (a)** $5 \cdot 18 \cdot 25$ **(b)** (5)(18)(25) **37.** 108 **38.** 720 **39.** 960 **40.** 7,770 **41.** 108 **42.** 300 **43.** 1,296 **44.** 55,776 **45. (a)** 250 **(b)** 2,500 **(c)** 25,000 **46. (a)** 1,390 **(b)** 1,390,000 **47.** $7,600 **48.** $882 **49.** 650 miles **50.** 2,790,000,000 miles **51.** $10,887,908 **52.** 55,942,455 gallons

1.6 DIVISION (÷)

Helen wishes to purchase 40 postage stamps at 20 cents each and she knows that a dollar contains 100 cents. How many dollars does she need to pay for these stamps? This question will be answered in this section.

Single-Digit Division

Division may be thought of as the process that is the opposite of multiplication. There are three common ways of expressing a division of two numbers. For example, 24 divided by 3 may be expressed as $24 \div 3$, or $\frac{24}{3}$, or $3\overline{)24}$. Similarly 56 divided by 7 may be written $56 \div 7$, or $\frac{56}{7}$, or $7\overline{)56}$.

The parts of a division are also named and it is helpful to know these names.

$$\begin{array}{rrl} & 8 & \leftarrow \textbf{quotient} \\ \textbf{divisor} \rightarrow & 3\overline{)25} & \leftarrow \textbf{dividend} \\ & \underline{24} & \\ & 1 & \leftarrow \textbf{remainder} \end{array}$$

Example 1 Find $168 \div 8$, or $\frac{168}{8}$, or $8\overline{)168}$

Solution step 1:

$$\begin{array}{r} 2 \\ 8\overline{)1\,6\,8} \\ \underline{\nearrow 1\,6\downarrow} \\ 0 \end{array}$$

8 is contained twice in 16. Bring down the 8 in the dividend.

step 2:

```
     2 1
  8)1 6 8
    1 6
    ---
      8
      8
      -
      0
```

8 is contained once in 8. The quotient is 21 and the remainder is 0.

Example 2 Find 4,195 ÷ 5, or $\frac{4{,}195}{5}$, or 5)4,195.

Solution

```
     8 3 9          8         3         9
  5)4, 1 9 5     5)4 1     5)1 9     5)4 5
    4 0 ↓          4 0       1 5       4 5
    ---            ---       ---       ---
      1 9            1         4         0
      1 5 ↓
      ---
        4 5
        4 5
        ---
          0
```

Answer: The quotient is 839 and the remainder is 0.

Example 3 Find 195 ÷ 7.

Solution step 1:

```
     2
  7)1 9 5
    1 4 ↓
    ---
      5
```

7 is contained twice in 19. Bring down the 5 in the dividend.

step 2:

```
     2 7
  7)1 9 5
    1 4
    ---
      5 5
      4 9
      ---
        6
```

7 is contained 7 times in 55.

Answer: The quotient is 27 and the remainder is 6.

Multiple-Digit Division When you divide whole numbers, keep in mind the following points, which we are about to illustrate.

1. The first digit in the quotient should be placed directly above the right most digit of the part of the dividend into which the divisor divides. Consider the problem of dividing 2,465 by 47.

```
           5
  4 7)2, 4 6 5
```

47 cannot be divided into 2. 47 cannot be divided into 24. 47 can be divided 5 times into 246. Place the 5 above the 6 in your dividend.

2. Multiply the divisor, 47, by the digit 5 in the quotient: 47 × 5. This partial product, 235, is placed beneath the first three numbers in your dividend. Subtract this partial product, 235, from the first three numbers, 246, in your dividend. The number resulting from this subtraction should always be less than the divisor.

```
           5
  4 7)2, 4 6 5
       2 3 5
       -----
         1 1
```

Observe that 11 is less than 47, or $11 < 47$.

3. Continue this process by bringing down the next digit of the dividend. The process stops when the dividend can no longer contain the number 47. Your remainder may be expressed by placing "$R =$ " immediately after the quotient

```
          5 2
    47)2, 4 6 5        R = 21        Answer: 52, R = 21
       2 3 5
         1 1 5
           9 4
           2 1
```

If the divisor will not divide into the dividend after one digit (the digits are 0, 1, 2, 3, . . ., 9) has been brought down, place a zero in the quotient and bring down the next digit. Here is an illustration of such a situation.

```
                    1 0 8
               58)6, 2 9 4       R = 30
                  5 8
58 can't be         4 9 4
divided into 49 →   4 6 4
                      3 0
```

Example 4 Find 33,712 ÷ 56.

Solution step 1:

```
          6
    56)3 3, 7 1 2
       3 3 6
           1
```

step 2:

```
          6 0
    56)3 3, 7 1 2
       3 3 6
           1 1
             0
           1 1
```

step 3:

```
          6 0 2
    56)3 3, 7 1 2
       3 3 6
           1 1
             0
           1 1 2
           1 1 2
               0
```

To check a division, multiply the quotient by the divisor and add the remainder, if any, to this product. Your answer should be the dividend.

(Quotient × Divisor) − Remainder = Dividend

Let's do the check for example 4.

$$\begin{array}{r} 602 \\ \times\ 56 \\ \hline 3\,612 \\ 30\,10 \\ \hline 33{,}712 \\ +\quad 0 \\ \hline 33{,}712 \end{array}$$

Example 5a Complete the division $146\overline{)5{,}473}$

Solution

$$\begin{array}{r} 37 \\ 146\overline{)5{,}473} \\ 4\,38 \\ \hline 1\,093 \\ 1\,022 \\ \hline 71 \end{array} \quad R = 71$$

Example 5b Check the work of Example 5a.

Solution Recall: (Quotient × Divisor) + Remainder = Dividend.
(146 × 37) + 71 should equal 5,473

$$\begin{array}{rl} 146 & \leftarrow \text{divisor} \\ \times\ 37 & \leftarrow \text{quotient} \\ \hline 1022 & \\ 438 & \\ \hline 5402 & \\ +\ 71 & \leftarrow \text{remainder} \\ \hline 5473 & \leftarrow \text{dividend} \end{array}$$

Example 6 Alex's take-home pay for any four-week month is $1,280, and this amount is paid to him in equal amounts. It happens this week that he has the following bills to pay: $46, $32, and $85. Assuming he pays these bills, how much of his week's pay remains?

Solution weekly pay = $1,280 ÷ 4 = $320
his bills total: $46 + $32 + $85 = $163
amount left over: $320 − $163 = $157

Example 7 Helen wishes to purchase 40 postage stamps at 20 cents each and she knows that a dollar contains 100 cents. How many dollars does she need to pay for these stamps?

Solution 40 stamps × 20 cents/stamp = 800 cents
800 cents ÷ 100 cents = 8 dollars

Example 8 The yearly rainfall (rounded to the nearest inch) for four consecutive years in Tifton was 22 inches, 28 inches, 31 inches, and 19 inches. Find the average rainfall.

Solution To find the average of a collection of numbers, find the sum and divide it by the number of items.

$$\text{the average} = \frac{22+28+31+19}{4} = \frac{100}{4} = 25 \text{ inches}$$

Division Involving Zero

When checking a division problem, remember to observe the following: *quotient × divisor = dividend.*

Example 9 $8\overline{)0}$

Solution $\begin{array}{r} 0 \\ 8\overline{)0} \end{array}$ check: $0 \times 8 = 0$

Example 10 $0\overline{)0}$

Solution (a) $\begin{array}{r} 1 \\ 0\overline{)0} \end{array}$ check: $1 \times 0 = 0$

(b) $\begin{array}{r} 5 \\ 0\overline{)0} \end{array}$ check: $5 \times 0 = 0$

(c) $\begin{array}{r} 11 \\ 0\overline{)\ 0} \end{array}$ check: $11 \times 0 = 0$

All of these answers are correct. Which one is the answer we are looking for? *We don't know.*

Example 11 $0\overline{)6}$

Solution $\begin{array}{r} \text{quotient} \\ 0\overline{)6\qquad\quad} \end{array}$ check: quotient $\times\ 0 = 0$, and $0 \neq 6$

No matter what number the quotient is, when multiplied by 0 the answer will always be 0.

Examples 10 and 11 show us that there is no unique answer to a problem in which you are dividing by zero. Thus division by 0 (using 0 as a divisor) cannot be done. When you are asked to do a problem such as $\frac{8}{0}$, $\frac{0}{0}$, $7 \div 0$, or $0 \div 0$, write "no solution."

1.6 EXERCISES

In problems 1–15 perform the indicated divisions, if possible.

1. $2\overline{)846}$ **2.** $3\overline{)963}$ **3.** $6\overline{)834}$

4. $5\overline{)505}$ **5.** $7\overline{)0}$ **6.** $0\overline{)12}$

7. $9\overline{)0}$ **8.** $7\overline{)392}$ **9.** $0\overline{)11}$

10. $8\overline{)5,016}$ **11.** $4\overline{)3,068}$ **12.** $9\overline{)40,325}$

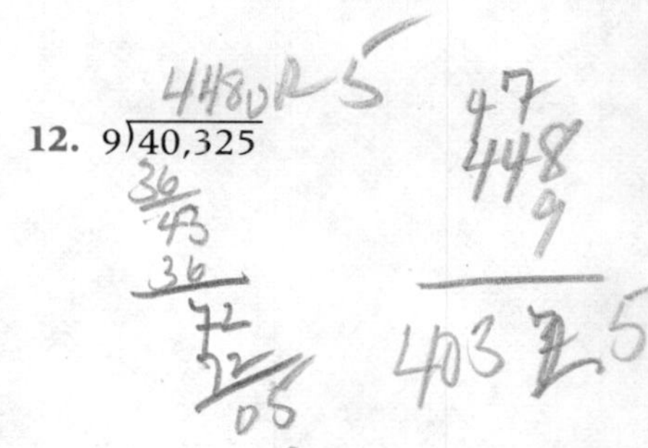

13. $5\overline{)61,312}$ **14.** $26\overline{)1,404}$ **15.** $18\overline{)3,240}$

Check your answers to problems 1–15. If you did not get at least 12 problems correct, turn to Practice Problem Set 4A in the Appendix.

In problems 16–36 perform the indicated divisions, if possible.

16. $35\overline{)4,410}$ **17.** $76\overline{)38,228}$ **18.** $84\overline{)18,984}$

19. $72\overline{)6,290}$ **20.** $63\overline{)3,486}$ **21.** $92\overline{)7,643}$

22. $37\overline{)9,447}$ **23.** $68\overline{)4,498}$ **24.** $25\overline{)2,055}$

25. $214\overline{)5,778}$ **26.** $354\overline{)9,912}$ **27.** $912\overline{)81,168}$

28. $342\overline{)9,747}$ **29.** $789\overline{)83,092}$ **30.** $547\overline{)57,435}$

31. $802\overline{)329{,}622}$ **32.** $321\overline{)129{,}042}$ **33.** $360\overline{)304{,}002}$

34. $304\overline{)305{,}216}$ **35.** $292\overline{)105{,}218}$ **36.** $702\overline{)591{,}066}$

37. Hi-fi Stereo Corporation purchased 24 stereos for $14,352. How much did the company pay for each stereo?

____

38. Lois received the following grades on her mathematics exams: 92, 76, 100, 84 and 93. What was her average for those exams?

____

39. A baseball player's batting average can be computed by multiplying the number of hits by 1,000 and dividing this product by the number of times he was at bat. If a player had 144 hits in 450 times at bat, what was his batting average?

____

40. During a four-week period a company has gross sales of $568,748 and returns and allowances of $78,960. What is the net sales average per week?

____

41. A woman agreed to pay \$7,920 on a car loan by making 48 equal monthly payments. How much must she pay each month to pay off her loan?

____

42. William earns \$30,420 per year. If he draws his pay in 52 equal weekly paychecks, how much does he receive each week?

____

ANSWERS

1. 423 **2.** 321 **3.** 139 **4.** 101 **5.** 0 **6.** no solution **7.** 0 **8.** 56
9. no solution **10.** 627 **11.** 767 **12.** 4,480, $R = 5$ **13.** 12,262, $R = 2$
14. 54 **15.** 180 **16.** 126 **17.** 503 **18.** 226 **19.** 87, $R = 26$ **20.** 55, $R = 21$
21. 83, $R = 7$ **22.** 255, $R = 12$ **23.** 66, $R = 10$ **24.** 82, $R = 5$ **25.** 27
26. 28 **27.** 89 **28.** 28, $R = 171$ **29.** 105, $R = 247$ **30.** 105 **31.** 411
32. 402 **33.** 844, $R = 162$ **34.** 1,004 **35.** 360, $R = 98$ **36.** 841, $R = 684$
37. \$598 **38.** 89 **39.** 320 **40.** \$122,447 **41.** \$165 **42.** \$585

1.7 POWERS OF WHOLE NUMBERS

Exponents When you multiply the same number again and again, such as $7 \times 7 \times 7 \times 7 \times 7$, the following shorthand notation is convenient: Write the repeating number, 7, and place a 5 (for the number of 7's being multiplied) above and to the right of the 7. The result is 7^5 or $7 \times 7 \times 7 \times 7 \times 7 = 7^5$. In this shorthand notation, the number 7 is called the "base" number and the number 5 is called the "exponent." The expression $4 \times 4 \times 4$ has the number 4 repeated 3 times; this may be shortened to 4^3. Here the 4 is the base and the 3 is the exponent.

If a is any whole number and n is a natural number, then:

$$\text{base} \rightarrow a^{n} = \underbrace{a \times a \times a \times \cdots \times a}_{\text{There are } n\ a\text{'s}} \quad (n \text{ is the exponent})$$

In the expression a^n, a is the **base** number and n is the **exponent.** The number n tells us how many times to write the base number a with multiplication between successive a's. The expression a^n is often read "a to the nth power." Hence, 2^7 is read "two to the seventh power," and 8^4 is read "eight to the fourth power."

If your exponent is one, you need not write it for it is understood. Observe, $2^1 = 2$, $3^1 = 3$, $4^1 = 4$, and so on.

The second power of a base number is often called the "square of the number." Thus, 5^2 is read "five squared" and 19^2 is read "nineteen squared." The third power of a base number is often called the "cube of the number." Thus, 7^3 is read "seven cubed" and 14^3 is read "fourteen cubed."

Example 1 Discuss the exponential expression 8^1.

Solution $8^1 = 8$. (The exponent of one need not be written.)

Example 2 Express $9 \cdot 9 \cdot 9 \cdot 9$ in exponential notation.

Solution 9^4

Example 3 Express (2)(2)(2)(2)(2)(2)(2) in exponential notation.

Solution 2^7

Example 4 Evaluate 4^3.

Solution 4^3 means $4 \times 4 \times 4 = 64$.

Example 5 Evaluate 6^5.

Solution 6^5 means $6 \times 6 \times 6 \times 6 \times 6 = 7{,}776$.

Example 6 (a) Write the expression "five squared" in exponential notation.
(b) Write the expression 2^3 in words.

Solution (a) 5^2
(b) The expression 2^3 can be written either "two cubed" or "two to the third power."

Order of Operations Given $5 \cdot 4 - 2$, do you multiply first or do you subtract first? How do you evaluate an expression such as $(9)(3) + 5^2 - 16 \div 2$? We will answer these questions in this section.

When addition, subtraction, multiplication, division, and powers are written in the same expression, we perform the operations in the following order:

> *Order of Operations*
> **1.** Powers of numbers are evaluated in any order.
> **2.** Multiplications and divisions are done in order from *left to right.*
> **3.** Additions and subtractions are done in order from *left to right.*

Example 7 Evaluate $6 \times 3 \div 2$.

Solution Proceed from left to right recognizing that multiplication and division are operations of equal priority.

Thus $\underbrace{6 \times 3}_{} \div 2 =$

$\underbrace{18 \div 2}_{} =$

9

Example 8 Evaluate $5 \cdot 4 - 2$.

Solution Multiplication proceeds subtraction.

Thus $5 \cdot 4 - 2 =$

$20 - 2 =$

18

Example 9 Evaluate $5 + 16 - 2^3$.

Solution $5 + 16 - 2^3 =$

$5 + 16 - 8 =$

$21 - 8 =$

13

Example 10 Evaluate $2^5 + 40 \div 10 - 3^2$.

Solution

$$\begin{aligned} 2^5 + 40 \div 10 - 3^2 &= 32 + 40 \div 10 - 9 \\ &= 32 + 4 - 9 \\ &= 27 \end{aligned}$$

Example 11 Evaluate $(9)(3) + 5^2 - 16 \div 2$.

Solution

$$\begin{aligned} (9)(3) + 5^2 - 16 \div 2 &= (9)(3) + 25 - 16 \div 2 \\ &= 27 + 25 - 8 \\ &= 52 - 8 \\ &= 44 \end{aligned}$$

1.7 EXERCISES

1. Given 7^4, what is the base? What is the exponent? ______________

2. Given 6^5, what is the base? What is the exponent? ______________

3. Express $13 \cdot 13 \cdot 13 \cdot 13$ in exponential form. ______________

4. Express $11 \cdot 11 \cdot 11$ in exponential form. ______________

5. Express $(4)(4)(4)(4)(4)(4)$ in exponential form. ______________

6. Express $(8)(8)(8)(8)(8)$ in exponential form. ______________

In problems 7–16, evaluate the given expression.

7. $2^6 =$ ______________ **8.** $3^3 =$ ______________

9. $5^3 =$ ______________ **10.** $6^2 =$ ______________

11. $10^5 =$ ______________ **12.** $10^4 =$ ______________

13. $11^2 =$ ______________ **14.** $8^3 =$ ______________

15. $2^7 =$ ______________ **16.** $3^5 =$ ______________

Express the written statements in problems 17–22 in exponential form.

17. twenty-six squared ________________

18. seventeen cubed ________________

19. thirteen cubed ________________

20. ten squared ________________

21. ten cubed ________________

22. twenty-five squared ________________

In problems 23–50 evaluate the given expression.

23. $6 - 2 + 3$ ________________

24. $9 \div 3 \cdot 5$ ________________

25. $18 - 7 + 3$ ________________

26. $26 + 8 - 15$ ________________

27. $100 - 2 \cdot 9$ ________________

28. $50 \div 2 + 5$ ________________

29. $2^4 - 10$ ________________

30. $15 \div 3 + 3^3$ ________________

31. $2 + 3(100) \div 25$ ________________

32. $28 \div 4 \cdot 2 + 6$ ________________

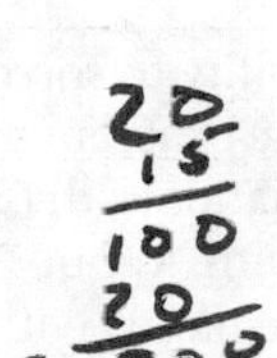

33. $(20)(15) \div 5$ ________________

34. $10 \div (2)(8)$ ________________

35. $3(2^4)$ ________________

36. $8(3^2)$ ________________

37. $4 + 77 \div 11 \cdot 2 - 5$ ________________

38. $31 - 60 \div 12 + 10$ ________________

39. $3^2 + (5)(5) - 8$ ________________

40. $90 \div 10 + 4^3 - 1$ ________________

41. $2^4 - 3^2 + 9 \div 3$ ________________

42. $36 \div 9 + 10^2 - 50$ ________________

43. $6 \cdot 3 \cdot 4 - 7^2 - 9$ ________________

44. $10^3 - 56 \cdot 8 + 2^5$ ________________

45. $18 \cdot 10^2 - 6^4 + 27^1$ ________________

46. $2^8 + 4^3 - (26)(12)$ ________________

47. $8^4 - 800 \div 8 - 3{,}000$ ________________

48. $37 + 216 \div 27 + 13^2 - 5^3$ ________________

49. $25{,}000 - 10^4 + 45 \div 9 \cdot 19$ ________________

50. $7{,}500 - (52)(86) + 3^6 \cdot 10^1$ ________________

ANSWERS

1. 7; 4 **2.** 6; 5 **3.** 13^4 **4.** 11^3 **5.** 4^6 **6.** 8^5 **7.** 64 **8.** 27 **9.** 125 **10.** 36 **11.** 100,000 **12.** 10,000 **13.** 121 **14.** 512 **15.** 128 **16.** 243 **17.** 26^2 **18.** 17^3 **19.** 13^3 **20.** 10^2 **21.** 10^3 **22.** 25^2 **23.** 7 **24.** 15 **25.** 14 **26.** 19 **27.** 82 **28.** 30 **29.** 6 **30.** 32 **31.** 14 **32.** 20 **33.** 60 **34.** 40 **35.** 48 **36.** 72 **37.** 13 **38.** 36 **39.** 26 **40.** 72 **41.** 10 **42.** 54 **43.** 14 **44.** 584 **45.** 531 **46.** 8 **47.** 996 **48.** 89 **49.** 15,095 **50.** 10,318

1.8 PROBLEM SOLVING

Four businessmen (A, B, C, and D) entered a meeting room. To become acquainted each person shook hands just once with everyone else. How many handshakes took place?

This is a problem expressed in words, or a "word problem." The ability to solve word problems is an important and often neglected skill. Let's discuss the problem of our four businessmen. If A shakes hands with everyone else, he shakes hands with B, C, and D. Here we have considered the case where B shakes hands with A, but B must still shake hands with C and D. There is one case left, C must shake hands with D. The answer to our introductory problem is that six handshakes took place.

A with B
A with C
A with D
B with C
B with D
C with D

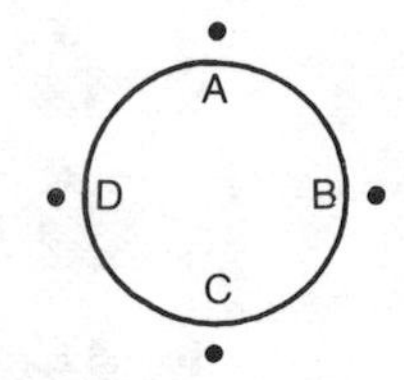

How does one become an effective problem solver? Generally this skill is gained through the effort involved in solving many problems. It is helpful to have a plan for attacking word problems. For some people this will be instinctive, but for other people this ability will be gained only through hard work. We cannot help the reader to gain skill, but we can help by suggesting a plan for attacking word problems. This is our next topic.

SOLVING WORD PROBLEMS

Step 1: Understand the problem.
- (a) Read all written parts of the problem until they are clearly understood. This has been accomplished if you can state the problem in your own words.
- (b) Do you understand what information is given?
- (c) Do you understand what information is missing or needed?

At this point, if you feel you are ready to solve the problem go to step 3. If you still feel that you do not have a grip on the problem go to step 2.

Step 2: Analyze the problem.
- (a) Draw a picture, a table, or a diagram of the problem.
- (b) Try a simpler problem.
- (c) Try a special case of the problem.
- (d) Do not be afraid to use trial and error.
- (e) Look for a pattern. Is our problem similar to another problem that we have already solved? Is it similar to a text example?

We realize that all of the suggestions in step 2 do not apply to every problem. In the beginning, if we are stuck we may use these suggestions as a checklist and proceed through them systematically.

Step 3: Try to solve the problem. This means do all the necessary computations.

Step 4: Check your answer.
- (a) Does it seem reasonable?
- (b) Does it check numerically?
- (c) Estimate numerical calculations as a check on your work.

Do not be afraid to try a problem again. Many problems are not solved on the first try. It is an amazing experience to find an insurmountable problem attacked on Monday suddenly become clear on Wednesday. The first two of the next examples show solutions to problems from earlier exercise sets.

Example 1 Raymond drove from Detroit to St. Louis for a sales meeting. On the way he stopped to see friends in Chicago. The distance from Detroit to Chicago is 435 km and from Chicago to St. Louis is 470 km. On the return trip he drove directly home and covered 819 km. How many kilometers did he drive round trip? How much shorter was the return trip?

Solution A little diagram is helpful.

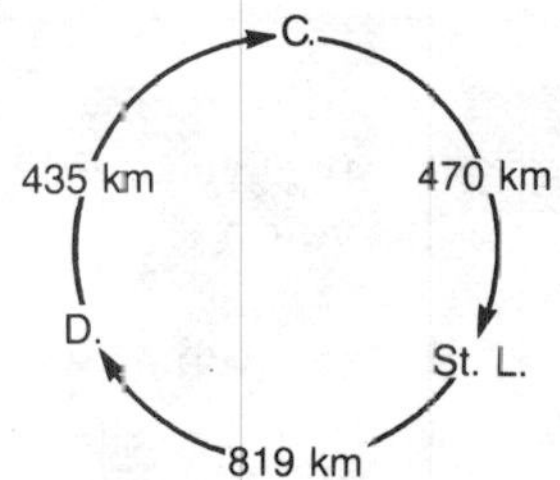

Round trip is 435 km + 470 km + 819 km = 1,724 km. From Detroit to Chicago to St. Louis is 435 km + 470 km = 905 km. The return trip is shorter by 905 km − 819 km = 86 km.

Example 2 In a particular year a company pays its employees an average salary of \$23,516. If the company has 463 employees, what is the total salary paid?

Solution Let's think about a simple analogy first. If the average of 4 mathematics tests is 70, what is the total of the 4 test grades? Answer: 4×70 or $(4)(70) = 280$. Now let's carry the thinking of this familiar little problem over into our present problem. The solution must be (463)(\$23,516) = \$10,887, 908.

Example 3 If a jogger, who runs at a constant speed, can travel 3 miles in 27 minutes, how long will it take him to travel 5 miles?

Solution Below is a figure to help visualize the problem.

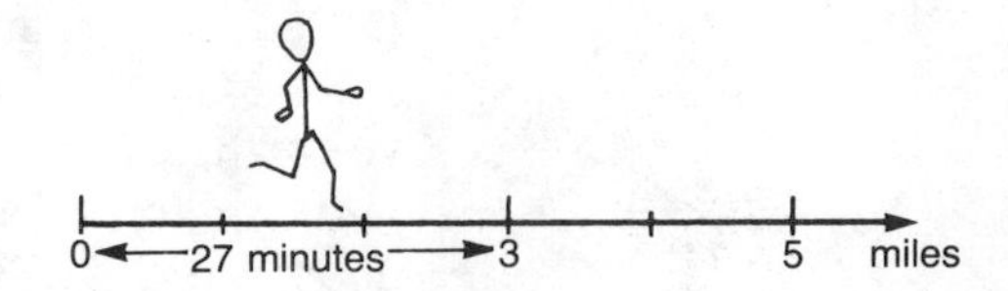

Let's estimate our answer. Double 3 miles gives 6 miles and double 27 minutes gives 54 minutes. Our answer for 5 miles must be less than 54 minutes. How long does it take him to run one mile? He can cover 1 mile in 27 minutes $\div$ 3 or in 9 minutes. Five miles must take 5×9 minutes or 45 minutes.

Example 4 A minibus has a load capacity of 2,000 pounds. On one trip there were five passengers, weighing 260 pounds, 132 pounds, 98 pounds, 290 pounds, and 155 pounds, respectively. Their luggage weighed 615 pounds. If the driver weighed 195 pounds, how much was the bus over or under capacity?

Solution We must add all of the passengers weights plus the luggage weight plus the driver's weight. Thus
$260 + 132 + 98 + 290 + 155 + 615 + 195 = 1,745$ pounds
The bus is then under capacity by $2,000 - 1,745 = 255$ pounds. How can we check this answer? If we again add all of the weights plus 255 pounds, the total should be 2,000 pounds. Try it!

1.8 Exercises

1. Allison, Joe, and Matthew ran for moderator of their town meeting. Allison received 1,709 votes, Joe received 1,682 votes, and Matthew received 1,438 votes.

   (a) What was the total number of votes cast? ______________

   (b) How many more votes did Allison receive than Matthew? ______________

2. If the population of Oklahoma was 3,219,786 in 1980 and 2,498,378 in 1970, what was the increase in population?

3. A clerk sold 39 tacks priced at 19 cents each. What was the total amount of this sale, in cents?

4. In a month's time (20 working days) a saleswoman's expenses were $3 per day for meals, $5 per day for gasoline, and a total monthly figure of $27 for miscellaneous expenses. What was the total expense bill for this saleswomen?

5. Alice has the following five history grades: 88, 92, 64, 80, and 76. What is her average grade?

6. Steve has a truck that can hold a load of 10,000 lb. He wants to carry crates of 3,217 lb, 5,618 lb, and 2,445 lb.
   (a) Can his truck carry this load?

   (b) By how much is his truck under (over) the truck's capacity?

7. Clinton makes $235 per week, Helen makes $195 per week, and Miguel makes $525 per week.
   (a) How much does each one earn in a year's time?

(b) What is the total salary, per year, for these three people?

____

8. The Ajax Chemical Company wishes to place 7, 176 lb, in equal amounts in 78 containers. How many pounds should be placed in each container?

____

9. Nancy must undertake a 1,984-mile cross-country trip. If she decides to stop after each quarter of the trip is complete, how many miles does she travel before making her first stop? How many miles remain after her first stop?

____

10. Four workers for Transwork Dial, Inc., must produce 1,298 parts in a week's time. If the first worker produces 350 parts, the second worker 375 parts, and the third worker 416 parts, how many parts remain for the fourth worker?

____

ANSWERS

1. (a) 4,829 votes; **(b)** 271 votes **2.** 721,408 **3.** 741 cents **4.** $187 **5.** 80 **6. (a)** no; total of 11, 280 **(b)** 1,280 pounds over capacity **7. (a)** Clinton, $12,220; Helen, $10,140; Miguel, $27,300 **(b)** $49,660 **8.** 92 lb **9 (a)** 496 miles **(b)** 1,488 miles **10.** 157 parts

FOCUS ON PROBLEM SOLVING

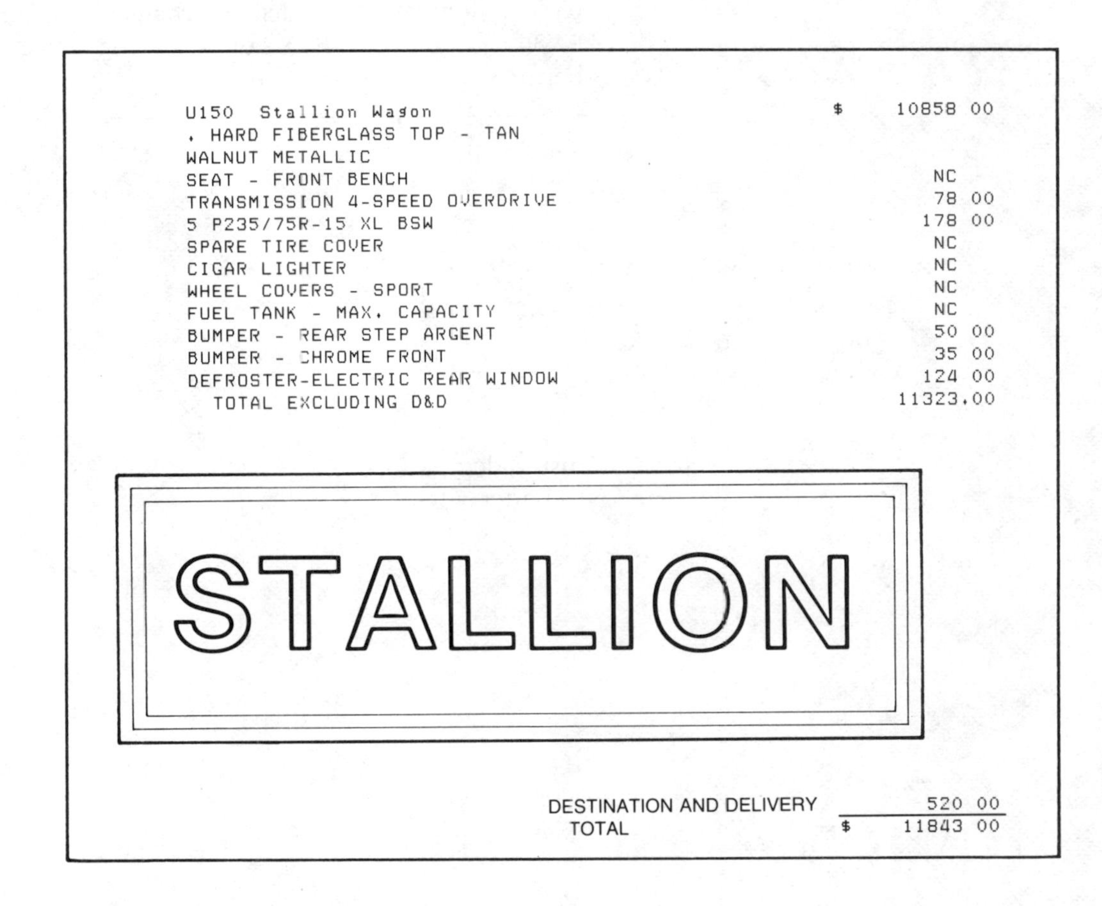

U150 Stallion Wagon	$	10858 00
. HARD FIBERGLASS TOP - TAN		
WALNUT METALLIC		
SEAT - FRONT BENCH		NC
TRANSMISSION 4-SPEED OVERDRIVE		78 00
5 P235/75R-15 XL BSW		178 00
SPARE TIRE COVER		NC
CIGAR LIGHTER		NC
WHEEL COVERS - SPORT		NC
FUEL TANK - MAX. CAPACITY		NC
BUMPER - REAR STEP ARGENT		50 00
BUMPER - CHROME FRONT		35 00
DEFROSTER-ELECTRIC REAR WINDOW		124 00
TOTAL EXCLUDING D&D		11323.00

STALLION

DESTINATION AND DELIVERY		520 00
TOTAL	$	11843 00

Two of the most important purchases many people make during their lifetime are the purchase of a home and an automobile. When a person decides to purchase an automobile, the first step is to choose a reputable dealer. The price of the automobile will be determined by its base price and the cost of any options the buyer may want plus destination and delivery charges. The total cost of the automobile will be indicated on a "sticker," as illustrated.

The base price of this Comet Stallion is $10,858. Standard equipment is indicated on the sticker by NC (no charge). Optional equipment is listed on the sticker with its cost. Destination and delivery charges are also indicated on the sticker ($520).

How much money must a potential buyer with $7,500 borrow to purchase this Comet Stallion with all of its options? How much must be borrowed if the dealer agrees to pay all destination and delivery charges?

Solution The buyer will look at the total sticker cost, which is determined by adding all of the sticker amounts:

$$\begin{array}{rr} & \$10,858 \\ & 78 \\ & 178 \\ & 50 \\ & 35 \\ & 124 \\ + & 520 \\ \hline \text{Total:} & \$11,843 \end{array}$$

The buyer must borrow \$11,843 − \$7,500 = \$4,343. For borrowing convenience, our purchaser borrows \$4,400. In the event the dealer pays all destination and delivery charges, our purchaser must borrow \$4,343 − \$520 = \$3,823. For borrowing convenience, our purchaser borrows \$3,900.

BUILDING YOUR MATH VOCABULARY

TERM	DEFINITION	EXAMPLE
Addend	A number being added in an addition problem.	8 ← addend + 6 ← addend 14
Base	The number being multiplied by itself in an exponential expression.	$3^5 = 243$ 3 is the base
Decimal System	This is our number system. It is based on expressing numbers in groups of 1: 10: 100: 1,000, and so on.	hundreds tens units 3 2 5
Difference	The answer to a subtraction problem.	25 − 10 15 ← difference
Digits	The first 10 whole numbers are called digits; they are 0, 1, 2, 3, 4, 5, 6, 7, 8, and 9.	30 is a number made up of two digits 3 and 0
Dividend	The number being divided by another number in a division problem.	3 $5\overline{)15}$ ← dividend
Divisor	The number doing the dividing in a division problem.	3 divisor → $5\overline{)15}$
Exponent	An exponent tells how many times another number called the base is to be multiplied by itself.	$3^5 = 243$ 5 is the exponent
Inequalities	$<$ is read "less than." $>$ is read "greater than." $\neq$ is read "not equal to."	$5<7$ 5 is less than 7 $9>4$ 9 is greater than 4. $6\neq5$ 6 is not equal to 5
Minuend	A number from which another number is subtracted.	25 ← minuend − 10 15
Multiplicand	A number being multiplied.	8 ← multiplicand × 7 56
Multiplier	A number doing the multiplying.	8 × 7 ← multiplier 56

TERM	DEFINITION	EXAMPLE
Natural or Counting Numbers	*N*: 1, 2, 3, 4, 5, . . .	12 is a natural number
Power	An exponential expression is called a power.	3^5 is read "3 to the fifth power"
Product	The answer to a multiplication problem.	$\begin{array}{r} 8 \\ \times\ 7 \\ \hline 56 \end{array} \leftarrow$ product
Quotient	The answer to a division problem.	$5\overline{)15}$ with 3 ← quotient
Remainder	The number left after you determine the maximum number of times the divisior is contained in the dividend.	$17\overline{)35}$ with quotient 2; 34; 1 ← remainder
Subtrahend	The number being taken away in a subtraction problem.	$\begin{array}{r} 25 \\ -\ 10 \\ \hline 15 \end{array}$ (10 ← subtrahend)
Sum	The answer to an addition problem.	$\begin{array}{r} 8 \\ +\ 6 \\ \hline 14 \end{array} \leftarrow$ sum
Whole Numbers	*W*: 0, 1, 2, 3, 4, 5, . . .	10 is a whole number

1

Chapter Review

In problems 1–4 express the number in written form.

1. 281 ____________

2. 76,599 ____________

3. 598,403 ____________

4. 2,073,547 ____________

In problems 5–8 express the number in numerical form.

5. seventy-one thousand, four hundred ninety-eight ____________

6. nine million, two hundred forty-eight thousand, nine hundred seven

____

7. eight hundred fifty-six ____________

8. fifty-five thousand, six hundred thirty-eight ____________

9. Multiply 53 by (a) 10, (b) 100, (c) 1,000. ____________

10. Multiply 618 by (a) 100, (b) 1,000, (c) 1,000,000. ____________

11. Multiply 29 by 10 and then multiply this product by 1,000. What is the final product?

____

In problems 12–27 perform the indicated operations.

12.
```
  7
  4
  9
  0
  2
+ 8
```

13.
```
  543
  784
  939
  618
+ 821
```

14.
```
  626
   72
  888
   63
+ 917
```

15.
```
  27,992
+ 26,084
```

16.
```
  716
-  98
```

17.
```
  43
- 15
```

18.
```
  478
- 342
```

19.
```
  6,702
- 5,944
```

20.
```
  703
×  20
```

21.
```
  9,623
×    25
```

22.
```
  287
×  36
```

23.
```
  515
× 487
```

24. $6\overline{)849}$ **25.** $85\overline{)5015}$ **26.** $48\overline{)9840}$

27. $762\overline{)85{,}598}$

In problems 28–31 write the given expression in exponential form.

28. 5×5 ________ **29.** $2 \times 2 \times 2 \times 2 \times 2$ ________

30. $7 \times 7 \times 7$ ________ **31.** $6 \times 6 \times 6 \times 6 \times 6 \times 6$ ________

In problems 32–35 evaluate the given expression.

32. 2^4 ________ **33.** 4^5 ________

34. 6^3 ________ **35.** 10^3 ________

In problems 36–43 evaluate the given expression.

36. $5 \times 2 - 6 + 2^2 =$ ________

37. $4 + 8 \div 2 - 5 =$ ________

38. $3^2 - 6 + 10 \div 5 \times 2 =$ ________

39. $50 \div 5^2 \times 6 - 3 =$ ________

40. $4^3 \div 8 - 4 \times 2 =$ ________

41. $36 - 9^2 \div 3^4 + 5 \times 3 =$ ________

42. $15 \div 5 \times 3 + 4^3 \times 10^2 =$ ________

43. $9 \times 3 + 5 - 2^3 \times 4 =$ ________

44. Roy Hagman has a record of the gasoline he used and the miles he drove his car the first year he owned it. His record showed 12,704 miles and 794 gallons of gasoline. How many miles did he average per gallon of gasoline?

45. Within 30 minutes a teller at Peoples Avenue Savings Bank took in deposits of $950, $50, $83, $529, and $200. In this same time frame she handled withdrawals of $80, $62, $465, and $93. What is the net amount that the bank took in (gave out)?

____

46. Last year Rose Marini paid $2,576 in real estate taxes; this year she is asked to pay $3,160. Find the amount of Rose's tax increase.

____

47. The Carter Paper Corporation has a monthly sales figure of $128,250. Their cost of materials for this month is $97,446, and their expenses are $5,608. If gross profit is sales less cost and net profit is gross profit less expenses, determine:

(a) the gross profit __________

(b) the net profit __________

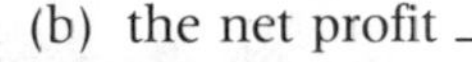

48. Lee Chang received the following grades in his high school physics course: 90, 86, 94, and 78. What is his grade average?

____

49. The quality control department of a large manufacturer checks parts on a daily basis and classifies them as good or bad. On Monday, 493 were found to be good and 2 bad; on Tuesday, 486 good and 14 bad; on Wednesday, 389 good and 11 bad; on Thursday 420 good and 18 bad; and on Friday 416 good and 27 bad.

(a) How many good parts were found during the week? __________

(b) How many bad parts were found during the week? __________

ANSWERS

1. two hundred eighty-one **2.** seventy-six thousand, five hundred ninety-nine **3.** five hundred ninety-eight thousand, four hundred three **4.** two million, seventy-three thousand, five hundred forty-seven **5.** 71,498 **6.** 9,248,9C7 **7.** 856 **8** 55,638 **9 (a)** 530 **(b)** 5,300 **(c)** 53,000 **10. (a)** 61,800 **(b)** 618,000 **(c)** 618,000,000 **11.** 290,000 **12.** 30 **13.** 3,705 **14.** 2,566 **15.** 54,076 **16.** 618 **17.** 28 **18.** 136 **19.** 758 **20.** 14,060 **21.** 240,575 **22.** 10,332 **23.** 250,805 **24.** 141, $R = 3$ **25.** 59 **26.** 205 **27.** 112, $R = 254$ **28.** 5^2 **29.** 2^5 **30.** 7^3 **31.** 6^6 **32.** 16 **33.** 1,024 **34.** 216 **35.** 1,000 **36.** 8 **37.** 3 **38.** 7 **39.** 9 **40** 0 **41.** 50 **42.** 6,409 **43.** 0 **44.** 16 **45.** took in $1,112 **46.** $584 **47. (a)** $30,804 **(b)** $25,196 **48.** 87 **49. (a)** 2,204 good parts **(b)** 72 bad parts

Warmup Test

Do not use a calculator.

1. Use the correct symbol ($<$ or $>$) to make each of the following statements true.

(a) 15 ____ 6

(b) 0 ____ 7

2. Express each number in written form.

(a) 15,012 ____________________

(b) 245,398 ____________________

3. Express each number in numerical form.

(a) sixty-five thousand, three hundred forty-six ____________

(b) three million, forty thousand, nine hundred seventy-eight ________

4. Round off each number to the indicated place.

(a) 4,298 to the nearest ten ____________

(b) 30,325 to the nearest thousand ____________

(c) 206,985 to the nearest ten thousand ____________

5. Add:

(a)
$$\begin{array}{r} 356 \\ 897 \\ +\ 688 \\ \hline \end{array}$$

(b)
$$\begin{array}{r} 95{,}406 \\ 2{,}976 \\ 112{,}689 \\ +\quad 4{,}008 \\ \hline \end{array}$$

(c) 1,046 + 812 + 92,687 + 476,350 + 78,908 = ____________

6. Subtract:

(a)
$$\begin{array}{r} 852 \\ -\ 648 \\ \hline \end{array}$$

(b)
$$\begin{array}{r} 47{,}061 \\ -\quad 9{,}846 \\ \hline \end{array}$$

(c)
$$\begin{array}{r} 64{,}000 \\ -\ 39{,}512 \\ \hline \end{array}$$

7. Multiply:

(a)
$$\begin{array}{r} 4{,}085 \\ \times\quad 29 \\ \hline \end{array}$$

(b)
$$\begin{array}{r} 32{,}561 \\ \times\quad 306 \\ \hline \end{array}$$

(c)
$$\begin{array}{r} 8{,}900 \\ \times\ 1{,}200 \\ \hline \end{array}$$

8. Divide:

(a) $56\overline{)2,688}$ (b) $67\overline{)13,936}$ (c) $254\overline{)18,669}$

9. Write each of the following in exponential form:

(a) $3 \times 3 \times 3$ ____ (b) $7 \times 7 \times 7 \times 7 \times 7 \times 7$ ____

10. Evaluate:

(a) 4^3 ____ (b) 2^5 ____

11. Evaluate:

(a) $2 \times 10 + 6 - 3^2$ ____ (b) $2^4 \times 3 + 12 \div 6$ ____

12. A salesman drove 160 miles on Monday, 212 miles on Tuesday, 89 miles on Wednesday, 157 miles on Thursday, and 197 miles on Friday. What was his average daily mileage for the week?

____

13. Tom makes 35 equal monthly payments of $145 on a 36-month car loan. The amount he must pay back is $5,200. How much does he pay the 36th month?

____

14. The balance in Debbie's bank account at the beginning of the month was $1,685. During the month she made withdrawals of $320, $462, $112, and $74 and made deposits of $228 and $65. What is her new balance?

____

15. Fuel oil produces 84,600 BTU per gallon. How many BTU are produced by 24 gallons of fuel oil?

____

Challenge Test

Do not use a calculator.

1. Use the correct symbol (< or >) to make each of the following statements true.

(a) 9 ____ 12 (b) 10 ____ 7

2. Express each number in written form.

(a) 9,461 ____________________

(b) 328,019 ____________________

3. Express each number in numerical form.

(a) one hundred twenty-eight thousand, four hundred sixty.

____

(b) five million, eight hundred fifty-six thousand, four hundred fifty-five

____

4. Round off each number to the indicated place.

(a) 6,738 to the nearest ten __________

(b) 55,801 to the nearest thousand __________

(c) 792,043 to the nearest ten thousand __________

5. Add:

(a)
$$\begin{array}{r} 276 \\ 981 \\ +\ 563 \\ \hline \end{array}$$

(b)
$$\begin{array}{r} 1{,}650 \\ 27{,}984 \\ 547{,}487 \\ +\ 319{,}056 \\ \hline \end{array}$$

(c) $776 + 1{,}804 + 263 + 196{,}762 + 87{,}705 =$ __________

6. Subtract:

(a)
$$\begin{array}{r} 912 \\ -\ 563 \\ \hline \end{array}$$

(b)
$$\begin{array}{r} 74{,}601 \\ -\ 38{,}590 \\ \hline \end{array}$$

(c)
$$\begin{array}{r} 55{,}000 \\ -\ 37{,}216 \\ \hline \end{array}$$

7. Multiply:

(a)
$$\begin{array}{r} 543 \\ \times\ 28 \\ \hline \end{array}$$

(b)
$$\begin{array}{r} 51{,}604 \\ \times\ 714 \\ \hline \end{array}$$

(c)
$$\begin{array}{r} 7{,}800 \\ \times\ 5{,}300 \\ \hline \end{array}$$

8. Divide:

(a) $72\overline{)6{,}120}$ (b) $48\overline{)19{,}440}$ (c) $245\overline{)31{,}916}$

9. Write each of the following expressions in exponential notation:

(a) $5 \cdot 5 \cdot 5 \cdot 5$ _____ (b) $(2)(2)(2)(2)(2)(2)(2)$ _____

10. Evaluate each of the following expressions:

(a) 3^4 _____ (b) 7^3 _____

11. Evaluate each of the following expressions:

(a) $9 \cdot 5 + 10 - 36 \div 3^2$ _____________

(b) $45 - 7 + 3^3 \div 9$ _____________

12. At the beginning of a trip the odometer of your car read 47,629 miles. At the end of the trip, the odometer read 48,793 miles. How many miles were traveled during the trip?

____

13. The history grades of four students are 76, 82, 91 and 67. What is the average grade?

____

14. A contest prize of $16,225 is divided equally among 5 people. What amount does each person receive?

____

15. A consumer makes a down payment on an item costing $4,210.
(a) If the down payment is $2,500, how much does she owe after making the down payment?

____

(b) Assuming that there are no interest charges, if she agrees to pay off the balance owed in 18 equal monthly payments, how much will she pay each month?

____

Fractions

CHAPTER OUTLINE

2–0 PRETEST

Solve the following problems. Do not use a calculator.

Problem	Section Reference	Learning Objectives
1. Classify each fraction as proper or improper. (a) $\frac{3}{2}$ __________ (b) $\frac{6}{7}$ __________ (c) $\frac{22}{25}$ __________ (d) $\frac{19}{14}$ __________	2.1	Recognize proper and improper fractions
2. Change these improper fractions to mixed numbers. (a) $\frac{17}{4}$ __________ (b) $\frac{20}{6}$ __________ (c) $\frac{100}{7}$ __________	2.2	Changing improper fractions to mixed numbers
3. Change these mixed numbers to improper fractions. (a) $6\frac{2}{3}$ __________ (b) $8\frac{4}{7}$ __________ (c) $15\frac{1}{4}$ __________	2.2	Changing mixed numbers to improper fractions
4. Reduce each of the following fractions to simplest or lowest terms. (a) $\frac{18}{45}$ ____ (b) $\frac{50}{35}$ ____ (c) $\frac{24}{44}$ ____	2.3	Reducing fractions to lowest terms
5. Raise each of the following fractions to higher terms. (a) $\frac{7}{8} = \frac{?}{32}$ ____ (b) $\frac{4}{3} = \frac{28}{?}$ ____ (c) $\frac{16}{42} = \frac{40}{?}$ ____	2.3	Raising fractions to higher terms
6. Add each of the following and write the answer in its simplest form. (a) $\frac{3}{8} + \frac{5}{6} =$ __________ (b) $\frac{1}{4} + \frac{3}{10} + \frac{4}{15} =$ __________ (c) $\frac{2}{3} + \frac{1}{7} + \frac{1}{12} =$ __________ (d) $3\frac{5}{6} + 7\frac{3}{10} =$ __________	2.4	Finding a least common denominator and adding fractions.

Problem	Section Reference	Learning Objectives
7. Subtract each of the following and write the answer in its simplest form. (a) $\frac{5}{6} - \frac{8}{15} =$ ____________ (b) $\frac{18}{35} - \frac{6}{20} =$ ____________ (c) $\frac{5}{12} - \frac{5}{28} =$ ____________ (d) $4\frac{1}{5} - 2\frac{2}{7} =$ ____________	2.5	Finding a least common denominator and subtracting fractions
8. Multiply each of the following and write the answer in its simplest form. (a) $\frac{7}{8} \times \frac{16}{35} =$ ____________ (b) $5\frac{1}{3} \times \frac{9}{32} =$ ____________ (c) $\frac{2}{5} \times \frac{10}{3} \times \frac{6}{14} =$ ____________ (d) $1\frac{2}{5} \times 3\frac{3}{4} =$ ____________	2.6	Multiplying fractions
9. Divide each of the following and write the answer in its simplest form. (a) $\frac{5}{6} \div \frac{25}{12} =$ ____________ (b) $4\frac{2}{7} \div 6 =$ ____________ (c) $3\frac{2}{3} \div 6\frac{3}{5} =$ ____________ (d) $\frac{\frac{3}{7}}{\frac{2}{5}} =$ ____________	2.7	Dividing fractions
10. When Mr. Perez started on a trip to visit his family the fuel tank of his car was filled with 24 gallons of gasoline. He used $\frac{3}{4}$ of a tank of gasoline. How many gallons of gasoline did he have left? ____________	2.8	Analyzing and solving fractional word problems
11. A salesman drove $3\frac{1}{2}$ hours on Monday, 4 hours on Tuesday, 5 hours on Wednesday, $5\frac{3}{4}$ hours on Thursday and $3\frac{3}{8}$ hours on Friday. How many hours did he drive? ____________	2.8	Analyzing and solving fractional word problems

Problem	Section Reference	Learning Objectives
12. What is the ratio of 24 square yards of lawn to 12 pounds of fertilizer? ______________	2–9	Writing ratios
13. If the ratio of men to women employed by a company is 7 to 9, how many men are employed if there are 126 women working for this company? ______________	2–9	Solving proportions

2.1 INTRODUCING FRACTIONS

You have polished one-half of your car, you are nine-tenths of the way through your history exam, or you have read five and one-third books. In each case you have completed only a portion of something. A bank raises its interest rate $\frac{1}{2}$ of 1%, a housewife needs $\frac{2}{3}$ of a cup of sugar, a football game is $\frac{3}{4}$ over, your car has depreciated $\frac{7}{8}$ of its original cost, or your local store is having a "$\frac{1}{4}$ off the original price" sale. These examples illustrate that fractions enter into our daily lives. Figure 2.1 depicts three different fractional parts of a circle.

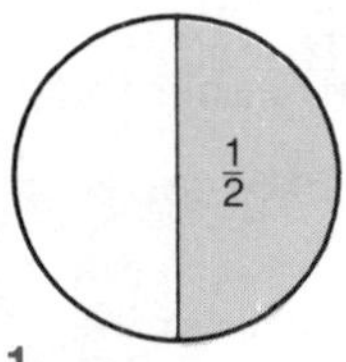

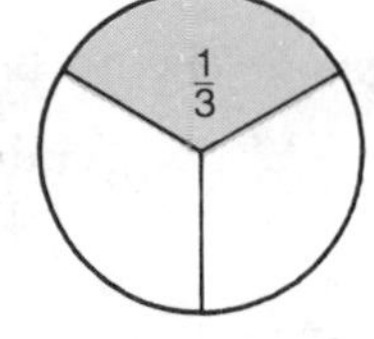

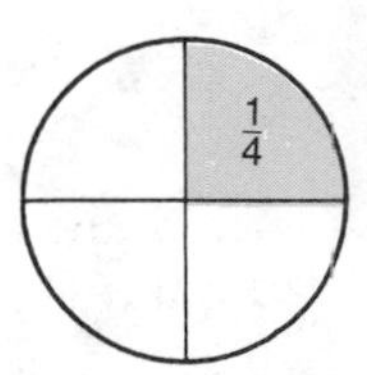

Figure 2.1

A **common fraction** is a quotient of two whole numbers. The number above the line "—" (a division bar or fraction line) is called the **numerator** and the number below the line is called the **denominator**.

$$\frac{5}{8} \begin{array}{l} \leftarrow \text{numerator} \\ \leftarrow \text{denominator} \end{array}$$

A **proper fraction** is one in which the numerator is less than the denominator. Some examples of proper fractions are:

$$\frac{1}{5}, \frac{3}{4}, \frac{5}{7}, \frac{11}{13}, \frac{7}{8}, \frac{241}{965}$$

An **improper fraction** is one in which the numerator is greater than or equal to the denominator. Some examples of improper fractions are:

$$\frac{3}{2}, \frac{8}{5}, \frac{6}{6}, \frac{11}{9}, \frac{7}{3}, \frac{50}{40}$$

Example 1 Can a number such as 8 be considered a fraction? Yes, because it can be expressed as $\frac{8}{1}$. Here 8 is the numerator and 1 is the denominator.

Because dividing a whole number by 1 does not change its value, it follows that all whole numbers are fractions.

Example 2 You are to take 375 objects and divide them into 15 equal piles. Express this division as a fraction. How many objects are in each pile?

Solution Our problem may be expressed by the fraction:

$$\frac{375}{15} = 375 \div 15 = 25 \text{ in each pile}$$

When you work with fractions, the denominator may refer to a number of equal parts that a quantity is divided into. In this case the numerator tells you how many of those equal parts you have. Thus, $\frac{3}{8}$ can mean that you have 3 of 8 equal parts.

Example 3 Each week the payroll clerk at Jenkins Auto Supply needs 8 hours to complete the company payroll. If the clerk works 3 hours on the payroll, $\frac{3}{8}$ of the payroll will be completed.

Fractions are a convenient way to express: (1) a comparison of one or more equal parts to a total number of equal parts; (2) a division between two whole numbers. In the case of division, it will always indicate that the numerator is to be divided by the denominator.

Example 4 The fraction $\frac{2}{7}$ means either that you have 2 of 7 equal parts or $2 \div 7$. Both interpretations are correct.

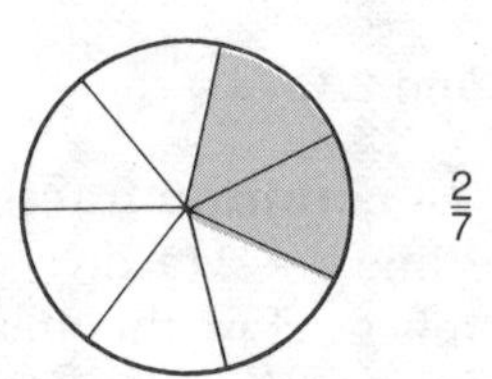

In the preceding chapter we explained how we can assign to each whole number a point on a number line. We continue the procedure by assigning a unique point on the number line to every fraction (proper or improper). Figure 2.2 shows this unique correspondence between some fractions and their corresponding points on the number line.

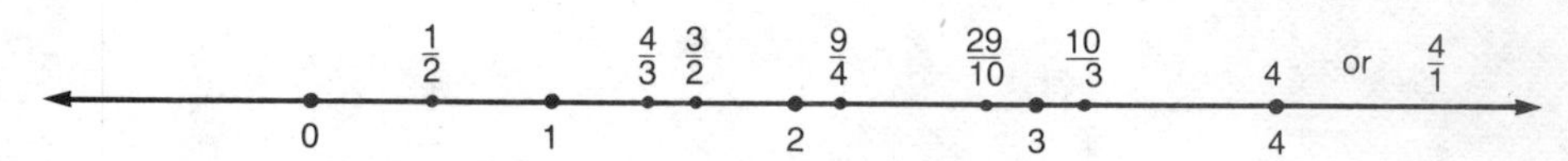

Figure 2.2

Given two fractions, or points, on the number line, the fraction farthest to the right is the largest. In Figure 2.2, observe that $\frac{3}{2} > \frac{4}{3}, \frac{29}{10} > \frac{1}{2}$ and $\frac{4}{1} > \frac{29}{10}$. You may write either $\frac{3}{2} > \frac{1}{2}$ or $\frac{1}{2} < \frac{3}{2}$.

The engineering feats of the ancient Egyptians clearly show their use of the mathematical tools we've been using in this chapter. *(Historical Pictures Services, Chicago)*

Historical records verify that the Egyptians worked with fractions. At first they used fractions with numerators of 1, such as $\frac{1}{2}$, $\frac{1}{3}$, and so forth. Later, other fractions were added to their set of numerals. The Babylonians used fractions with denominators of 60 or in powers of 60. Since our units of time were borrowed from the Babylonians, one minute is $\frac{1}{60}$ of an hour and one second is $\frac{1}{60}$ a minute or $\frac{1}{3{,}600}$ of an hour. The Romans made little use of fractions; most of their fractions had denominators of 12 and were associated with measurements. Many different notations have been used for fractions. The current notational form for fractions, such as two-thirds, $\frac{2}{3}$, came into general use in the sixteenth century.

Equality and Inequality

We can illustrate that $\frac{3}{4}$, $\frac{6}{8}$, and $\frac{12}{16}$ represent the same part of a whole and hence are equal fractions.

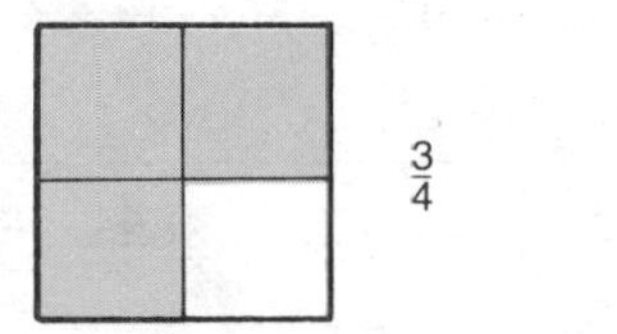

$\frac{3}{4}$

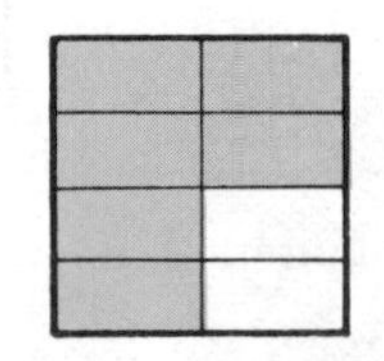

$\frac{6}{8}$

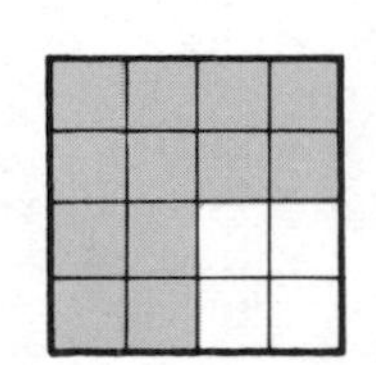

$\frac{12}{16}$

It is more difficult to visualize that $\frac{12}{18} = \frac{8}{12}$. Points representing $\frac{3}{7}$ and $\frac{6}{13}$ on the number line lie close together and a visual determination of inequality is difficult. Fortunately, we have an easy way to solve the problem of equality and inequality.

> Two fractions, $\frac{a}{b}$ and $\frac{c}{d}$, are equal whenever the products $a \cdot d$ and $b \cdot c$ are equal. If the product $a \cdot d$ is greater than the product $b \cdot c$, then $\frac{a}{b} > \frac{c}{d}$. If the product $a \cdot d$ is less than the product $b \cdot c$, then $\frac{a}{b} < \frac{c}{d}$.

Example 5 Does $\frac{3}{4} = \frac{1}{2}$?

Solution $a = 3$, $b = 4$, $c = 1$, and $d = 2$
$3 \cdot 2 \ ? \ 4 \cdot 1$
$6 > 4$
Thus the fractions are not equal; $\frac{3}{4} > \frac{1}{2}$.

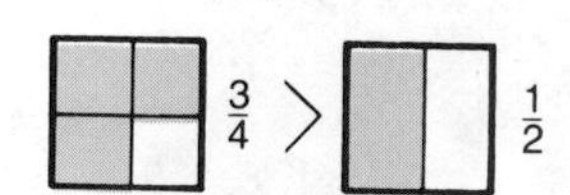

Example 6 Does $\frac{12}{18} = \frac{8}{12}$?

Solution $a = 12$, $b = 18$, $c = 8$, and $d = 12$
$12 \cdot 12 \ ? \ 18 \cdot 8$
$144 = 144$
Thus $\frac{12}{18} = \frac{8}{12}$.

Example 7 Does $\frac{3}{7} = \frac{6}{13}$?

Solution $a = 3$, $b = 7$, $c = 6$, and $d = 13$
$3 \cdot 13 \ ? \ 7 \cdot 6$
$39 < 42$
Thus the fractions are not equal; $\frac{3}{7} < \frac{6}{13}$.

Example 8 Does $\frac{15}{24} = \frac{30}{48}$?

Solution $a = 15$, $b = 24$, $c = 30$, and $d = 48$
$15 \cdot 48 \ ? \ 24 \cdot 30$
$720 = 720$
Thus $\frac{15}{24} = \frac{30}{48}$.

2.1 EXERCISES

1. Given the fraction $\frac{11}{13}$, identify the numerator and denominator.

___

2. Given the fraction $\frac{15}{17}$, identify the numerator and denominator.

___

3. In Figure 2.2, is $\frac{9}{4}$ greater than or less than $\frac{4}{3}$?

___

4. In Figure 2.2, is $\frac{29}{10}$ greater than or less than $\frac{10}{3}$?

___

Problems 5–8 deal with the points on the following number line. Insert the symbol > or <.

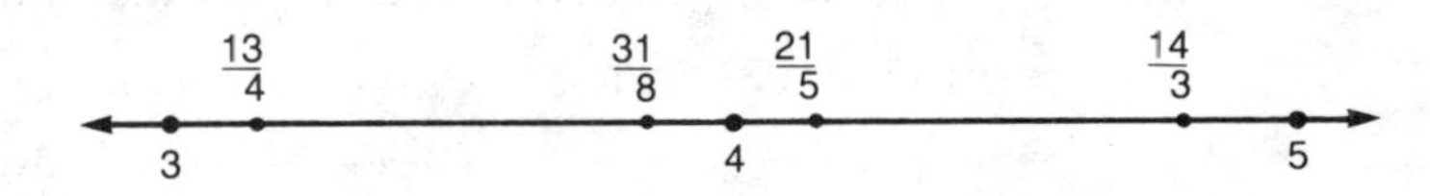

5. $\frac{13}{4}$ ____ $\frac{14}{3}$

6. $\frac{21}{5}$ ____ $\frac{31}{8}$

7. $\frac{14}{3}$ ____ $\frac{21}{5}$

8. $\frac{31}{8}$ ____ $\frac{14}{3}$

9. (true or false) If $\frac{1}{3} < \frac{5}{8}$, then $\frac{1}{3}$ must lie on the number line to the left of $\frac{5}{8}$.

____

10. (true or false) If $\frac{9}{10} > \frac{4}{5}$, then $\frac{9}{10}$ must lie on the number line to the left of $\frac{4}{5}$.

____

11. Express 7 as a fraction. What is the numerator? The denominator?

____

12. Express 4 as a fraction. What is the numerator? The denominator?

____

13. If $\frac{1}{2} < \frac{3}{4}$ and $\frac{3}{4} < \frac{7}{8}$, what is the relationship between $\frac{1}{2}$ and $\frac{7}{8}$?

____

14. If $\frac{2}{3} > \frac{2}{5}$ and $\frac{2}{5} > \frac{1}{4}$, what is the relationship between $\frac{2}{3}$ and $\frac{1}{4}$?

____

In problems 15–32 classify each fraction as proper or improper.

15. $\frac{5}{8}$ ____________

16. $\frac{7}{6}$ ____________

17. $\frac{4}{4}$ ____________

18. $\frac{98}{100}$ ____________

19. $\frac{100}{98}$ ____________

20. $\frac{8}{7}$ ____________

21. 2 ____________

22. $\frac{1}{3}$ ____________

23. $\frac{9}{7}$ ____________

24. $\frac{21}{25}$ ____________

25. $\frac{8}{11}$ ____________

26. 6 ____________

27. $\frac{103}{97}$ ____________

28. $\frac{4}{9}$ ____________

29. $\frac{51}{26}$ ____________

30. $\frac{37}{48}$ ____________

31. $\frac{1{,}017}{948}$ ____________

32. $\frac{767}{2{,}548}$ ____________

In problems 33–48 determine whether the given fractions are equal. If the fractions are not equal, state which one is larger.

33. $\frac{9}{12}, \frac{3}{4}$ ______ **34.** $\frac{5}{4}, \frac{10}{8}$ ______

35. $\frac{2}{5}, \frac{3}{7}$ ______ **36.** $\frac{5}{8}, \frac{15}{21}$ ______

37. $\frac{35}{40}, \frac{15}{18}$ ______ **38.** $\frac{9}{15}, \frac{21}{35}$ ______

39. $\frac{16}{10}, \frac{9}{6}$ ______ **40.** $\frac{6}{15}, \frac{10}{25}$ ______

41. $\frac{3}{4}, \frac{300}{400}$ ______ **42.** $\frac{22}{33}, \frac{44}{77}$ ______

43. $\frac{3}{9}, \frac{6}{18}$ ______ **44.** $\frac{4}{10}, \frac{3}{7}$ ______

45. $\frac{8}{12}, \frac{10}{18}$ ______ **46.** $\frac{14}{12}, \frac{8}{7}$ ______

47. $\frac{30}{35}, \frac{24}{28}$ ______ **48.** $\frac{24}{21}, \frac{72}{50}$ ______

ANSWERS

1. numerator = 11; denominator = 13 **2.** numerator = 15; denominator = 17 **3.** $\frac{9}{4} > \frac{4}{3}$ **4.** $\frac{29}{10} < \frac{10}{3}$ **5.** < **6.** > **7.** > **8.** < **9.** true **10.** false **11.** $\frac{7}{1}$; 7; 1 **12.** $\frac{4}{1}$; 4; 1 **13.** $\frac{1}{2} < \frac{7}{8}$ **14.** $\frac{2}{3} > \frac{1}{4}$ **15.** proper **16.** improper **17.** improper **18.** proper **19.** improper **20.** improper **21.** improper **22.** proper **23.** improper **24.** proper **25.** proper **26.** improper **27.** improper **28.** proper **29.** improper **30.** proper **31.** improper **32.** proper **33.** equal **34.** equal **35.** $\frac{2}{5} < \frac{3}{7}$ **36.** $\frac{5}{8} < \frac{15}{21}$ **37.** $\frac{35}{40} > \frac{15}{18}$ **38.** equal **39.** $\frac{16}{10} > \frac{9}{6}$ **40.** equal **41.** equal **42.** $\frac{22}{33} > \frac{44}{77}$ **43.** equal **44.** $\frac{4}{10} < \frac{3}{7}$ **45.** $\frac{8}{12} > \frac{10}{18}$ **46.** $\frac{14}{12} > \frac{8}{7}$ **47.** equal **48.** $\frac{24}{21} < \frac{72}{50}$

2.2 MIXED NUMBERS

Numbers such as $3\frac{1}{2}$, $9\frac{3}{5}$, and $2\frac{8}{13}$ are called **mixed numbers.** A mixed number consists of a whole number added to a proper fraction. The addition sign is understood but not written. Thus, $3\frac{1}{2}$ means $3 + \frac{1}{2}$. Further, $19\frac{3}{4}$ may be written as $19 + \frac{3}{4}$ and $6\frac{7}{8}$ may be written as $6 + \frac{7}{8}$.

Example 1 (a) $5 + \frac{2}{3}$ may be shortened to $5\frac{2}{3}$.
(b) $6\frac{4}{7}$ denotes $6 + \frac{4}{7}$.

Converting Mixed Numbers to Improper Fractions

We can always change a mixed number to an improper fraction greater than 1. To convert $3\frac{1}{2}$ to an improper fraction, multiply the whole number, 3, by the denominator, 2, and add the numerator, 1, to obtain the new numerator, 7. Place this numerator over the denominator, 2. Thus

$$3\frac{1}{2} = \frac{(3 \times 2) + 1}{2} = \frac{6 + 1}{2} = \frac{7}{2}$$

> *To convert a mixed number to an improper fraction,* multiply the whole number by the denominator of the fraction and add the numerator of the fraction. This will give you your new numerator; your denominator is the denominator of the fractional part of the mixed number.

Example 2 Convert $1\frac{3}{5}$ to an improper fraction.

Solution $\boxed{\frac{1}{5}\,\frac{1}{5}\,\frac{1}{5}\,\frac{1}{5}\,\frac{1}{5}} + \boxed{\frac{1}{5}\,\frac{1}{5}\,\frac{1}{5}} = \frac{8}{5}$

$$1\frac{3}{5} = \frac{(1 \times 5) + 3}{5} = \frac{5 + 3}{5} = \frac{8}{5}$$

Example 3 Convert $6\frac{3}{4}$ to an improper fraction.

Solution

$$6\frac{3}{4} = \frac{(6 \times 4) + 3}{4} = \frac{24 + 3}{4} = \frac{27}{4}$$

Example 4 Convert $12\frac{5}{8}$ to an improper fraction.

Solution

$$12\frac{5}{8} = \frac{(12 \times 8) + 5}{8} = \frac{96 + 5}{8} = \frac{101}{8}$$

Converting Improper Fractions to Mixed Numbers

To convert the improper fraction $\frac{27}{4}$ to a mixed number, divide the denominator into the numerator. This yields the whole number portion of your answer, 6. The remainder is the numerator of the fractional portion of your mixed number, and the denominator of the fractional portion is the same as the original denominator, 4. Thus $\frac{27}{4} = 6\frac{3}{4}$.

> *To convert an improper fraction to a mixed number,* divide the denominator into the numerator. This yields the whole number portion of your answer. The remainder, if any, is the numerator of your fraction, and the denominator of the improper fraction is the denominator of the fractional part of the mixed number.

Example 5 Convert $\frac{7}{4}$ to a mixed number.

Solution Carry out the indicated division.

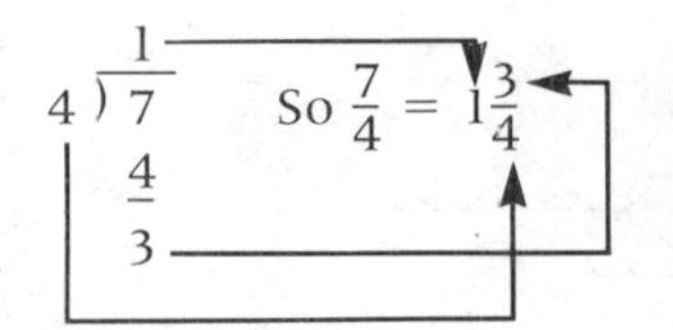

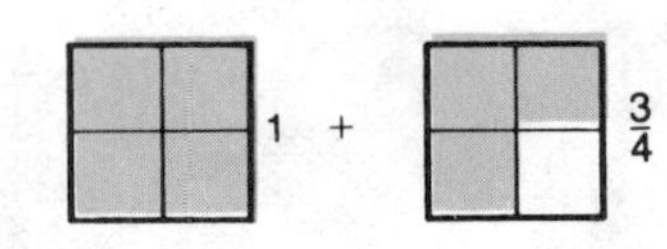

Example 6 Convert $\frac{39}{5}$ to a mixed number.

Solution

$$\begin{array}{r} 7 \\ 5\overline{)39} \\ \underline{35} \\ 4 \end{array} \qquad \text{So } \frac{39}{5} = 7\frac{4}{5}$$

Example 7 Convert $\frac{47}{8}$ to a mixed number.

Solution

$$\begin{array}{r} 5 \\ 8\overline{)47} \\ \underline{40} \\ 7 \end{array} \qquad \text{So } \frac{47}{8} = 5\frac{7}{8}$$

2.2 EXERCISES

In problems 1–4 identify the whole-number portion and the fractional portion of the given mixed number.

1. $4\frac{2}{5}$ __________ **2.** $6\frac{7}{8}$ __________

3. $9\frac{1}{10}$ __________ **4.** $5\frac{3}{7}$ __________

In problems 5–24 express the given fraction as a whole or a mixed number.

5. $\frac{15}{3}$ __________ **6.** $\frac{9}{2}$ __________

7. $\frac{7}{3}$ __________ **8.** $\frac{68}{3}$ __________

9. $\frac{45}{4}$ __________ **10.** $\frac{13}{5}$ __________

11. $\frac{17}{4}$ __________ **12.** $\frac{300}{100}$ __________

13. $\frac{14}{3}$ __________ **14.** $\frac{17}{5}$ __________

15. $\frac{19}{5}$ __________ **16.** $\frac{42}{8}$ __________

17. $\frac{8}{4}$ __________ **18.** $\frac{28}{7}$ __________

19. $\frac{15}{9}$ __________ **20.** $\frac{88}{15}$ __________

21. $\frac{172}{35}$ __________ **22.** $\frac{196}{42}$ __________

23. $\frac{206}{45}$ __________ **24.** $\frac{154}{38}$ __________

In problems 25–44 change each mixed number to an improper fraction.

25. $5\frac{1}{2}$ __________ **26.** $3\frac{2}{7}$ __________

27. $2\frac{1}{5}$ __________ **28.** $4\frac{3}{4}$ __________

29. $6\frac{6}{11}$ __________ **30.** $7\frac{2}{9}$ __________

31. $2\frac{5}{6}$ __________ **32.** $4\frac{1}{8}$ __________

33. $11\frac{5}{9}$ __________ **34.** $5\frac{5}{7}$ __________

35. $7\frac{1}{3}$ __________ 36. $11\frac{2}{7}$ __________

37. $7\frac{3}{4}$ __________ 38. $3\frac{5}{6}$ __________

39. $2\frac{3}{5}$ __________ 40. $9\frac{1}{2}$ __________

41. $10\frac{5}{6}$ __________ 42. $4\frac{5}{12}$ __________

43. $15\frac{7}{8}$ __________ 44. $16\frac{9}{10}$ __________

ANSWERS

1. $4; \frac{2}{5}$ **2.** $6; \frac{7}{8}$ **3.** $9; \frac{1}{10}$ **4.** $5; \frac{3}{7}$ **5.** 5 **6.** $4\frac{1}{2}$ **7.** $2\frac{1}{3}$ **8.** $22\frac{2}{3}$ **9.** $11\frac{1}{4}$
10. $2\frac{3}{5}$ **11.** $4\frac{1}{4}$ **12.** 3 **13.** $4\frac{2}{3}$ **14.** $3\frac{2}{5}$ **15.** $3\frac{4}{5}$ **16.** $5\frac{1}{4}$ **17.** 2 **18.** 4
19. $1\frac{2}{3}$ **20.** $5\frac{13}{15}$ **21.** $4\frac{32}{35}$ **22.** $4\frac{2}{3}$ **23.** $4\frac{26}{45}$ **24.** $4\frac{1}{19}$ **25.** $\frac{11}{2}$ **26.** $\frac{23}{7}$
27. $\frac{11}{5}$ **28.** $\frac{19}{4}$ **29.** $\frac{72}{11}$ **30.** $\frac{65}{9}$ **31.** $\frac{17}{6}$ **32.** $\frac{33}{8}$ **33.** $\frac{104}{9}$ **34.** $\frac{40}{7}$
35. $\frac{22}{3}$ **36.** $\frac{79}{7}$ **37.** $\frac{31}{4}$ **38.** $\frac{23}{6}$ **39.** $\frac{13}{5}$ **40.** $\frac{19}{2}$ **41.** $\frac{65}{6}$ **42.** $\frac{53}{12}$
43. $\frac{127}{8}$ **44.** $\frac{169}{10}$

2.3 REDUCING FRACTIONS TO LOWEST TERMS

A fraction is said to be in **simplest form** or in **lowest terms** when its numerator and denominator have no common divisor greater than 1. The fractions $\frac{3}{4}, \frac{5}{6},$ and $\frac{7}{8}$ are expressed in lowest terms. On the other hand, the fraction $\frac{6}{9}$ is not expressed in lowest terms because both its numerator and its denominator may be divided by 3. When $\frac{6}{9}$ is expressed in lowest terms we obtain $\frac{2}{3}$. See Figure 2.3.

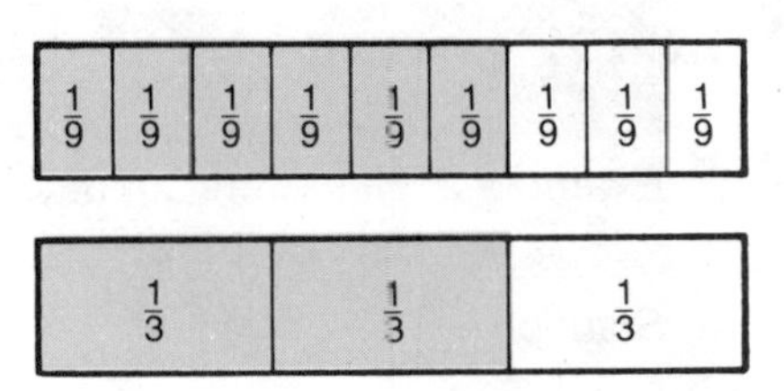

Figure 2.3

The process of expressing a fraction in lowest terms is called **reducing the fraction.** It is done by dividing both the numerator and denominator by the same nonzero number called a **common divisor.** This process can be simplified by determining the largest or **greatest common divisor,** the GCD, of the numerator and denominator. Finding the greatest common divisor can often be accomplished by inspection. When this is not possible, the following discussion will be helpful.

Prime Factors

A number greater than 1 that has no other divisors except itself and 1 is called a **prime number** or simply a "prime." From this definition, we see that 2 is the smallest prime number and the number 35 is not a prime because it is divisible by 5 and 7. The complete list of primes from 1 to 100 is:

2, 3, 5, 7, 11, 13, 17, 19, 23, 29, 31, 37, 41, 43, 47, 53, 59, 61, 67, 71, 73, 79, 83, 89, and 97

Numbers greater than 1 such as 30, 72, and 140, that are not prime possess prime divisors or factors. To find the smallest such prime divisor, divide your number successively by 2, 3, 5, 7, 11, and so on until you determine the first prime divisor. Additional prime divisors, if any, may be found by repeating this trial-and-error process using the quotient from the previous step. Let's illustrate this trial-and-error process for the number 30. Thirty is divisible by 2 ($30 \div 2 = 15$). In turn, 15 is divisible by 3 ($15 \div 3 = 5$). At this point, we observe that 5 is a prime. Hence, $30 = 2 \times 3 \times 5$. You have located your last prime factor when the quotient in this process is prime. As illustrated below, a **factor tree** is helpful in determining the prime divisors of a number.

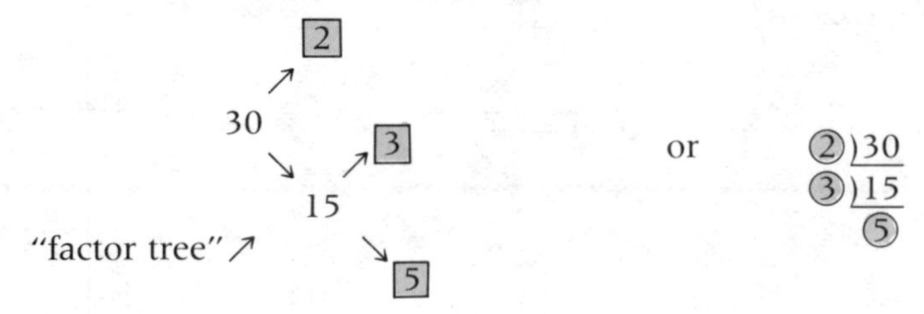

Thus $30 = 2 \times 3 \times 5$; prime factorization of 30.

Example 1 Determine all of the prime factors of 72. That is, determine its prime factorization.

Solution

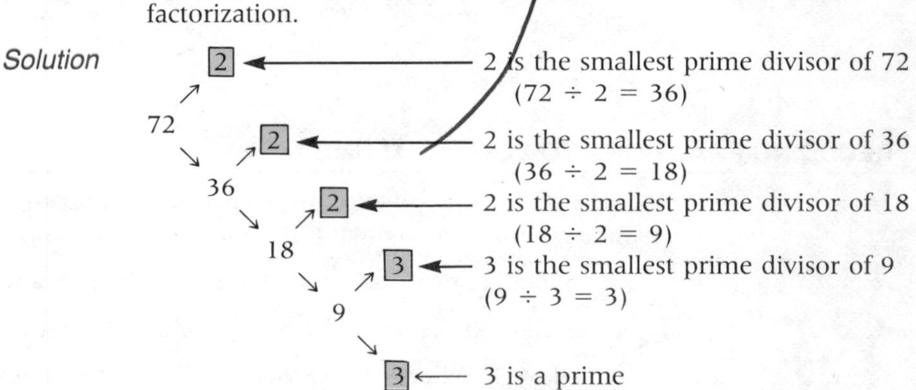

Thus $72 = 2 \times 2 \times 2 \times 3 \times 3$; prime factorization of 72

Example 2 Determine all of the prime factors of 140. That is, determine its prime factorization.

Solution

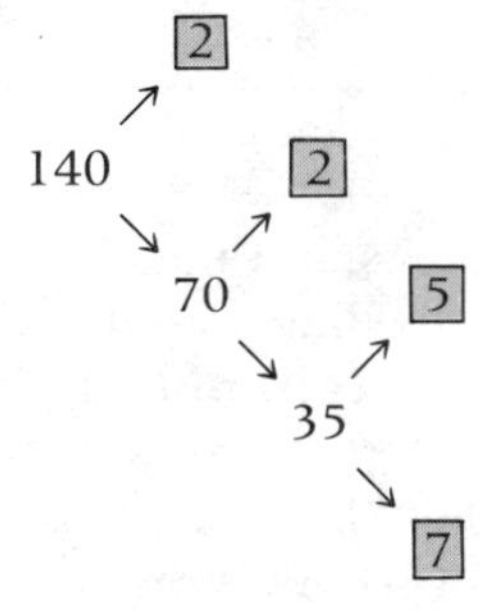

Thus $140 = 2 \times 2 \times 5 \times 7$; prime factorization of 140

Example 3 Find the prime factorization of 273.

Solution

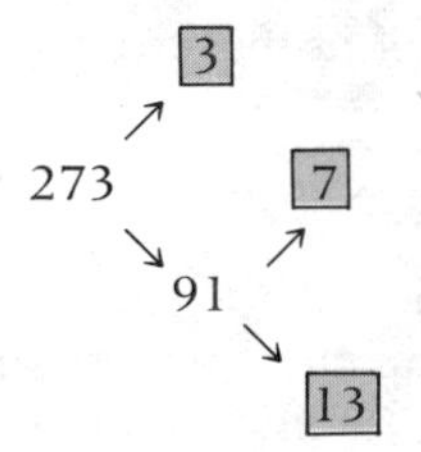

Thus $273 = 3 \times 7 \times 13$; prime factorization of 273.

The Greatest Common Divisor, GCD

Prime divisors, also called **prime factors,** are helpful in finding the greatest common divisor of a collection of whole numbers. *The GCD is the largest number that will divide into (remainder of 0) each number in the collection.*

To Find the GCD of Two or More Numbers

1. Write each number as a product of its prime factors.
2. List all prime factors **common** to the numbers.
3. Determine the **least** number of times each of the common factors is used in any one of the numbers.
4. The GCD is the product of all factors in step 3.

Example 4a Find the GCD of 18 and 24.

Solution
step 1: $18 = 2 \times 3 \times 3$
$24 = 2 \times 2 \times 2 \times 3$
step 2: 2 and 3
step 3: 2, 3
step 4: The GCD of 18 and 24 is $2 \times 3 = 6$

Example 4b Reduce $\frac{18}{24}$ to lowest terms

Solution The GCD of 18 and 24 is 6. Divide both numerator and denominator by 6 to obtain:

$$\frac{18}{24} = \frac{18 \div 6}{24 \div 6} = \frac{3}{4}$$

$\frac{18}{24}$ and $\frac{3}{4}$ are called **equivalent fractions.**

Equivalent fractions are fractions having the same value.

Example 5a Find the GCD of 48 and 60.

Solution
step 1: $48 = 2 \times 2 \times 2 \times 2 \times 3$
$60 = 2 \times 2 \times 3 \times 5$
step 2: 2 and 3
step 3: 2×2, 3
step 4: The GCD of 48 and 60 is $2 \times 2 \times 3 = 12$.

Example 5b Reduce $\frac{48}{60}$ to lowest terms.

Solution The GCD of 48 and 60 is 12. Divide both numerator and denominator by 12 to obtain:

$$\frac{48}{60} = \frac{48 \div 12}{60 \div 12} = \frac{4}{5}$$

$\frac{48}{60}$ and $\frac{4}{5}$ are equivalent fractions.

Example 6a Find the GCD of 210 and 945.

Solution As soon as you can condense steps, do so.
$210 = 2 \times 3 \times 5 \times 7$
$945 = 3 \times 3 \times 3 \times 5 \times 7$
The GCD of 210 and 945 is $3 \times 5 \times 7 = 105$.

Example 6b Reduce $\frac{210}{945}$ to lowest terms.

Solution $\frac{210}{945} = \frac{210 \div 105}{945 \div 105} = \frac{2}{9}$; $\frac{210}{945}$ and $\frac{2}{9}$ are equivalent fractions.

Example 7 Find the GCD of 6, 9, and 15.

Solution

step 1: $6 = 2 \times 3$
$9 = 3 \times 3$
$15 = 3 \times 5$

step 2: 3
step 3: 3
step 4: The GCD of 6, 9, and 15 is 3.

Raising a Fraction to Higher Terms

There are times when you will find it necessary to raise a fraction to "higher terms"—that is, multiply both the numerator and denominator by the same nonzero number. *Just as reducing a fraction to lowest terms does not change the value of a fraction, raising a fraction to higher terms does not change its value.*

Example 8 Convert $\frac{5}{8}$ to twenty-fourths. That is, find the missing numerator that will allow you to express $\frac{5}{8}$ as some number over 24.

Solution Observe that 8 times 3 gives 24. Hence the desired numerator must be 5 times 3 or 15.

Thus $\frac{5}{8} = \frac{15}{24}$.

Once again, $\frac{5}{8} = \frac{15}{24}$ are said to be equivalent fractions.

Example 9 Given $\frac{1}{2} = \frac{?}{8}$, what is the missing numerator?

Solution Observe that 2 times 4 gives 8. Hence the desired numerator must be 1×4 or 4.

Thus $\frac{1}{2} = \frac{4}{8}$.

Example 10 Given $\frac{3}{7} = \frac{9}{?}$, what is the missing denominator?

Solution Observe that $3 \times 3 = 9$. So the missing denominator must be 21.

Thus $\frac{3}{7} = \frac{9}{21}$.

Example 11 Given $\frac{7}{12} = \frac{70}{?}$, what is the missing denominator?

Solution Observe that $7 \times 10 = 70$. So the missing denominator must be $12 \times 10 = 120$.

Thus $\frac{7}{12} = \frac{70}{120}$.

> In summary, the value of a fraction is unchanged if both the numerator and the denominator are multiplied (or divided) by the same nonzero number.

2.3 EXERCISES

1. State all the primes between 40 and 50. ______

2. Why is 2 the only even prime number? ______

In problems 3–22 break the number down into a product of prime divisors (factors).

3. 44 ______
4. 36 ______
5. 12 ______
6. 40 ______
7. 42 ______
8. 55 ______
9. 20 ______
10. 22 ______
11. 21 ______
12. 33 ______
13. 63 ______
14. 120 ______
15. 72 ______
16. 81 ______
17. 418 ______
18. 396 ______
19. 570 ______
20. 1,875 ______
21. 1,596 ______
22. 2,584 ______

In problems 23–32 find the GCD for the given collection of numbers.

23. 10, 25 ________ **24.** 16, 20 ________

25. 36, 54 ________ **26.** 44, 66 ________

27. 40, 140 ________ **28.** 84, 210 ________

29. 90, 420 ________ **30.** 105, 462 ________

31. 49, 81 ________ **32.** 26, 77 ________

In problems 33–64 reduce the given fractions to lowest terms.

33. $\frac{4}{10}$ ________ **34.** $\frac{7}{21}$ ________

35. $\frac{8}{40}$ ________ **36.** $\frac{15}{20}$ ________

37. $\frac{36}{60}$ ________ **38.** $\frac{18}{48}$ ________

39. $\frac{17}{34}$ ________ **40.** $\frac{50}{70}$ ________

41. $\frac{42}{52}$ ________ **42.** $\frac{120}{330}$ ________

43. $\frac{26}{39}$ ________ **44.** $\frac{128}{260}$ ________

45. $\frac{100}{250}$ ________ **46.** $\frac{160}{450}$ ________

47. $\frac{25}{150}$ ________ **48.** $\frac{320}{480}$ ________

49. $\frac{20}{48}$ ________ **50.** $\frac{39}{65}$ ________

51. $\frac{66}{420}$ ________ **52.** $\frac{21}{70}$ ________

53. $\frac{60}{90}$ ________ **54.** $\frac{105}{120}$ ________

55. $\frac{77}{55}$ ________ **56.** $\frac{132}{60}$ ________

57. $\frac{18}{78}$ ________ **58.** $\frac{52}{103}$ ________

59. $\frac{9}{41}$ ________ **60.** $\frac{51}{119}$ ________

61. $\frac{45}{50}$ ________ **62.** $\frac{33}{18}$ ________

63. $\frac{64}{48}$ ________ **64.** $\frac{72}{100}$ ________

In problems 65–76 raise the fractions to higher terms by finding the missing numerator.

65. $\frac{1}{3} = \frac{?}{6}$ ________ **66.** $\frac{7}{8} = \frac{?}{24}$ ________

67. $\frac{5}{9} = \frac{?}{18}$ ________ **68.** $\frac{3}{8} = \frac{?}{40}$ ________

69. $\frac{3}{5} = \frac{?}{100}$ ________ **70.** $\frac{4}{5} = \frac{?}{30}$ ________

71. $\frac{4}{5} = \frac{?}{10}$ ________ **72.** $\frac{6}{7} = \frac{?}{56}$ ________

73. $\frac{3}{8} = \frac{?}{40}$ ________________ **74.** $\frac{2}{11} = \frac{?}{55}$ ________________

75. $\frac{9}{4} = \frac{?}{20}$ ________________ **76.** $\frac{8}{7} = \frac{?}{63}$ ________________

In problems 77–84 raise the fractions to higher terms by finding the missing denominator.

77. $\frac{2}{3} = \frac{6}{?}$ ________________ **78.** $\frac{4}{7} = \frac{16}{?}$ ________________

79. $\frac{1}{2} = \frac{5}{?}$ ________________ **80.** $\frac{5}{6} = \frac{10}{?}$ ________________

81. $\frac{10}{13} = \frac{30}{?}$ ________________ **82.** $\frac{15}{17} = \frac{45}{?}$ ________________

83. $\frac{8}{15} = \frac{32}{?}$ ________________ **84.** $\frac{11}{24} = \frac{44}{?}$ ________________

ANSWERS

1. 41, 43, 47 **2.** Because every other even number is divisible by 2. **3.** $2 \times 2 \times 11$
4. $2 \times 2 \times 3 \times 3$ **5.** $2 \times 2 \times 3$ **6.** $2 \times 2 \times 2 \times 5$ **7.** $2 \times 3 \times 7$ **8.** 5×11
9. $2 \times 2 \times 5$ **10.** 2×11 **11.** 3×7 **12.** 3×11 **13.** $3 \times 3 \times 7$
14. $2 \times 2 \times 2 \times 3 \times 5$ **15.** $2 \times 2 \times 2 \times 3 \times 3$ **16.** $3 \times 3 \times 3 \times 3$ **17.** $2 \times 11 \times 19$
18. $2 \times 2 \times 3 \times 3 \times 11$ **19.** $2 \times 3 \times 5 \times 19$ **20.** $3 \times 5 \times 5 \times 5 \times 5$
21. $2 \times 2 \times 3 \times 7 \times 19$ **22.** $2 \times 2 \times 2 \times 17 \times 19$ **23.** 5 **24.** 4 **25.** 18
26. 22 **27.** 20 **28.** 42 **29.** 30 **30.** 21 **31.** 1 **32.** 1 **33.** $\frac{2}{5}$ **34.** $\frac{1}{3}$
35. $\frac{1}{5}$ **36.** $\frac{3}{4}$ **37.** $\frac{3}{5}$ **38.** $\frac{3}{8}$ **39.** $\frac{1}{2}$ **40.** $\frac{5}{7}$ **41.** $\frac{21}{26}$ **42.** $\frac{4}{11}$ **43.** $\frac{2}{3}$
44. $\frac{32}{65}$ **45.** $\frac{2}{5}$ **46.** $\frac{16}{45}$ **47.** $\frac{1}{6}$ **48.** $\frac{2}{3}$ **49.** $\frac{5}{12}$ **50.** $\frac{3}{5}$ **51.** $\frac{11}{70}$ **52.** $\frac{3}{10}$
53. $\frac{2}{3}$ **54.** $\frac{7}{8}$ **55.** $\frac{7}{5}$ **56.** $\frac{11}{5}$ **57.** $\frac{3}{13}$ **58.** $\frac{52}{103}$ **59.** $\frac{9}{41}$ **60.** $\frac{51}{119}$ **61.** $\frac{9}{10}$
62. $\frac{11}{6}$ **63.** $\frac{4}{3}$ **64.** $\frac{18}{25}$ **65.** 2 **66.** 21 **67.** 10 **68.** 15 **69.** 60
70. 24 **71.** 8 **72.** 48 **73.** 15 **74.** 10 **75.** 45 **76.** 72 **77.** 9 **78.** 28
79. 10 **80.** 12 **81.** 39 **82.** 51 **83.** 60 **84.** 96

2.4 ADDITION (+)

How much is $\frac{2}{9} + \frac{5}{9}$? If you add $\frac{1}{3}, \frac{2}{4}$, and $\frac{3}{5}$, what is the sum? Albert has two cans of antifreeze, one $\frac{3}{7}$ full and the other $\frac{4}{9}$ full. When the two cans are combined, will Albert have a full can of antifreeze? These questions will be answered in this section.

Adding Fractions with Common Denominators

Below is the method for adding fractions with common denominators.

> *To add two or more fractions with common denominators,* add the numerators and place the total over the common denominator. Check to see whether your answer is in its simplest form.

Example 1 Add $\frac{2}{8} + \frac{5}{8}$

Solution $\frac{2}{8} + \frac{5}{8} = \frac{2 + 5}{8} = \frac{7}{8}$

$\frac{2}{8}$ +

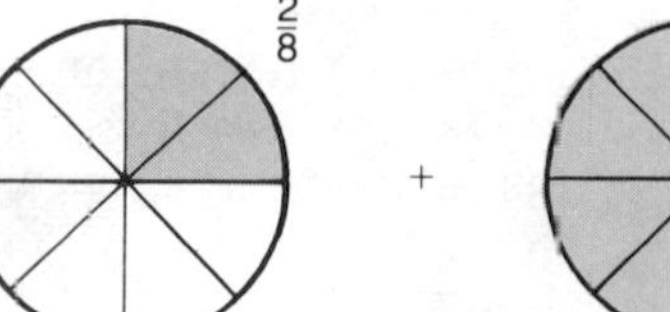

=

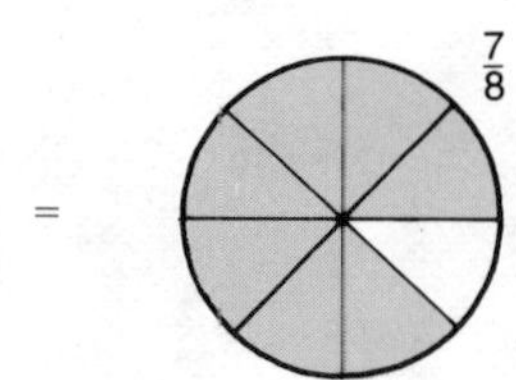

Example 2 Add $\frac{7}{9} + \frac{5}{9}$

Solution $\frac{7}{9} + \frac{5}{9} = \frac{7+5}{9} = \frac{12}{9} = 1\frac{3}{9} = 1\frac{1}{3}$

When fractions do not have a common denominator one must be found. The following discussion is helpful in enabling us to achieve this goal.

The Least Common Multiple, LCM

Prime divisors (or factors) are helpful in finding the **least common multiple,** LCM, of a collection of whole numbers. *The LCM is defined as the smallest nonzero whole number that is divisible by each number in the collection.*

To Find the Least Common Multiple (LCM)

1. Write each number as a product of its prime factors.
2. List each of the **different** prime factors.
3. Determine the **greatest** number of times each of the prime factors is used in any of the numbers.
4. The LCM is the product of all the factors in step 3.

Example 3 Find the LCM of 6 and 21.

Solution
step 1: $6 = 2 \times 3$
$21 = 3 \times 7$
step 2: 2, 3, and 7
step 3: 2, 3, 7
step 4: LCM of 6 and 21 is $2 \times 3 \times 7 = 42$.

Example 4 Find the LCM of 20 and 30.

Solution
step 1: $20 = 2 \times 2 \times 5$
$30 = 2 \times 3 \times 5$
step 2: 2, 3, and 5
step 3: 2×2, 3, 5
step 4: LCM of 20 and 30 is $2 \times 2 \times 3 \times 5 = 60$.

Example 5 Find the LCM of 20, 25, and 30.

Solution
step 1: $20 = 2 \times 2 \times 5$
$25 = 5 \times 5$
$30 = 2 \times 3 \times 5$
step 2: 2, 3, and 5
step 3: 2×2, 3, 5×5
step 4: LCM of 20, 25 and 30 is $2 \times 2 \times 3 \times 5 \times 5 = 300$.

Example 6 Find the LCM of 90, 84, and 105.

Solution As soon as you can condense steps, do so.

$$\begin{aligned} 90 &= 2 \times 3 \times 3 \times 5 \\ 84 &= 2 \times 2 \times 3 \times 7 \\ 105 &= 3 \times 5 \times 7 \\ \hline \text{LCM} &= 2 \times 2 \times 3 \times 3 \times 5 \times 7 = 1{,}260 \end{aligned}$$

Adding Fractions with Unlike Denominators

To add fractions with unlike denominators, convert the given fractions to equivalent fractions with a common denominator and then add. The common denominator should be the **least common denominator, LCD.** Sometimes the least common denominator can be found by inspection. When this is not possible, *it is important to understand that the least common denominator, LCD, is the LCM of the given denominators.*

Here is a step-by-step procedure for adding fractions with unlike denominators.

> **To Add Fractions with Unlike Denominators**
>
> **1.** Find the least common denominator, LCD. This is the least common multiple, LCM, of the given denominators.
> **2.** Convert each of the given fractions to an equivalent fraction whose denominator is the LCD.
> **3.** Add the fractions and reduce the sum to lowest terms.

Example 7 Add $\frac{3}{7} + \frac{1}{21}$.

Solution step 1: Find the LCM of 7 and 21.

$7 = 1 \times \quad 7$

$21 = \quad 3 \times 7$

LCM of 7 and 21 is $3 \times 7 = 21$.

Thus the least common denominator is 21.

(Remember: if you can find the least common denominator by inspection, do so.)

step 2: $\frac{3}{7} = \frac{9}{21}; \frac{1}{21} = \frac{1}{21}$

step 3: $\frac{3}{7} + \frac{1}{21} = \frac{9}{21} + \frac{1}{21} = \frac{9+1}{21} = \frac{10}{21}$

Example 8 Add $\frac{1}{4} + \frac{3}{5}$.

Solution step 1: Find the LCM of 4 and 5.

$4 = \quad 2 \times 2$

$5 = 1 \times \quad 5$

LCM of 4 and 5 is $2 \times 2 \times 5 = 20$.

Thus the least common denominator is 20.

step 2: $\frac{1}{4} = \frac{5}{20}; \frac{3}{5} = \frac{12}{20}$

step 3: $\frac{1}{4} + \frac{3}{5} = \frac{5}{20} + \frac{12}{20} = \frac{5+12}{20} = \frac{17}{20}$

Example 9 Add $\frac{1}{8} + \frac{2}{9} + \frac{5}{12}$.

Solution step 1: Find the LCM of 8, 9, and 12.

$8 = 2 \times 2 \times 2$

$9 = \quad 3 \times 3$

$12 = 2 \times 2 \times \quad 3$

LCM of 8, 9 and 12 is $2 \times 2 \times 2 \times 3 \times 3 = 72$.

Thus the least common denominator is 72.

step 2: $\frac{1}{8} = \frac{9}{72}; \frac{2}{9} = \frac{16}{72}; \frac{5}{12} = \frac{30}{72}$

step 3: $\frac{1}{8} + \frac{2}{9} + \frac{5}{12} = \frac{9}{72} + \frac{16}{72} + \frac{30}{72} = \frac{9+16+30}{72} = \frac{55}{72}$

Example 10 Add $\frac{1}{6} + \frac{3}{10} + \frac{1}{15}$.

Solution step 1: Find the LCM of 6, 10, and 15.

$6 = 2 \times 3$
$10 = 2 \times 5$
$15 = 3 \times 5$

LCM of 6, 10, and 15 is $2 \times 3 \times 5 = 30$.
Thus the least common denominator is 30.

step 2: $\frac{1}{6} = \frac{5}{30}$; $\frac{3}{10} = \frac{9}{30}$; $\frac{1}{15} = \frac{2}{30}$

step 3: $\frac{1}{6} + \frac{3}{10} + \frac{1}{15} = \frac{5}{30} + \frac{9}{30} + \frac{2}{30} = \frac{5+9+2}{30} = \frac{16}{30} = \frac{8}{15}$

Example 11 Add $\frac{1}{18} + \frac{7}{30} + \frac{1}{36}$.

Solution step 1: Find the LCM of 18, 30 and 36.

$18 = 2 \times 3 \times 3$
$30 = 2 \times 3 \times 5$
$36 = 2 \times 2 \times 3 \times 3$

LCM of 18, 30, and 36 is $2 \times 2 \times 3 \times 3 \times 5 = 180$.
Thus the least common denominator is 180.

step 2: $\frac{1}{18} = \frac{10}{180}$, $\frac{7}{30} = \frac{42}{180}$, $\frac{1}{36} = \frac{5}{180}$

step 3: $\frac{1}{18} + \frac{7}{30} + \frac{1}{36} = \frac{10}{180} + \frac{42}{180} + \frac{5}{180} = \frac{10+42+5}{180} = \frac{57}{180} = \frac{19}{60}$

Example 12 Albert has two cans of antifreeze, one $\frac{3}{7}$ full and the other $\frac{4}{9}$ full. When these two cans are combined, will Albert have a full can?

Solution The least common denominator is 63. So

$\frac{4}{9} + \frac{3}{7} = \frac{28}{63} + \frac{27}{63} = \frac{55}{63} < 1$

He does not have a full can.

Adding Mixed Numbers Here is a step-by-step procedure for adding mixed numbers.

To Add Mixed Numbers

1. If necessary, change fractional parts to equivalent fractions with least common denominators.
2. Add the fractional parts.
3. Add the whole numbers.
4. Simplify and reduce the answer to lowest terms.

Example 13

$$\begin{array}{r} 26\frac{4}{15} \\ +\ 42\frac{5}{15} \\ \hline 68\frac{9}{15} \end{array} = 68\frac{3}{5}$$

Example 14

$$\begin{array}{rcr} 3\frac{1}{5} & = & 3\frac{2}{10} \\ +\ 2\frac{7}{10} & = & +\ 2\frac{7}{10} \\ \hline & & 5\frac{9}{10} \end{array}$$

Example 15

$$\begin{array}{rcr} 6\frac{1}{4} & = & 6\frac{3}{12} \\ +\ 3\frac{1}{3} & = & +\ 3\frac{4}{12} \\ \hline & & 9\frac{7}{12} \end{array}$$

Example 16

$$\begin{array}{rcr} 5\frac{7}{8} & = & 5\frac{21}{24} \\ +\ 3\frac{5}{12} & = & +\ 3\frac{10}{24} \\ \hline & & 8\frac{31}{24} \end{array} = 8 + 1 + \frac{7}{24} = 9\frac{7}{24}$$

Because $\frac{31}{24}$ is an improper fraction, we converted it to a mixed number.

Example 17 Add the following numbers: $12\frac{5}{6}$, $7\frac{1}{4}$, and $29\frac{9}{16}$.

$$\begin{array}{rcr} 12\frac{5}{6} & = & 12\frac{40}{48} \\ 7\frac{1}{4} & = & 7\frac{12}{48} \\ +\ 29\frac{9}{16} & = & +\ 29\frac{27}{48} \\ \hline & & 48\frac{79}{48} \end{array} = 48 + 1 + \frac{31}{48} = 49\frac{31}{48}$$

Convert the improper fraction $\frac{79}{48}$ to $1\frac{31}{48}$ and then add this to the 48.

2.4 EXERCISES

In problems 1–20 find the LCM for the given collection of numbers.

1. 3, 12 ____ **2.** 4, 20 ____ **3.** 9, 27 ____

4. 6, 7 ____ **5.** 4, 9 ____ **6.** 7, 11 ____

7. 6, 8 ____ **8.** 14, 21 ____ **9.** 18, 45 ____

10. 16, 24 ____ **11.** 3, 6, 12 ____ **12.** 5, 10, 30 ____

13. 8, 12, 24 ____ **14.** 2, 5, 7 ____ **15.** 4, 5, 9 ____

16. 5, 8, 10 ____ **17.** 9, 12, 15 ____ **18.** 10, 12, 16 ____

19. 8, 12, 18 ____ **20.** 15, 24, 30 ____

In problems 21–37 state your answer as a fraction in simplest form or as a mixed number in simplest form.

21. $\frac{1}{3} + \frac{1}{2}$ ____ 22. $\frac{4}{5} + \frac{7}{10}$ ____ 23. $\frac{3}{4} + \frac{1}{8}$ ____

24. $\frac{1}{2} + \frac{5}{16}$ ____ 25. $\frac{3}{4} + \frac{7}{8}$ ____ 26. $\frac{2}{5} + \frac{7}{20}$ ____

27. $\frac{1}{6} + \frac{2}{15}$ ____ 28. $\frac{7}{12} + \frac{5}{8}$ ____ 29. $\frac{7}{10} + \frac{5}{18}$ ____

30. $\frac{3}{4} + \frac{2}{9}$ 31. $\frac{7}{8} + \frac{5}{16}$ 32. $\frac{4}{5} + \frac{2}{7}$ 33. $\frac{1}{3} + \frac{5}{8}$

34. $\frac{1}{10} + \frac{5}{6}$ 35. $\frac{17}{20} + \frac{9}{10}$ 36. $\frac{5}{7} + \frac{30}{49}$ 37. $\frac{1}{9} + \frac{5}{6}$

Check your answers to problems 21–35. If you did not get at least 12 problems correct, turn to Practice Problem Set 5A in the Appendix.

In problems 38–58 state your answer as a fraction in simplest form or as a mixed number in simplest form.

38. $\frac{1}{3} + \frac{5}{6} + \frac{7}{12}$ ____ 39. $\frac{3}{4} + \frac{5}{16} + \frac{1}{8} + \frac{5}{6}$ ____

40. $\frac{3}{5} + \frac{2}{3} + \frac{11}{15}$ ____ 41. $\frac{2}{7} + \frac{1}{4} + \frac{3}{14}$ ____

42. $\frac{1}{12} + \frac{11}{18} + \frac{5}{24}$ ____ 43. $\frac{1}{3} + \frac{2}{5} + \frac{1}{6} + \frac{7}{10}$ ____

44. $\frac{3}{7} + \frac{2}{5} + \frac{5}{21} + \frac{6}{35}$ ____ 45. $\frac{4}{3} + \frac{7}{12} + \frac{8}{15}$ ____

46. $\frac{5}{21} + \frac{1}{3} + \frac{9}{14} + \frac{2}{7}$ ____ 47. $\frac{21}{16} + \frac{4}{5} + \frac{7}{20} + \frac{11}{10}$ ____

48. $\frac{9}{28} + \frac{5}{16} + \frac{8}{21}$ ____ 49. $4\frac{7}{12} + 7\frac{2}{3}$ ____

50. $3\frac{2}{5} + 2\frac{3}{4}$ 51. $11\frac{7}{9} + 8\frac{2}{3}$ 52. $24\frac{3}{8} + 12\frac{4}{7}$ 53. $19\frac{5}{6} + 37\frac{3}{5}$

54. $9\frac{1}{2} + 6\frac{4}{5}$ ____ 55. $4\frac{2}{3} + 1\frac{3}{4} + 3\frac{1}{5}$ ____

56. $38\frac{5}{6} + 29\frac{2}{3} + 46\frac{1}{2}$ ______

57. $72\frac{3}{7} + 41\frac{3}{14} + 21\frac{5}{8}$ ______

58. $23\frac{3}{11} + 75\frac{5}{12} + 28\frac{5}{6}$ ______

59. Show that $\frac{1}{2} + \frac{1}{3} \neq \frac{1}{5}$. (*Hint:* $\neq$ is the symbol denoting "not equal to.")

____

60. Frank did $\frac{1}{2}$ of his homework one day and $\frac{1}{5}$ of his homework the next day. How much of his homework was completed in two days?

____

61. A contractor moved back and forth between two jobs. He did $\frac{1}{10}$ of a specific job the first week, $\frac{1}{4}$ of the job the second week, and $\frac{2}{5}$ of the job the third week. How much of this job was completed at the end of the third week?

____

62. A painter used $2\frac{3}{4}$ gallons of paint to paint a deck, $1\frac{5}{8}$ gallons to paint a tool shed and $2\frac{1}{2}$ gallons to paint the trim of a house. How many gallons of paint did he use in all?

____

63. A jogger ran $4\frac{7}{10}$ miles, $3\frac{5}{12}$ miles, $5\frac{1}{4}$ miles and $6\frac{7}{15}$ miles. How many miles did he run in all?

____

64. A Greyhound bus made the following trips: 165 miles, $212\frac{1}{2}$ miles, $87\frac{7}{10}$ miles, and 92 miles. What is the total mileage?

____

ANSWERS

1. 12 **2.** 20 **3.** 27 **4.** 42 **5.** 36 **6.** 77 **7.** 24 **8.** 42 **9.** 90 **10.** 48 **11.** 12 **12.** 30 **13.** 24 **14.** 70 **15.** 180 **16.** 40 **17.** 180 **18.** 240 **19.** 72 **20.** 120 **21.** $\frac{5}{6}$ **22.** $1\frac{1}{2}$ **23.** $\frac{7}{8}$ **24.** $\frac{13}{16}$ **25.** $1\frac{5}{8}$ **26.** $\frac{3}{4}$ **27.** $\frac{3}{10}$ **28.** $1\frac{5}{24}$ **29.** $\frac{44}{45}$ **30.** $\frac{35}{36}$ **31.** $1\frac{3}{16}$ **32.** $1\frac{3}{35}$ **33.** $\frac{23}{24}$ **34.** $\frac{14}{15}$ **35.** $1\frac{3}{4}$ **36.** $1\frac{16}{49}$ **37.** $\frac{17}{18}$ **38.** $1\frac{3}{4}$ **39.** $2\frac{1}{48}$ **40.** 2 **41.** $\frac{3}{4}$ **42.** $\frac{65}{72}$ **43.** $1\frac{3}{5}$ **44.** $1\frac{5}{21}$ **45.** $2\frac{9}{20}$ **46.** $1\frac{1}{2}$ **47.** $3\frac{9}{16}$ **48.** $1\frac{5}{336}$ **49.** $12\frac{1}{4}$ **50.** $6\frac{3}{20}$ **51.** $20\frac{4}{9}$ **52.** $36\frac{53}{56}$ **53.** $57\frac{13}{30}$ **54.** $16\frac{3}{10}$ **55.** $9\frac{37}{60}$ **56.** 115 **57.** $135\frac{15}{56}$ **58.** $127\frac{23}{44}$ **59.** $\frac{1}{2} + \frac{1}{3} = \frac{5}{6} \neq \frac{1}{5}$ **60.** $\frac{7}{10}$ completed **61.** $\frac{3}{4}$ completed **62.** $6\frac{7}{8}$ gallons **63.** $19\frac{5}{6}$ miles **64.** $557\frac{1}{5}$ miles

2.5 SUBTRACTION (−)

How much is $\frac{7}{4} - \frac{2}{5}$? If you start with 15 and take away $8\frac{3}{4}$, how much remains? A biochemist has one bottle of insulin and it is $\frac{3}{4}$ full. If the biochemist uses $\frac{1}{5}$ of a bottle of insulin for an experiment, how much remains? Later that day, the biochemist must conduct an experiment for which she needs $\frac{1}{2}$ of a bottle of insulin. Will she be able to conduct this experiment? These questions will be answered in this section.

Subtracting Fractions with Common Denominators

Below is the method for subtracting fractions with common denominators.

> *To subtract two fractions with common denominators,* subtract the numerators and place the difference over the common denominator. Check to see whether your answer is in simplest form.

Example 1 Subtract $\frac{9}{12} - \frac{5}{12}$.

Solution $\frac{9}{12} - \frac{5}{12} = \frac{9-5}{12} = \frac{4}{12} = \frac{1}{3}$

Example 2 Subtract $\frac{17}{24} - \frac{11}{24}$.

Solution $\frac{17}{24} - \frac{11}{24} = \frac{17-11}{24} = \frac{6}{24} = \frac{1}{4}$

Subtracting Fractions with Unlike Denominators

Here is a step-by-step procedure for subtracting fractions with unlike denominators.

> **To Subtract Fractions with Unlike Denominators**
>
> **1.** Find the least common denominator, LCD. This is the least common multiple, LCM, of the given denominators.
> **2.** Convert each of the given fractions to an equivalent fraction whose denominator is the least common denominator.
> **3.** Subtract the fractions and reduce the difference to lowest terms.

Example 3 Subtract $\frac{11}{12} - \frac{2}{3}$.

Solution step 1: The least common denominator is 12.
(Remember, if you can find the LCD by inspection, do so.)

step 2: $\frac{11}{12}$; $\frac{2}{3} = \frac{8}{12}$

step 3: $\frac{11}{12} - \frac{8}{12} = \frac{11-8}{12} = \frac{3}{12} = \frac{1}{4}$

Example 4 Subtract $\frac{7}{4} - \frac{2}{5}$.

Solution step 1: The least common denominator is 20.

step 2: $\frac{7}{4} = \frac{35}{20}$; $\frac{2}{5} = \frac{8}{20}$

step 3: $\frac{7}{4} - \frac{2}{5} = \frac{35}{20} - \frac{8}{20} = \frac{27}{20} = 1\frac{7}{20}$

Example 5 Subtract $\frac{1}{15}$ from $\frac{5}{18}$.

Solution step 1: Find the LCM of 15 and 18.

$15 = \quad\;\; 3 \times 5$

$18 = 2 \times 3 \times 3$

LCM of 15 and 18 is $2 \times 3 \times 3 \times 5 = 90$.

Thus the least common denominator is 90.

step 2: $\frac{5}{18} = \frac{25}{90}$; $\frac{1}{15} = \frac{6}{90}$

step 3: $\frac{5}{18} - \frac{1}{15} = \frac{25}{90} - \frac{6}{90} = \frac{19}{90}$

Example 6 $\frac{17}{21} - \frac{1}{15} - \frac{2}{35}$.

Solution step 1: Find the LCM of 21, 15, and 35.

$15 = 3 \times 5$

$21 = 3 \times \quad 7$

$35 = \quad\; 5 \times 7$

LCM of 21, 15, and 35 is $3 \times 5 \times 7 = 105$.

Thus the least common denominator is 105.

step 2: $\frac{17}{21} = \frac{85}{105}$; $\frac{1}{15} = \frac{7}{105}$; $\frac{2}{35} = \frac{6}{105}$

step 3: $\frac{17}{21} - \frac{1}{15} - \frac{2}{35} = \frac{85}{105} - \frac{7}{105} - \frac{6}{105} = \frac{85 - 7 - 6}{105}$

$= \frac{72}{105} = \frac{24}{35}$

Example 7 Subtract $1 - \frac{3}{8}$.

Solution Least common denominator is 8.

$1 - \frac{3}{8} = \frac{8}{8} - \frac{3}{8} = \frac{8 - 3}{8} = \frac{5}{8}$

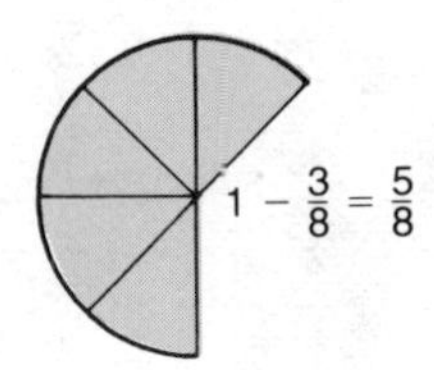

Example 8 A biochemist has one bottle of insulin and it is $\frac{3}{4}$ full. If the biochemist uses $\frac{1}{5}$ of a bottle of insulin for an experiment, how much remains? Later that day, the biochemist must conduct an experiment for which she needs $\frac{1}{2}$ of a bottle of insulin. Will she be able to conduct this experiment?

Solution $\frac{3}{4} - \frac{1}{5} = \frac{15}{20} - \frac{4}{20} = \frac{11}{20}$ remains

Because $\frac{11}{20} > \frac{1}{2}$, she will be able to conduct the second experiment.

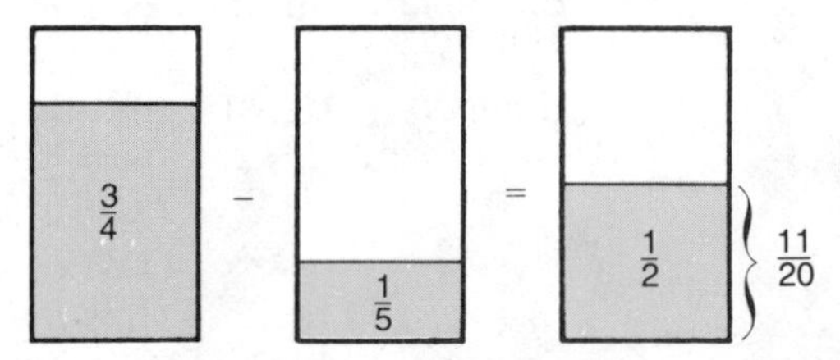

Subtracting Mixed Numbers

Here is the step-by-step procedure for subtracting mixed numbers.

To Subtract Mixed Numbers

1. If necessary, change the fractional parts to equivalent fractions with least common denominator.
2. Subtract the fractional parts. To accomplish this, you may have to convert a whole number, in the minuend, into its fractional parts of one. This is called "borrowing."
3. Subtract the whole numbers.
4. Simplify and reduce the answer to lowest terms.

Example 9 Subtract $7\frac{9}{10} - 3$.

Solution

$$\begin{array}{r} 7\frac{9}{10} \\ -\ 3 \\ \hline 4\frac{9}{10} \end{array}$$

Example 10 Subtract $47\frac{5}{8} - 29\frac{2}{8}$

Solution

$$\begin{array}{r} 47\frac{5}{8} \\ -\ 29\frac{2}{8} \\ \hline 18\frac{3}{8} \end{array}$$

Example 11 Subtract $12\frac{5}{6} - 7\frac{1}{2}$

Solution

$$\begin{array}{rcr} 12\frac{5}{6} & = & 12\frac{5}{6} \\ -\ 7\frac{1}{2} & = & -\ 7\frac{3}{6} \\ \hline & & 5\frac{2}{6} = 5\frac{1}{3} \end{array}$$

Example 12 Subtract $34\frac{2}{5} - 10\frac{1}{6}$

Solution

$$\begin{array}{rcr} 34\frac{2}{5} & = & 34\frac{12}{30} \\ -\ 10\frac{1}{6} & = & -\ 10\frac{5}{30} \\ \hline & & 24\frac{7}{30} \end{array}$$

Example 13 Subtract $1\frac{3}{4}$ from $5\frac{1}{3}$.

Solution

$$\begin{array}{rcrcr} 5\frac{1}{3} & = & 5\frac{4}{12} & = & 4\frac{16}{12} \\ -\ 1\frac{3}{4} & = & 1\frac{9}{12} & = & 1\frac{9}{12} \\ \hline & & & & 3\frac{7}{12} \end{array}$$

Notice when we wished to subtract the fractional parts, the bottom fraction was greater than the top fraction $\left(\frac{9}{12} > \frac{4}{12}\right)$. This required us to "borrow" one from the 5 and convert it into $\frac{12}{12}$. Then $\frac{12}{12}$ was added to $\frac{4}{12}$ to yield $\frac{16}{12}$. At this point we were able to subtract the fractional and whole number portions of our mixed numbers.

Example 14 Subtract $3\frac{7}{8}$ from $6\frac{1}{6}$.

Solution

$$\begin{array}{rcrcr} 6\frac{1}{6} & = & 6\frac{4}{24} & = & 5\frac{28}{24} \\ -\ 3\frac{7}{8} & = & 3\frac{21}{24} & = & 3\frac{21}{24} \\ \hline & & & & 2\frac{7}{24} \end{array}$$

2.5 EXERCISES

In problems 1–15 state your answer as a fraction in simplest form or as a mixed number in simplest form.

1. $\begin{array}{r} \frac{9}{10} \\ -\ \frac{2}{10} \\ \hline \end{array}$ **2.** $\begin{array}{r} \frac{14}{20} \\ -\ \frac{9}{20} \\ \hline \end{array}$ **3.** $\begin{array}{r} \frac{12}{18} \\ -\ \frac{5}{18} \\ \hline \end{array}$ **4.** $\begin{array}{r} \frac{152}{160} \\ -\ \frac{105}{160} \\ \hline \end{array}$

5. $\frac{90}{14} - \frac{5}{14}$ ____ **6.** $\frac{5}{8} - \frac{1}{2}$ ____ **7.** $\frac{7}{10} - \frac{2}{5}$ ____

8. $\frac{13}{18} - \frac{2}{3}$ ____ **9.** $\frac{6}{7} - \frac{1}{3}$ ____ **10.** $\frac{7}{8} - \frac{1}{3}$ ____

11. $\frac{1}{6} - \frac{1}{7}$ ____ **12.** $\frac{4}{5} - \frac{1}{4}$ ____ **13.** $\frac{7}{9} - \frac{1}{2}$ ____

14. $\frac{5}{8} - \frac{2}{5}$ ______

15. $\begin{array}{r} 1 \\ -\ \frac{2}{9} \\ \hline \end{array}$

Check your answers to problems 1–15. If you did not get at least 12 problems correct, turn to Practice Problem Set 6A in the Appendix.

In exercises 16–53 state your answer as a fraction in simplest form or as a mixed number in simplest form.

16. $\begin{array}{r} 2 \\ -\ 1\frac{3}{4} \\ \hline \end{array}$

17. $\begin{array}{r} \frac{5}{6} \\ -\ \frac{1}{3} \\ \hline \end{array}$

18. $\begin{array}{r} \frac{7}{8} \\ -\ \frac{3}{4} \\ \hline \end{array}$

19. $\begin{array}{r} \frac{4}{5} \\ -\ \frac{2}{3} \\ \hline \end{array}$

20. $\begin{array}{r} \frac{9}{10} \\ -\ \frac{2}{5} \\ \hline \end{array}$

21. $\begin{array}{r} \frac{7}{8} \\ -\ \frac{5}{6} \\ \hline \end{array}$

22. $\begin{array}{r} \frac{3}{4} \\ -\ \frac{3}{5} \\ \hline \end{array}$

23. $\frac{9}{10} - \frac{3}{4}$ ______

24. $\frac{16}{21} - \frac{4}{15}$ ______

25. $\frac{8}{11} - \frac{1}{4}$ ______

26. $\frac{11}{24} - \frac{5}{36}$ ______

27. $\frac{11}{18} - \frac{7}{15}$ ______

28. $\frac{7}{12} - \frac{5}{18}$ ______

29. $\frac{9}{16} - \frac{5}{12}$ ______

30. $\frac{18}{35} - \frac{1}{20}$ ______

31. $\frac{16}{25} - \frac{4}{15}$ ______

32. $\frac{13}{18} - \frac{11}{20}$ ______

33. $\frac{10}{21} - \frac{5}{14}$ ______

34. $\frac{21}{40} - \frac{8}{25}$ ______

35. $\frac{17}{20} - \frac{13}{30}$ ______

36. $\frac{10}{21} - \frac{5}{28}$ ______

Hint: Work from left to right in problems 37, 39, and 40.

37. $\frac{15}{16} - \frac{1}{4} - \frac{1}{2}$ ______

38. $\frac{18}{33} - \frac{9}{22}$ ______

39. $\frac{7}{8} + \frac{2}{3} - \frac{3}{4}$ ______

40. $\frac{4}{12} + \frac{1}{8} - \frac{1}{3}$ ______

41. $9\frac{2}{3} - 4\frac{2}{5}$ ______

42. $4\frac{3}{4} - 2\frac{7}{8}$ ______

43. $15\frac{1}{2} - 4\frac{3}{4}$ ______

44. $7\frac{3}{7} - 5\frac{2}{3}$ ______

45. $5\frac{5}{18} - 2\frac{7}{12}$ ________________

46. $8 - 5\frac{3}{7}$

47. $10 - 4\frac{9}{17}$

48. $9\frac{3}{5} - 2\frac{3}{7}$

49. $6\frac{5}{9} - 2\frac{1}{3}$

50. $25\frac{1}{6} - 11\frac{1}{2}$

51. $47\frac{3}{8} - 17\frac{4}{5}$

52. $19\frac{1}{4} - 7\frac{2}{5}$

53. $28\frac{1}{3} - 7\frac{3}{4}$

54. A glazier begins work with $8\frac{4}{5}$ lb of glazing compound. If he uses $4\frac{2}{3}$ lb, how much remains?

____

55. On a particular day Weymouth Cleaners receives 210 gallons of number-one cleaning solution and uses $27\frac{3}{8}$ gallons of this solution. How much of the day's shipment remains in stock?

____

56. Miguel went on a $55\frac{1}{2}$-km hike. He traveled $19\frac{3}{8}$ km the first day and $24\frac{1}{3}$ km the second day. How many km has he traveled after two days? Assuming that he must complete the trip in three days, how many km remain to be traveled on day three?

____

57. A $3\frac{7}{9}$ foot post was set $1\frac{1}{4}$ feet into the ground. What was the exact length of the post above ground?

____

58. Henri and Jacques painted $\frac{2}{5}$ and $\frac{1}{3}$ of their home, respectively. How much of the house did they paint? How much is left to be painted?

____

59. A salesperson drove $184\frac{7}{8}$, $216\frac{3}{5}$, and $210\frac{1}{10}$ miles. How many miles were driven in all? Assuming that this salesperson is allowed expense money for 800 miles of driving, how many of the 800 expense-paid miles remain?

____

60. A hardware store received a $55\frac{3}{4}$-gallon shipment of a particular brand of paint followed by a $62\frac{1}{2}$-gallon shipment of paint. How much paint was received? If three contractors purchase $27\frac{1}{4}$ gallons, 19 gallons, and $21\frac{3}{8}$ gallons from the hardware store, how much of the two shipments remains?

____

ANSWERS

1. $\frac{7}{10}$ **2.** $\frac{1}{4}$ **3.** $\frac{7}{18}$ **4.** $\frac{47}{160}$ **5.** $6\frac{1}{14}$ **6.** $\frac{1}{8}$ **7.** $\frac{3}{10}$ **8.** $\frac{1}{18}$ **9.** $\frac{11}{21}$
10. $\frac{13}{24}$ **11.** $\frac{1}{42}$ **12.** $\frac{11}{20}$ **13.** $\frac{5}{18}$ **14.** $\frac{9}{40}$ **15.** $\frac{7}{9}$ **16.** $\frac{1}{4}$ **17.** $\frac{1}{2}$ **18.** $\frac{1}{8}$
19. $\frac{2}{15}$ **20.** $\frac{1}{2}$ **21.** $\frac{1}{24}$ **22.** $\frac{3}{20}$ **23.** $\frac{3}{20}$ **24.** $\frac{52}{105}$ **25.** $\frac{21}{44}$ **26.** $\frac{23}{72}$
27. $\frac{13}{90}$ **28.** $\frac{11}{36}$ **29.** $\frac{7}{48}$ **30.** $\frac{13}{28}$ **31.** $\frac{28}{75}$ **32.** $\frac{31}{180}$ **33.** $\frac{5}{42}$ **34.** $\frac{41}{200}$
35. $\frac{5}{12}$ **36.** $\frac{25}{84}$ **37.** $\frac{3}{16}$ **38.** $\frac{3}{22}$ **39.** $\frac{19}{24}$ **40.** $\frac{1}{8}$ **41.** $5\frac{4}{15}$ **42.** $1\frac{7}{8}$
43. $10\frac{3}{4}$ **44.** $1\frac{16}{21}$ **45.** $2\frac{25}{36}$ **46.** $2\frac{4}{7}$ **47.** $5\frac{8}{17}$ **48.** $7\frac{6}{35}$ **49.** $4\frac{2}{9}$ **50.** $13\frac{2}{3}$
51. $29\frac{23}{40}$ **52.** $11\frac{17}{20}$ **53.** $20\frac{7}{12}$ **54.** $4\frac{2}{15}$ **55.** $182\frac{5}{8}$ **56.** $43\frac{17}{24}$; $11\frac{19}{24}$ **57.** $2\frac{19}{36}$
58. $\frac{11}{15}$; $\frac{4}{15}$ **59.** $611\frac{23}{40}$ in all; $188\frac{17}{40}$ miles remain **60.** $118\frac{1}{4}$; $50\frac{5}{8}$ gallons are left

2.6 MULTIPLICATION (×)

How much is $\frac{3}{2} \times \frac{8}{5}$? You must give a person $\frac{2}{5}$ of \$2,020. How much do you give this person? Elma Lewis invested \$18,900 in her business last year. This year she invested $2\frac{1}{2}$ times as much in her business. How much did she invest this year? These questions will be answered in this section.

To answer these questions, you must be familiar with the following method for multiplying fractions.

> *To multiply two or more fractions,* multiply the numerators to obtain the numerator of the answer and multiply the denominators to obtain the denominator of the answer. Simplify and reduce your answer to lowest terms.

Example 1 Multiply $6 \times \frac{5}{7}$

Solution $6 \times \frac{5}{7} = \frac{6}{1} \times \frac{5}{7} = \frac{6 \times 5}{1 \times 7} = \frac{30}{7}$ or $4\frac{2}{7}$

Example 2 Multiply $\frac{2}{3} \times \frac{1}{5} \times \frac{9}{4}$

Solution $\frac{2}{3} \times \frac{1}{5} \times \frac{9}{4} = \frac{2 \times 1 \times 9}{3 \times 5 \times 4} = \frac{18}{60} = \frac{3}{10}$

Example 3a Multiply $\frac{3}{2} \times \frac{8}{5}$

Solution $\frac{3}{2} \times \frac{8}{5} = \frac{3 \times 8}{2 \times 5} = \frac{24}{10} = 2\frac{4}{10} = 2\frac{2}{5}$

Example 3b Return to Example 3a and solve the problem by **canceling** a common divisor (factor). Observe that 2 may be divided into the denominator of $\frac{3}{2}$ and into the numerator of $\frac{8}{5}$. When we divide each of these numbers by 2, called "canceling," and complete the multiplication we obtain:
$\frac{3}{\cancel{2}_1} \times \frac{\cancel{8}^4}{5} = \frac{12}{5} = 2\frac{2}{5}$

Canceling in multiplication involves dividing the numerator and denominator of two different fractions, or the numerator and denominator of the same fraction, by the same number. Why does canceling work in the preceding example? The answer lies in our method of multiplying fractions. From this method, it follows that $\frac{3}{2} \times \frac{8}{5} = \frac{3}{5} \times \frac{8}{2}$. When $\frac{8}{2}$ is expressed in lowest terms, we obtain $\frac{4}{1}$ or 4. *Thus, canceling in a multiplication problem is really a reduction of a fraction to its lowest terms.* Cancel whenever possible, for it will make your numbers more manageable.

Return to Example 2. Two cancellations are possible in this example. One cancellation involves dividing the numbers 2 and 4 by 2. Can you find the other cancellation? Below are additional examples of canceling in multiplication.

Example 4 $\frac{3}{14} \cdot \frac{21}{5}$ means $\frac{3}{\cancel{14}_2} \times \frac{\cancel{21}^3}{5} = \frac{9}{10}$

Example 5 $\left(\frac{10}{6}\right)\left(\frac{14}{15}\right)$ means $\frac{{}^2\cancel{10}}{\cancel{6}_3} \times \frac{\cancel{14}^7}{\cancel{15}_3} = \frac{14}{9}$, or $1\frac{5}{9}$

Example 6 $\frac{15}{20} \times \frac{7}{3} \times \frac{6}{5} = \frac{21}{10}$, or $2\frac{1}{10}$ The reader is encouraged to complete the cancellations in this example.

Example 7
$\frac{2}{5}$ of \$2,020 $= \frac{2}{\cancel{5}_1} \times \frac{\overset{404}{\cancel{\$2,020}}}{1} = \$808$
Notice how the word "of" signals a multiplication.

Example 8 How much is $\frac{1}{3}$ of $\frac{1}{2}$?

Solution To visualize our solution, we have drawn a circle in which *ACB* represents $\frac{1}{2}$ of the circle. We must determine $\frac{1}{3}$ of *ACB*. Clearly our answer must be less than $\frac{1}{2}$ (of the circle).
$\frac{1}{3} \times \frac{1}{2} = \frac{1}{6}$; shaded area

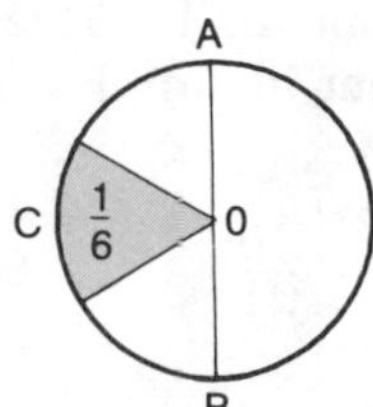

When you multiply mixed numbers, always begin by converting them to improper fractions. This will be illustrated in the next examples.

Example 9 Multiply $6\frac{2}{3}$ by $4\frac{1}{5}$.

Solution When you multiply mixed numbers, you should always begin by converting them to improper fractions. Look for possible cancellations, and then complete the multiplication.

$$6\frac{2}{3} \times 4\frac{1}{5} = \frac{\overset{4}{\cancel{20}}}{\underset{1}{\cancel{3}}} \times \frac{\overset{7}{\cancel{21}}}{\underset{1}{\cancel{5}}} = 28$$

Example 10 $3\frac{1}{4} \cdot 1\frac{1}{2} = \frac{13}{4} \cdot \frac{3}{2} = \frac{39}{8} = 4\frac{7}{8}$

Example 11 Evaluate $\left(\frac{2}{5} + \frac{1}{3}\right) \times \frac{9}{22}$.

Solution Here "evaluate" means to do the indicated addition and multiplication. The parentheses indicate that we are to do the addition first. Thus:

$$\left(\frac{2}{5} + \frac{1}{3}\right) \times \frac{9}{22} = \left(\frac{6+5}{15}\right) \times \frac{9}{22}$$

$$= \frac{\overset{1}{\cancel{11}}}{\underset{5}{\cancel{15}}} \times \frac{\overset{3}{\cancel{9}}}{\underset{2}{\cancel{22}}} = \frac{3}{10}$$

Example 12 A certain mixture requires at most $\frac{2}{3}$ of 36 available grams of a particular ingredient. What is the largest possible amount of this ingredient that is required?

Solution $\frac{2}{\underset{1}{\cancel{3}}} \times \overset{12}{\cancel{36}}$ grams $= 24$ grams

Example 13 Elma Lewis invested $18,900 in her business last year. This year she invested $2\frac{1}{2}$ times as much in her business. How much did she invest this year?

Solution step 1: $2\frac{1}{2} = \frac{5}{2}$

step 2: $\frac{5}{\underset{1}{\cancel{2}}} \times \$\overset{9,450}{\cancel{18,900}} = 5 \times \$9,450 = \$47,250$

2.6 EXERCISES

In problems 1–15 state your answer as a fraction in simplest form or as a mixed number in simplest form.

1. $\frac{3}{5} \times \frac{10}{21}$ ________________

2. $\frac{6}{7} \times \frac{14}{15}$ ________________

3. $\frac{3}{8} \times \frac{5}{4}$ ________________

4. $\frac{9}{20} \times \frac{7}{5}$ ________________

5. $\frac{3}{4} \times \frac{1}{6}$ ______

6. $8 \times \frac{3}{4}$ ______

7. $\frac{6}{7} \times \frac{21}{10}$ ______

8. $16 \times \frac{5}{8}$ ______

9. $\frac{16}{25} \times \frac{15}{32}$ ______

10. $\frac{30}{44} \times \frac{33}{10}$ ______

11. $\frac{7}{2} \times \frac{3}{4}$ ______

12. $\frac{6}{5} \times \frac{3}{5}$ ______

13. $\frac{3}{4} \cdot \frac{7}{9}$ ______

14. $\frac{4}{1} \cdot \frac{5}{3}$ ______

15. $\frac{5}{1} \cdot \frac{9}{4}$ ______

Check your answers to problems 1–15. If you did not get at least 12 problems correct, turn to Practice Problem Set 7A in the Appendix.

In problems 16–55 state your answer as a fraction in simplest form or as a mixed number in simplest form.

16. $\frac{2}{5} \cdot \frac{3}{6}$ ______

17. $\frac{21}{4} \cdot \frac{16}{7}$ ______

18. $\frac{32}{5} \cdot \frac{25}{12}$ ______

19. $\frac{14}{5} \cdot \frac{3}{7} \cdot \frac{15}{4}$ ______

20. $\frac{16}{5} \cdot \frac{35}{6} \cdot \frac{3}{4}$ ______

21. $\frac{3}{7} \times \frac{14}{15} \times \frac{5}{2}$ ______

22. $\frac{2}{3} \times \frac{8}{15} \times \frac{25}{32}$ ______

23. $\frac{7}{12} \times \frac{9}{4} \times \frac{24}{49}$ ______

24. $\frac{9}{30} \times \frac{10}{27} \times 6$ ______

25. $\frac{5}{2} \times \frac{7}{10} \times \frac{4}{21}$ ______

26. $\frac{12}{17} \times 2 \times \frac{51}{18}$ ______

27. $5\frac{2}{3} \times 3\frac{3}{8}$ ______

28. $3\frac{2}{5} \times 7$ ______

29. $3\frac{3}{4} \times \frac{8}{15}$ ______

30. $3\frac{1}{2} \times \frac{5}{14} \times \frac{4}{15}$ ______

31. $2\frac{1}{4} \times 3\frac{2}{3} \times \frac{8}{33}$ ______

32. $1\frac{2}{3} \times 2\frac{3}{4} \times 1\frac{4}{5}$ ______

33. $\left(4\frac{1}{2}\right)\left(\frac{10}{21}\right)$ ______

34. $\left(3\frac{1}{16}\right)\left(\frac{2}{7}\right)$ ______

35. $\left(6\frac{2}{5}\right)\left(2\frac{1}{12}\right)$ ______

36. $\left(4\frac{1}{6}\right)\left(2\frac{4}{5}\right)$ ______

37. $\left(\frac{2}{3}\right)\left(6\frac{3}{4}\right)$ ______

38. $\left(10\frac{2}{5}\right)\left(1\frac{2}{13}\right)$ ______________

39. $\left(1\frac{2}{7}\right)\left(\frac{2}{3}\right)\left(4\frac{2}{3}\right)$ ______________

40. $\left(2\frac{1}{2}\right)\left(1\frac{1}{15}\right)\left(2\frac{1}{4}\right)$ ______________

41. Find $\frac{4}{9}$ of 72. ______________

42. Find $\frac{3}{8}$ of 32. ______________

43. Find $\frac{2}{9}$ of 16. ______________

44. Find $\frac{5}{3}$ of 20. ______________

45. Find $\frac{7}{8}$ of 96. ______________

46. Find $\frac{4}{5}$ of 120. ______________

47. Find $\frac{2}{3}$ of 150. ______________

48. Find $\frac{9}{10}$ of 180. ______________

49. $\left(\frac{1}{4} + \frac{2}{7}\right) \times \frac{8}{3}$ ______________

50. $\left(\frac{5}{8} + \frac{5}{6}\right) \times \frac{4}{7}$ ______________

51. $\left(\frac{2}{3} + \frac{1}{2}\right) \times \frac{12}{5}$ ______________

52. $\left(\frac{3}{10} + \frac{2}{5}\right) \times \frac{10}{12}$ ______________

53. $\left(\frac{10}{7} + \frac{2}{3}\right) \times \frac{3}{11}$ ______________

54. $\left(\frac{5}{8} + \frac{2}{7}\right) \times \left(\frac{1}{3} + \frac{1}{9}\right)$ ______________

55. $\left(3\frac{1}{4} + \frac{3}{5}\right) \times \left(3\frac{1}{3} + \frac{5}{11}\right)$ ______________

56. Aspirin is available in $\frac{3}{10}$ gram tablets. How many grams will $2\frac{1}{2}$ aspirin tablets make?

____

57. Maria estimates that a pair of nylons last her 6 months. She believes that by purchasing two pairs at a time and rotating them, she can extend the life of a pair of nylons by $\frac{1}{3}$. If she is correct, how long will a new pair of nylons last her?

____

58. Sidney paints his house every four years. He does the work himself at odd times. He estimates that this saves him $8\frac{1}{4}$ times the amount spent on materials, which cost him \$180. How much has he saved by painting his own house?

____

59. Mrs. Walker has saved material to make a hooked rug. She estimates that it will cost her \$18 to make this rug in spite of the material saved. However, she knows that the finished rug is worth $7\frac{3}{4}$ times her cost. What is the rug worth? How much has she saved?

____

60. Andrew, John, and James were to help their father wash the family car. Andrew washed $\frac{1}{2}$ of the car. John washed $\frac{1}{2}$ of what Andrew left, and James washed $\frac{1}{4}$ of the part left after both Andrew and John had finished. How much of the car was washed? How much remains to be washed?

____

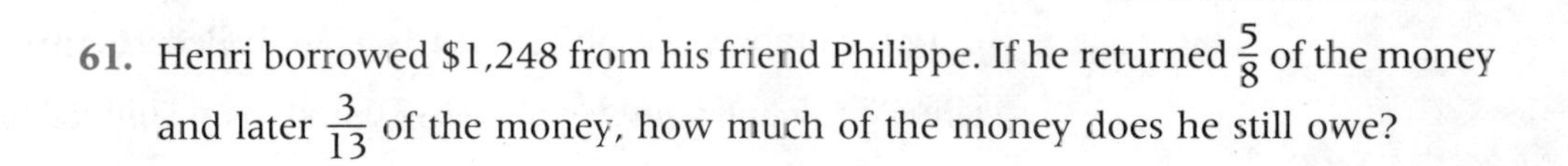

61. Henri borrowed \$1,248 from his friend Philippe. If he returned $\frac{5}{8}$ of the money and later $\frac{3}{13}$ of the money, how much of the money does he still owe?

____

62. Ira and Laura have 15 grams of pure gold. Ira wants to make a 9-gram golf ball that will be $\frac{1}{3}$ gold, and Laura wants to make a 15-gram brooch that will be $\frac{2}{3}$ gold. Do they have enough gold to make both of these items? How much is left over (or short)?

____

63. If a share of Exxon stock costs \$$56\frac{7}{8}$, how much do 24 shares cost?

____

64. Alice worked $22\frac{1}{2}$ hours on a job, and Bob worked $\frac{5}{8}$ as long as Alice. How long did Bob work?

____

65. A carpenter wanted three pieces of wood each $1\frac{5}{8}$ feet long. If he planned to cut them from a single piece of wood, how long would that piece of wood have to be?

____

ANSWERS

1. $\frac{2}{7}$ **2.** $\frac{4}{5}$ **3.** $\frac{15}{32}$ **4.** $\frac{63}{100}$ **5.** $\frac{1}{8}$ **6.** 6 **7.** $1\frac{4}{5}$ **8.** 10 **9.** $\frac{3}{10}$ **10.** $2\frac{1}{4}$
11. $2\frac{5}{8}$ **12.** $\frac{18}{25}$ **13.** $\frac{7}{12}$ **14.** $6\frac{2}{3}$ **15.** $11\frac{1}{4}$ **16.** $\frac{1}{5}$ **17.** 12 **18.** $13\frac{1}{3}$
19. $4\frac{1}{2}$ **20.** 14 **21.** 1 **22.** $\frac{5}{18}$ **23.** $\frac{9}{14}$ **24.** $\frac{2}{3}$ **25.** $\frac{1}{3}$ **26.** 4 **27.** $19\frac{1}{8}$
28. $23\frac{4}{5}$ **29.** 2 **30.** $\frac{1}{3}$ **31.** 2 **32.** $8\frac{1}{4}$ **33.** $2\frac{1}{7}$ **34.** $\frac{7}{8}$ **35.** $13\frac{1}{3}$
36. $11\frac{2}{3}$ **37.** $4\frac{1}{2}$ **38.** 12 **39.** 4 **40.** 6 **41.** 32 **42.** 12 **43.** $3\frac{5}{9}$
44. $33\frac{1}{3}$ **45.** 84 **46.** 96 **47.** 100 **48.** 162 **49.** $1\frac{3}{7}$ **50.** $\frac{5}{6}$ **51.** $2\frac{4}{5}$
52. $\frac{7}{12}$ **53.** $\frac{4}{7}$ **54.** $\frac{17}{42}$ **55.** $14\frac{7}{12}$ **56.** $\frac{3}{4}$ **57.** 8 months **58.** \$1,485
59. \$$139\frac{1}{2}$; \$$121\frac{1}{2}$ **60.** $\frac{13}{16}$; $\frac{3}{16}$ **61.** \$180 still owed **62.** yes; 2 grams left
63. \$1,365 **64.** $14\frac{1}{16}$ hours **65.** $4\frac{7}{8}$ feet long

2.7 DIVISION (÷)

How much is $\frac{4}{7} \div \frac{8}{3}$? What number is the reciprocal of $\frac{4}{5}$? What does it mean to say that two numbers are reciprocals? In the course of a year, Nicole's Candy shipped 19,272 pounds of hard candy in containers that hold $48\frac{2}{3}$ pounds each. How many containers of candy were shipped? We will answer these questions in this section.

Two numbers whose product is 1 are called **reciprocals.** Thus $\frac{7}{8}$ and $\frac{8}{7}$ are reciprocals, 5 and $\frac{1}{5}$ are reciprocals. To find the reciprocal of a fraction, interchange the numerator and the denominator.

Example 1 (a) Find the reciprocal of $\frac{4}{5}$.

(b) Find the reciprocal of $\frac{7}{6}$.

(c) Find the reciprocal of $\frac{1}{3}$.

Solution (a) $\frac{5}{4}$ (b) $\frac{6}{7}$ (c) $\frac{3}{1}$ or 3

Is it clear that to find the reciprocal of a fraction you have only to "invert" it?

Example 2 Find the reciprocal of 9.

Solution 9 can be written as $\frac{9}{1}$. Hence the reciprocal of 9 is $\frac{1}{9}$.

Before we state the method used to divide fractions, remember when a division of fractions is written horizontally the first fraction (on the left of ÷) is the dividend and the second fraction (on the right of ÷) is the divisor.

> *To divide two fractions,* multiply the first fraction by the reciprocal of the second fraction. Cancel whenever possible and leave your answer in simplest form.

Example 3 Divide $\frac{5}{7} \div \frac{8}{3}$

Solution $\frac{5}{7} \div \frac{8}{3} = \frac{5}{7} \times \frac{3}{8} = \frac{15}{56}$

Example 4 Divide $\frac{6}{11} \div 2$

Solution

$$\frac{6}{11} \div 2 = \frac{\overset{3}{\cancel{6}}}{11} \times \frac{1}{\underset{1}{\cancel{2}}} = \frac{3}{11}$$

When you divide mixed numbers, always begin by converting them to improper fractions. This will be illustrated in the next examples.

Example 5 Divide $1\frac{3}{5}$ by $2\frac{2}{7}$.

Solution When you divide mixed numbers, you should always begin by converting them to improper fractions. Look for possible cancellations and then complete the division.

$$1\frac{3}{5} \div 2\frac{2}{7} = \frac{8}{5} \div \frac{16}{7} = \frac{\overset{1}{\cancel{8}}}{5} \times \frac{7}{\underset{2}{\cancel{16}}} = \frac{7}{10}$$

Example 6 Divide $6\frac{2}{3} \div \frac{5}{18}$

Solution

$$6\frac{2}{3} \div \frac{5}{18} = \frac{20}{3} \div \frac{5}{18} = \frac{\overset{4}{\cancel{20}}}{\underset{1}{\cancel{3}}} \times \frac{\overset{6}{\cancel{18}}}{\underset{1}{\cancel{5}}} = 24$$

Example 7 Divide $1\frac{3}{5} \div 2\frac{1}{3}$

Solution $1\frac{3}{5} \div 2\frac{1}{3} = \frac{8}{5} \div \frac{7}{3} = \frac{8}{5} \times \frac{3}{7} = \frac{24}{35}$

Example 8 Find the value of $4 \div \frac{1}{2}$. This may also be expressed, "How many times is $\frac{1}{2}$ contained in 4?"

Solution 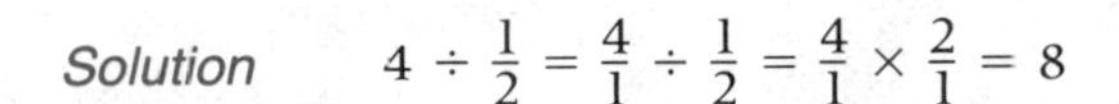

$4 \div \frac{1}{2} = \frac{4}{1} \div \frac{1}{2} = \frac{4}{1} \times \frac{2}{1} = 8$

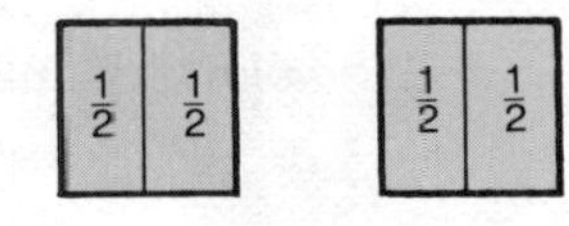

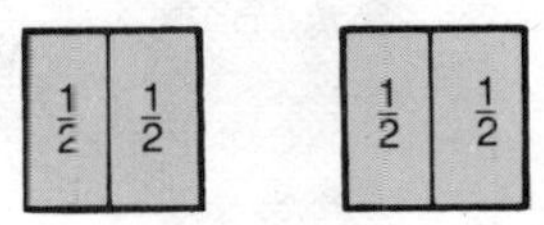

Example 9 A carpenter has a piece of molding $2\frac{3}{4}$ feet long and he must divide it into 4 equal parts. How long should each piece of molding be?

Solution $2\frac{3}{4} \div 4 = \frac{11}{4} \times \frac{1}{4} = \frac{11}{16}$ of a foot

Example 10 In the course of a year, Nicole's Candy shipped 19,272 pounds of hard candy in containers that hold $48\frac{2}{3}$ pounds each. How many containers of candy were shipped?

Solution $19{,}272 \div 48\frac{2}{3} = 19{,}272 - \frac{146}{3} = 19{,}272 \times \frac{3}{146} = 396$ containers

Complex Fractions A fraction that has a number other than a whole number in its numerator or denominator or both is called a **complex fraction.** Some complex fractions are:

$$\frac{\frac{2}{3}}{\frac{7}{8}}, \quad \frac{5}{\frac{1}{2}}, \quad \frac{\frac{3}{4}}{9}, \text{ and } \frac{\frac{3}{5}}{\frac{4}{10}}$$

How do we simplify these types of complex fractions? We do the indicated division. Here is an illustration:

$$\frac{\frac{2}{3}}{\frac{7}{8}}$$

denotes $\frac{2}{3} \div \frac{7}{8}$. Observe the longer fraction line used to emphasize this division.

$$\frac{\frac{2}{3}}{\frac{7}{8}} = \frac{2}{3} \div \frac{7}{8} = \frac{2}{3} \times \frac{8}{7} = \frac{16}{21}$$

Example 11 Simplify $\frac{\frac{5}{8}}{\frac{10}{24}}$.

Solution

$$\frac{\frac{5}{8}}{\frac{10}{24}} = \frac{5}{8} \div \frac{10}{24} = \frac{\overset{1}{\cancel{5}}}{\underset{1}{\cancel{8}}} \times \frac{\overset{3}{\cancel{24}}}{\underset{2}{\cancel{10}}} = \frac{3}{2} = 1\frac{1}{2}$$

Example 12 Simplify $\frac{\frac{9}{10}}{\frac{27}{25}}$.

Solution

$$\frac{\frac{9}{10}}{\frac{27}{25}} = \frac{9}{10} \div \frac{27}{25} = \frac{\overset{1}{\cancel{9}}}{\underset{2}{\cancel{10}}} \times \frac{\overset{5}{\cancel{25}}}{\underset{3}{\cancel{27}}} = \frac{5}{6}$$

Example 13 Simplify $\frac{\frac{3}{4}}{7}$.

Solution $\frac{3}{4} \div 7 = \frac{3}{4} \times \frac{1}{7} = \frac{3}{28}$

2.7 EXERCISES

In problems 1–14 find the reciprocal of the given number.

1. $\frac{6}{7}$ ____________ **2.** $\frac{2}{5}$ ____________

3. $\frac{3}{11}$ __________ 4. 5 __________

5. $\frac{1}{8}$ __________ 6. $\frac{7}{4}$ __________

7. $\frac{2}{9}$ __________ 8. $\frac{11}{6}$ __________

9. 3 __________ 10. 8 __________

11. $\frac{10}{3}$ __________ 12. $\frac{5}{16}$ __________

13. $\frac{1}{9}$ __________ 14. $\frac{1}{12}$ __________

15. What is the product of $\frac{3}{4}$ times its reciprocal? __________

16. What is the product of $\frac{7}{8}$ times its reciprocal? __________

In problems 17–36 state your answer as a fraction in simplest form or as a mixed number in simplest form.

17. $4 \div \frac{1}{3}$ __________ 18. $5 \div \frac{1}{5}$ __________

19. $\frac{2}{7} \div 3$ __________ 20. $\frac{3}{4} \div 5$ __________

21. $6 \div \frac{1}{4}$ __________ 22. $\frac{1}{4} \div 6$ __________

23. $\frac{1}{2} \div 10$ __________ 24. $10 \div \frac{1}{2}$ __________

25. $\frac{3}{4} \div \frac{7}{3}$ __________ 26. $\frac{7}{9} \div \frac{2}{3}$ __________

27. $\frac{3}{5} \div \frac{5}{6}$ __________ 28. $\frac{10}{21} \div \frac{20}{7}$ __________

29. $\frac{8}{11} \div \frac{7}{8}$ __________ 30. $\frac{7}{10} \div \frac{5}{3}$ __________

31. $\frac{15}{33} \div \frac{3}{11}$ __________ 32. $\frac{15}{24} \div \frac{5}{16}$ __________

33. $\frac{11}{6} \div \frac{22}{3}$ __________ 34. $\frac{3}{7} \div \frac{3}{7}$ __________

35. $\frac{8}{15} \div \frac{24}{25}$ __________ 36. $8 \div \frac{2}{3}$ __________

Check your answers to problems 17–36. If you did not get at least 16 problems correct turn to Practice Problem Set 8A in the Appendix.

In problems 37–56 state your answer as a fraction in simplest form or as a mixed number in simplest form.

37. $\frac{5}{8} \div 5$ __________ 38. $2 \div \frac{2}{5}$ __________

39. $2\frac{1}{4} \div \frac{15}{8}$ __________ 40. $2\frac{5}{8} \div \frac{3}{8}$ __________

41. $5\frac{2}{9} \div 4\frac{5}{6}$ __________

42. $\frac{10}{21} \div 4\frac{1}{6}$ __________

43. $3\frac{1}{5} \div \frac{4}{15}$ __________

44. $35 \div \frac{7}{5}$ __________

45. $\frac{9}{4} \div 18$ __________

46. $\frac{100}{63} \div \frac{10}{9}$ __________

47. $\frac{225}{81} \div \frac{25}{27}$ __________

48. $\frac{225}{17} \div \frac{15}{34}$ __________

49. $4\frac{1}{2} \div 2\frac{1}{10}$ __________

50. $3\frac{1}{16} \div 6\frac{1}{2}$ __________

51. $6\frac{2}{5} \div 1\frac{4}{10}$ __________

52. $3\frac{3}{4} \div 5\frac{7}{8}$ __________

53. $5\frac{1}{4} \div 2\frac{1}{4}$ __________

54. $1\frac{5}{7} \div 4\frac{5}{7}$ __________

55. $2\frac{7}{8} \div 1\frac{1}{2}$ __________

56. $5\frac{3}{10} \div 2\frac{1}{5}$ __________

In problems 57–72 simplify the given complex fraction.

57. $\dfrac{\frac{3}{7}}{\frac{9}{14}}$ __________

58. $\dfrac{\frac{8}{9}}{16}$ __________

59. $\dfrac{\frac{4}{10}}{\frac{2}{25}}$ __________

60. $\dfrac{\frac{3}{10}}{\frac{5}{7}}$ __________

61. $\dfrac{\frac{2}{3}}{\frac{7}{8}}$ __________

62. $\dfrac{\frac{5}{7}}{10}$ __________

63. $\dfrac{7}{\frac{14}{5}}$ __________

64. $\dfrac{\frac{63}{81}}{\frac{77}{90}}$ __________

65. $\dfrac{\frac{14}{3}}{8}$ __________

66. $\dfrac{4}{\frac{16}{15}}$ __________

67. $\dfrac{\frac{20}{30}}{\frac{10}{15}}$ __________

68. $\dfrac{\frac{28}{105}}{\frac{35}{125}}$ __________

69. $\dfrac{\frac{4}{5}}{\frac{8}{10}}$ __________

70. $\dfrac{\frac{2}{9}}{\frac{6}{12}}$ __________

71. $\dfrac{\frac{50}{100}}{2}$ ____________ **72.** $\dfrac{\frac{22}{6}}{\frac{33}{5}}$ ____________

73. How many pieces of rope, each $2\frac{3}{5}$ inches long, can be cut from a 65-inch piece of rope?

____

74. Claire Hendrickson paid \$6,435 for stock costing $\$12\frac{3}{8}$ per share. How many shares did she buy?

____

75. A farmer packs peaches in baskets that contain $\frac{2}{3}$ of a bushel. If he picks 14 full bushels, how many baskets will he fill?

____

76. A board is to be cut into $2\frac{1}{2}$-foot lengths. How many pieces can be cut from a board 15 feet long?

____

77. A car averages $27\frac{1}{2}$ miles per gallon of gasoline on a trip. How many gallons of gasoline does the car use on a 440-mile trip?

____

78. Joni runs a $6\frac{1}{5}$-mile race in 31 minutes. What is her average rate of speed per minute?

____

79. Alex owns a $12\frac{2}{3}$-acre tract of land, and he plans to subdivide this tract into $\frac{1}{4}$-acre lots. If he allows one-sixth of the land for roads, how many lots will the tract yield?

____

ANSWERS

1. $\frac{7}{6}$ **2.** $\frac{5}{2}$ **3.** $\frac{11}{3}$ **4.** $\frac{1}{5}$ **5.** 8 **6.** $\frac{4}{7}$ **7.** $\frac{9}{2}$ **8.** $\frac{6}{11}$ **9.** $\frac{1}{3}$ **10.** $\frac{1}{8}$
11. $\frac{3}{10}$ **12.** $\frac{16}{5}$ **13.** 9 **14.** 12 **15.** 1 **16.** 1 **17.** 12 **18.** 25 **19.** $\frac{2}{21}$
20. $\frac{3}{20}$ **21.** 24 **22.** $\frac{1}{24}$ **23.** $\frac{1}{20}$ **24.** 20 **25.** $\frac{9}{28}$ **26.** $\frac{7}{6}$ **27.** $\frac{18}{25}$
28. $\frac{1}{6}$ **29.** $\frac{64}{77}$ **30.** $\frac{21}{50}$ **31.** $\frac{5}{3}$ **32.** 2 **33.** $\frac{1}{4}$ **34.** 1 **35.** $\frac{5}{9}$ **36.** 12
37. $\frac{1}{8}$ **38.** 5 **39.** $1\frac{1}{5}$ **40.** 7 **41.** $1\frac{7}{87}$ **42.** $\frac{4}{35}$ **43.** 12 **44.** 25 **45.** $\frac{1}{8}$
46. $1\frac{3}{7}$ **47.** 3 **48.** 30 **49.** $2\frac{1}{7}$ **50.** $\frac{49}{104}$ **51.** $4\frac{4}{7}$ **52.** $\frac{30}{47}$ **53.** $2\frac{1}{3}$ **54.** $\frac{4}{11}$
55. $1\frac{11}{12}$ **56.** $2\frac{9}{22}$ **57.** $\frac{2}{3}$ **58.** $\frac{1}{18}$ **59.** 5 **60.** $\frac{21}{50}$ **61.** $\frac{16}{21}$ **62.** $\frac{1}{14}$ **63.** $2\frac{1}{2}$
64. $\frac{10}{11}$ **65.** $\frac{7}{12}$ **66.** $3\frac{3}{4}$ **67.** 1 **68.** $\frac{20}{21}$ **69.** 1 **70.** $\frac{4}{9}$ **71.** $\frac{1}{4}$ **72.** $\frac{5}{9}$
73. 25 pieces **74.** 520 shares **75.** 21 baskets **76.** 6 pieces **77.** 16 gallons
78. $\frac{1}{5}$ mile per minute **79.** 42 lots

2.8 COMMON FRACTIONAL PROBLEMS

How much is $\frac{3}{4}$ of \$40? If $\frac{3}{8}$ of a cheesecake weighs $2\frac{7}{10}$ ounces, how much does the entire cake weigh? If $\frac{3}{4}$ of a quantity costs \$9, how much will $\frac{7}{8}$ of the same quantity cost?

These questions represent three different common fractional problems: (1) finding a part of a whole quantity; (2) finding the whole quantity given a part; and (3) given one part of a whole quantity, finding another part of the same whole quantity. In our next examples we will solve each of these three common types of fractional problems.

Example 1 How much is $\frac{3}{4}$ of \$40?

Solution To find a fractional part of a whole quantity, multiply the whole quantity by the specified fractional part. Thus:

$$\frac{3}{4} \times \$40 = \$30$$

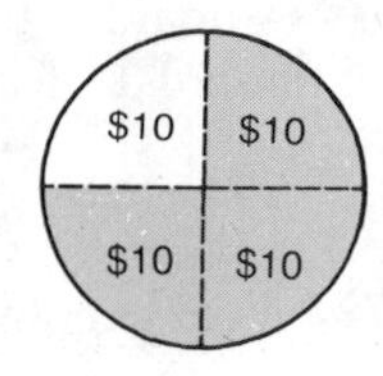

Example 2 Find $\frac{4}{5}$ of 70.

Solution

$$\frac{4}{5} \times 70 = \frac{4}{\cancel{5}_{1}} \times \frac{\cancel{70}^{14}}{1} = 56$$

Once again, notice that the word "of" signals a multiplication.

Example 3 $\frac{3}{5}$ of what quantity equals 150?

Solution To find a whole quantity given a fractional amount of this quantity, divide the fractional amount by the fraction. Thus:

$$150 \div \frac{3}{5} = \cancel{150}^{50} \times \frac{5}{\cancel{3}_{1}} = 250$$

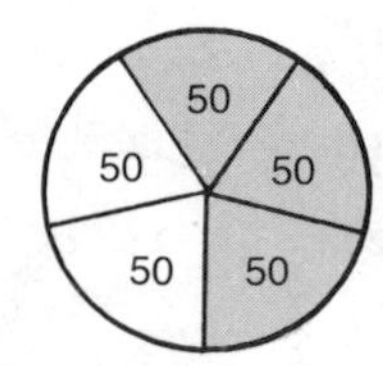

Example 4 If $\frac{3}{8}$ of a cheesecake weighs $2\frac{7}{10}$ ounces, how much does the entire cake weigh?

Solution Once again, we are asked to find a whole quantity given a fractional part of this quantity.

$$2\frac{7}{10} \div \frac{3}{8} = 2\frac{7}{10} \times \frac{8}{3} = \frac{\cancel{27}^{9}}{\cancel{10}_{5}} \times \frac{\cancel{8}^{4}}{\cancel{3}_{1}} = \frac{36}{5} = 7\frac{1}{5} \text{ ounces}$$

Example 5 If $\frac{2}{5}$ of an electric range costs \$172, what is the cost of the range?

Solution

$$\$344 \div \frac{2}{5} = \$\overset{172}{\cancel{344}} \times \frac{5}{\underset{1}{\cancel{2}}} = \$860$$

Example 6 If $\frac{3}{4}$ of a quantity costs \$12, how much does $\frac{7}{8}$ of the same quantity cost?

Solution To solve this problem we combine the techniques learned in the previous examples. First, given that $\frac{3}{4}$ costs \$12, we must determine the value of the whole quantity. Second, once we know the value of the whole quantity, we must determine $\frac{7}{8}$ of it. This is a two-step process as follows:

step 1: $\$12 \div \frac{3}{4} = \$12 \times \frac{4}{3} = \$16$ (whole quantity)

step 2: $\frac{7}{8} \times \$16 = \$14$ (answer)

Example 7 Ira has saved \$60 for a piece of stereo equipment. This amount represents $\frac{2}{5}$ of the cost of the equipment. His father tells Ira that if he can save $\frac{9}{10}$ of the cost he will pay the remaining $\frac{1}{10}$. What is the total amount that Ira must save? How much money is Ira's father willing to pay?

Solution step 1: $\$60 \div \frac{2}{5} = \$60 \times \frac{5}{2} = \$150$ (total cost)

step 2: $\frac{9}{10} \times \$150 = \$135$ (amount Ira must save)

His father will pay $\$150 - \$135 = \$15$.

2.8 EXERCISES

1. Find $\frac{3}{8}$ of \$16. __________

2. Find $\frac{7}{11}$ of \$99. __________

3. Find $\frac{2}{5}$ of \$75. __________

4. Find $\frac{4}{5}$ of \$120. __________

5. Find $\frac{7}{8}$ of \$96. __________

6. Find $\frac{5}{6}$ of \$72. __________

7. $\frac{2}{3}$ of what quantity equals 12? __________

8. $\frac{7}{8}$ of what quantity equals 49? __________

9. $\frac{3}{4}$ of what quantity equals 270? __________

10. $\frac{5}{6}$ of what quantity equals 700? ________________

11. $\frac{4}{11}$ of what quantity equals 44? ________________

12. $\frac{2}{7}$ of what quantity equals 100? ________________

13. If $\frac{5}{6}$ of a quantity equals 40, how much is $\frac{1}{3}$ of this quantity?

____

14. If $\frac{2}{3}$ of a quantity equals 72, how much is double the whole quantity?

____

15. If $\frac{3}{4}$ of a quantity equals \$6, how much is double the whole quantity?

____

16. If $\frac{7}{8}$ of a quantity equals \$28, how much is $\frac{1}{8}$ of the whole quantity?

____

17. If $\frac{5}{9}$ of a quantity equals 45, how much is $\frac{1}{3}$ of the whole quantity?

____

18. If $\frac{4}{7}$ of a quantity equals 36, how much is $\frac{1}{3}$ of the whole quantity?

____

19. A woman drove $75\frac{6}{10}$ miles, which was $\frac{3}{15}$ of the distance to her destination. Find the total distance that she planned to travel.

____

20. If $\frac{2}{21}$ of a professional basketball player's annual salary is \$8,148, what is his annual salary?

____

21. If $\frac{3}{48}$ of the total cost of an automobile is tax and delivery charge in the amount of $\$638\frac{2}{5}$, find the total cost of the automobile. What is the total cost excluding tax and delivery charge?

____

22. A book shelf is 42 inches long. How many books that are $\frac{3}{4}$ inch thick can be stocked on this shelf?

____

23. Steven eats $\frac{1}{4}$ of a pie. Later that day Brian eats $\frac{1}{2}$ of what is left of the pie. How much pie has been eaten? How much of the pie is left?

____

24. Paula takes 2 hours to do a job and Pat takes 3 hours to do the same job. How many jobs (including parts) are completed if each works 5 hours.

____

25. A committee is $\frac{2}{3}$ female. One-fourth of the male members are married. If there are 9 bachelors on this committee, how many members are there altogether?

____

26. If $p = \frac{3}{5}q$ and q is 1, 2, 3, . . . , which of the following could be a value of p: 2, 3, 4, 5, 8?

____

ANSWERS

1. \$6 **2.** \$63 **3.** \$30 **4.** \$96 **5.** \$84 **6.** \$60 **7.** 18 **8.** 56 **9.** 360 **10.** 840 **11.** 121 **12.** 350 **13.** 16 **14.** 216 **15.** \$16 **16.** \$4 **17.** 27

18. 21 **19.** 378 miles **20.** \$85.554 **21.** \$10,214$\frac{2}{5}$; \$9,576 **22.** 56 books

23. $\frac{5}{8}$; $\frac{3}{8}$ **24.** $4\frac{1}{6}$ **25.** 36 **26.** 3

2.9 RATIO AND PROPORTION

To compare the daily output of two oil wells, one producing 750 barrels of oil and the other 250 barrels of oil, we compute the quotient $\frac{750}{250} = 3$ or 3 to 1. This tells us that one well is producing three times what the other well is producing. That is, the well with the larger output is producing three barrels of oil for each barrel produced by the other well.

Ratio

The **ratio** of two numbers is the quotient of the first number divided by the second number. The ratio of 5 to 4 may be expressed in three ways:

1. by using a division sign; $5 \div 4$
2. by using a ratio sign, ":"; 5:4
3. by using a fraction; $\frac{5}{4}$

Example 1a A company has \$7,250,000 in assets and \$2,500,000 in liabilities. Compute the ratio of assets to liabilities for this company.

Solution Order is important in writing a ratio. We want the ratio of assets to liabilities. Thus we write the assets number in the numerator or first:

$$\frac{\$7{,}250{,}000}{\$2{,}500{,}000} = \frac{7{,}250}{2{,}500} = \frac{29}{10}, \quad \text{or} \quad 29{:}10$$

Example 1b Return to Example 1a and compute the ratio of liabilities to assets.

Solution

$$\frac{\$2{,}500{,}000}{\$7{,}250{,}000} = \frac{2{,}500}{7{,}250} = \frac{10}{29}, \quad \text{or} \quad 10{:}29$$

The ratio of 2 pounds to 4 pounds is $\frac{2}{4}$, which is equivalent to $\frac{1}{2}$. This ratio may also be written 2:4 or 1:2. Similarly, the ratio of 15 apples to 20 apples is 15:20 or 3:4. A ratio has no units, such as feet, pounds, or grams.

A ratio such as 5:4 compares not only 5 to 4 but 10 to 8, 15 to 12, 25 to 20 and so on. *A ratio is usually a comparison of two like quantities.* This allows us to obtain a unique answer. For example, if you wish to compare 2 feet to 4 inches (2 feet = 24 inches), you have the ratio of 24 inches to 4 inches or 6:1.

Example 2 What is the ratio of 14 yards to 63 feet?

Solution In order to compare like measures, we must change yards to feet or feet to yards. If we change 63 feet to 21 yards, we obtain 14:21, or 2:3, as our ratio.

However, there are applied instances when it is convenient to compare "unlike quantities," such as a student-teacher ratio of 25 to 1, or 25:1.

Example 3 On a particular map 1 inch represents 5 miles. What is the ratio of inches to miles?

Solution 1 to 5, or 1:5

Proportion

One year of a dog's life is equivalent to 7 years of man's life, 1:7. How many years of a dog's life, ? , are equivalent to 35 years of a man's life? Here we have two equal ratios, $\frac{1}{7}$ and $\frac{?}{35}$. That is, $\frac{1}{7} = \frac{?}{35}$. By inspecting the two equal fractions, you can see that the ? must be 5 or 5 years.

A **proportion** is a statement that two ratios are equal. Some examples of proportions are: $\frac{2}{3} = \frac{10}{15}$, $\frac{1}{4} = \frac{25}{100}$, and $\frac{35}{40} = \frac{7}{8}$. These same proportions may also be expressed 2:3 = 10:15, 1:4 = 25:100, and 35:40 = 7:8. *In general, a proportion is of the form* $\frac{a}{b} = \frac{c}{d}$, but it is sometimes written in the form $a{:}b = c{:}d$. The four numbers, a, b, c, and d, are called the *terms of the proportion.* The first and last terms, a and d, are called the **extremes;** the second and third terms, b and c, are called the **means.**

$$\begin{array}{l} \text{extreme} \rightarrow \\ \text{mean} \rightarrow \end{array} \frac{a}{b} = \frac{c}{d} \begin{array}{l} \leftarrow \text{mean} \\ \leftarrow \text{extreme} \end{array}$$

Rule for Proportions

In a proportion, the product of the means equals the product of the extremes: $a \cdot d = b \cdot c$ or $ad = bc$.

Example 4 Given the proportion $\frac{45}{65} = \frac{9}{13}$, verify that the product of the means equals the product of the extremes.

Solution The means are 65 and 9. Their product is $(65)(9) = 585$.
The extremes are 45 and 13. Their product is $(45)(13) = 585$.

Problems are frequently encountered in nursing, chemistry, business and so on in which three of the four terms of a proportion are given and you are to find the fourth term or "solve the proportion" for the missing term. It is convenient to name the missing term by a letter, such as x. That is, x is the value of the unknown term. This convention allows us to discuss the missing term prior to determining its actual value. *The key to solving a proportion is remembering that the product of the means equals the product of the extremes (Rule for Proportions).* This is illustrated in the next examples.

Example 5 If $\frac{x}{8} = \frac{15}{24}$, find the value of x.

Solution

$$24 \cdot x = 8 \cdot 15 \quad \text{(Rule for Proportions)}$$

$$24 \cdot x = 120$$

$$\frac{\overset{1}{\cancel{24}} \cdot x}{\underset{1}{\cancel{24}}} = \frac{\overset{5}{\cancel{120}}}{\underset{1}{\cancel{24}}} \quad \text{(divide each side by 24)}$$

$$x = 5$$

Check Replace x by 5 to obtain $\frac{5}{8}$. Can you see that $\frac{15}{24}$ is equivalent to $\frac{5}{8}$? Because this implies two equal ratios, the solution $x = 5$ is correct. As an alternative, you may wish to observe that the product of the means, $(8)(15) = 120$ and the product of the extremes $(5)(24) = 120$. Because the Rule for Proportions is satisfied, you have a correct answer.

Example 6 If $\frac{12}{8} = \frac{3}{x}$, find the value of x.

Solution

$$12 \cdot x = 8 \cdot 3 \quad \text{(Rule for Proportions)}$$

$$12 \cdot x = 24$$

$$\frac{\overset{1}{\cancel{12}} \cdot x}{\underset{1}{\cancel{12}}} = \frac{\overset{2}{\cancel{24}}}{\underset{1}{\cancel{12}}} \quad \text{(divide each side by 12)}$$

$$x = 2$$

Example 7 If $\frac{5}{7} = \frac{x}{11}$, find the value of x.

Solution

$$5 \cdot 11 = 7 \cdot x \quad \text{(Rule for Proportions)}$$

$$55 = 7x$$

$$\frac{55}{7} = \frac{7x}{7} \quad \text{(divide each side by 7)}$$

$$7\tfrac{6}{7} = x$$

Example 8 Proportions are helpful in solving scale problems. Given

$$\frac{1 \text{ inch}}{x \text{ inches}} = \frac{30 \text{ blocks}}{135 \text{ blocks}}, \text{ find the value of } x.$$

Solution

$$30 \cdot x = 135 \quad \text{(Rule for Proportions)}$$

$$\frac{\overset{1}{\cancel{30}} \cdot x}{\underset{1}{\cancel{30}}} = \frac{\overset{9}{\cancel{135}}}{\underset{2}{\cancel{30}}} \quad \text{(divide by 30)}$$

$$x = \frac{9}{2}, \text{ or } 4\frac{1}{2}$$

Comparing Unlike Quantities In any proportion, due to the Rule for Proportions, you may interchange either the means or the extremes. The result of doing this is a different but true proportion. Let's illustrate how this may be done with the familiar concepts of feet and yards.

Given: $\frac{3 \text{ feet}}{12 \text{ feet}} = \frac{1 \text{ yard}}{4 \text{ yards}}$

When the means are interchanged we get: $\frac{3 \text{ feet}}{\boxed{1 \text{ yard}}} = \frac{\boxed{12 \text{ feet}}}{4 \text{ yards}}$

When the extremes are interchanged we get: $\frac{\boxed{4 \text{ yards}}}{12 \text{ feet}} = \frac{1 \text{ yard}}{\boxed{3 \text{ feet}}}$

Observe, in each case your ratios are comparisons of unlike quantities.

Applications of Proportions Setting up a proportion in which the ratios are comparisons of unlike quantities is useful when we are solving problems. Because proportions are helpful in solving a variety of applied problems, we will (1) offer a step-by-step process for solving problems involving proportions, and (2) present a sequence of applied examples.

Solving Problems Involving Proportions

1. Identify the unknown and label it x.
2. Make a proportion by setting up two equal ratios. Frequently, your ratios will involve comparisons of unlike quantities, such as feet and yards.
3. Use the Rule for Proportions (product of the means equals the product of the extremes).
4. Divide each side of your statement by the number multiplying x.

Example 9 Children's medicine is administered by dosage per weight. If 2 milliliters of a liquid medicine is given for every 15 pounds, how many milliliters should be given to a child weighing 25 pounds?

Solution step 1: Let x be the dosage for a 25-pound child.

step 2: $\frac{2 \text{ milliliters}}{15 \text{ pounds}} = \frac{x \text{ milliliters}}{25 \text{ pounds}}$

step 3: $15 \cdot x = 2 \cdot 25$

step 4: $\frac{\overset{1}{\cancel{15}} \cdot x}{\underset{1}{\cancel{15}}} = \frac{\overset{10}{\cancel{50}}}{\underset{3}{\cancel{15}}}$

$x = 3\frac{1}{3}$ milliliters

Example 10 A flagpole casts a shadow 40 feet long. Next to it, a 3-foot ruler casts a shadow 2 feet long. How tall is the flagpole?

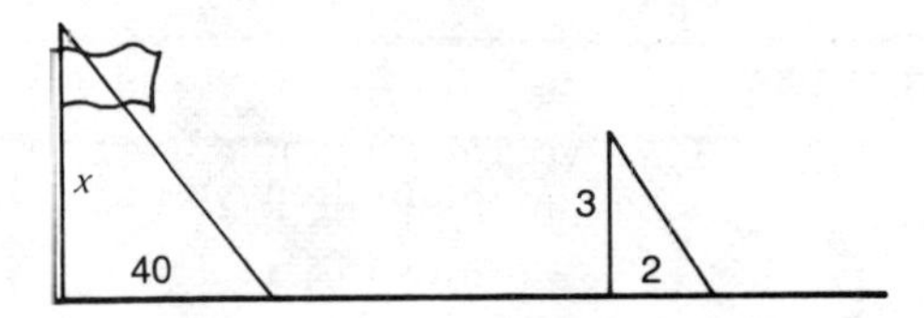

Solution Since the triangles in our figure formed by the shadows have the same shape, we can assume that the sides are proportional: height to shadow. (This fact can be proven in geometry.)

step 1: Let x be the height of the flagpole.

step 2: $\frac{x \text{ (height)}}{40 \text{ (shadow)}} = \frac{3 \text{ (height)}}{2 \text{ (shadow)}}$

step 3: $2 \cdot x = 3 \cdot 40$

step 4: $\frac{\overset{1}{\cancel{2}} \cdot x}{\underset{1}{\cancel{2}}} = \frac{\overset{60}{\cancel{120}}}{\underset{1}{\cancel{2}}}$

$x = 60$ feet

Example 11 If a baseball player hits 10 home runs in the first 45 games of the season, how many home runs can he expect to hit in 162 games?

Solution Let x be the number of home runs in 162 games

$$\frac{x \text{ home runs}}{162 \text{ games}} = \frac{10 \text{ home runs}}{45 \text{ games}}$$

$45 \cdot x = 10 \cdot 162$

$x = 36$

Example 12 If 1 meter is approximately equal to $3\frac{1}{2}$ feet, how many feet are approximately equal to 8 meters?

Solution Let x be the number of feet equivalent to 8 meters.

$$\frac{x \text{ feet}}{8 \text{ meters}} = \frac{\frac{10}{3} \text{ feet}}{1 \text{ meter}}$$

$x = (8)\left(\frac{10}{3}\right)$

$x = \frac{80}{3}$ feet $= 26\frac{2}{3}$ feet

Example 13 Fred drove 450 miles on 25 gallons of gasoline. How far can he drive on 60 gallons of gasoline?

Solution Let x be the distance traveled using 60 gallons.

$$\frac{x \text{ miles}}{60 \text{ gallons}} = \frac{450 \text{ miles}}{25 \text{ gallons}}$$

$$25 \cdot x = (60)(450)$$

$$x = 1{,}080 \text{ miles}$$

2.9 EXERCISES

1. State the given ratio in simplest (reduced) form.

(a) 4:8 ______ (b) 5:15 ______ (c) 4 meters:16 meters ______

2. State the given ratio in simplest form.

(a) 36:16 ______ (b) 20:25 ______ (c) \$0.35:\$0.85 ______

3. State the given ratio in simplest form.

(a) 1 pound:1 ounce ______ (b) 12 months:2 years ______

(c) 3 feet:40 inches ______

4. A company has assets of \$8,400,000 and liabilities of \$2,800,000.

(a) Compute the ratio of assets to liabilities. ______

(b) Compute the ratio of liabilities to assets. ______

5. A profit of \$360 is realized on a coat that sold for \$1,200.

(a) Compute the ratio of profit to selling price. ______

(b) Compute the ratio of profit to cost. ______

6. Given $\frac{3}{4} = \frac{15}{20}$. (a) State the means. ______ (b) State the extremes. ______

7. Express the ratio of 3 to 2 in three ways. ______

8. Express the ratio of 6 to 5 in three ways. ______

9. What is the ratio of 20 quarters to 4 quarters? ______

10. What is the ratio of 50 pennies to 40 pennies? ______

11. Given $\frac{a}{b} = \frac{c}{d}$. (a) State the means. ______ (b) State the extremes. ______

12. In a class of 30 students, 6 students received honor grades. What is the ratio of honor students to total enrollment?

______

In problems 13–21 solve each proportion for the unknown.

13. $\frac{7}{8} = \frac{x}{40}$ __________

14. $\frac{9}{2} = \frac{63}{y}$ __________

15. $\frac{11}{x} = \frac{132}{24}$ __________

16. $\frac{2}{3} = \frac{10}{n}$ __________

17. $\frac{5}{13} = \frac{x}{65}$ __________

18. $\frac{9}{8} = \frac{y}{32}$ __________

19. $\frac{7}{x} = \frac{1}{9}$ __________

20. $\frac{12}{5} = \frac{60}{x}$ __________

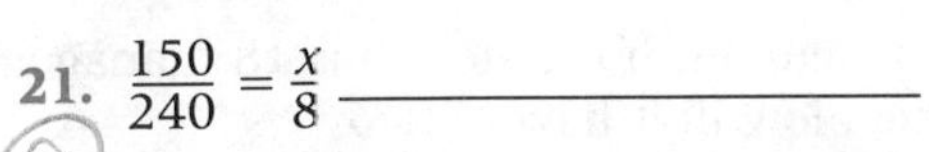

21. $\frac{150}{240} = \frac{x}{8}$ __________

22. If the ratio of a to b is 3 to 1, what is the value of a if $b = 24$?

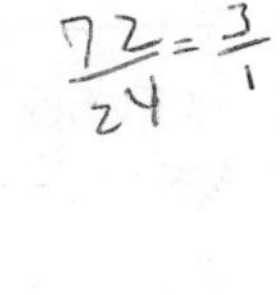

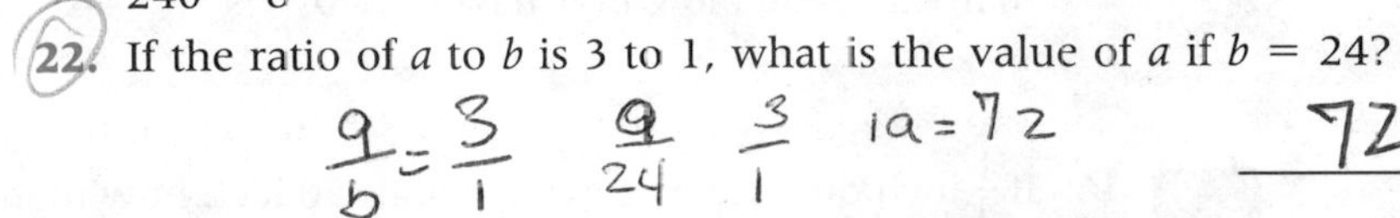

____

23. If the ratio of a to b is 3 to 4, what is the value of a if $b = 60$?

____

24. If three oranges cost 90 cents, how much will 12 oranges cost?

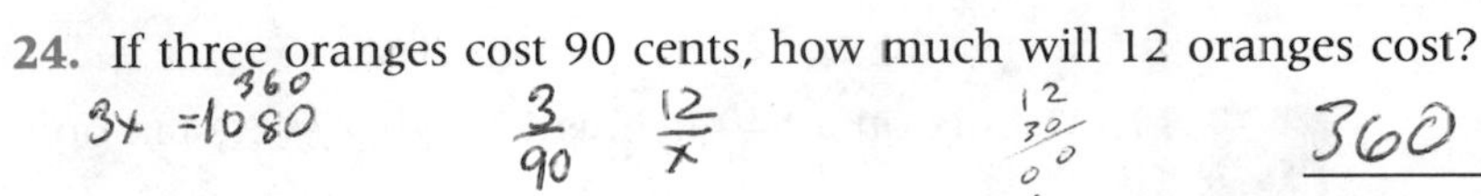

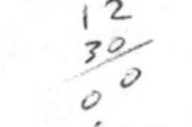

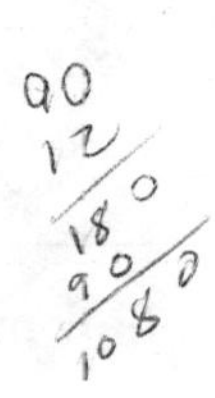

____

25. If one inch equals $2\frac{1}{2}$ centimeters, how many inches equal 100 centimeters?

____

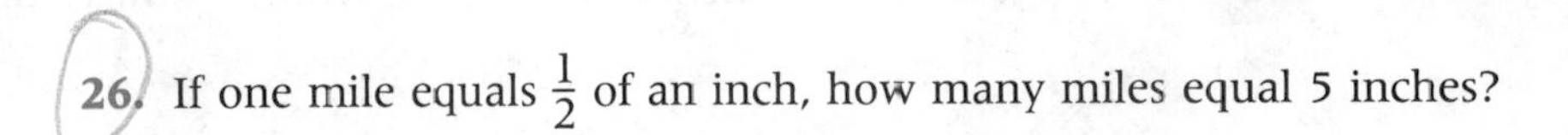

26. If one mile equals $\frac{1}{2}$ of an inch, how many miles equal 5 inches?

____

27. If 90 feet of wire weighs 18 pounds, what will 110 feet of the same wire weigh?

____

28. What actual distance is represented on a map by $2\frac{1}{4}$ inches if 60 miles is represented by $1\frac{1}{2}$ inches?

____

29. If $\frac{70}{n} = \frac{5}{7}$, find the value of n. __________

30. If $\frac{n}{48} = \frac{5}{3}$, find the value of n. __________

31. Brian is the clean-up hitter on the Braintree Bombers. At present he has 75 hits in 110 times at bat. At this rate, how many hits (to the nearest whole number) will he get in 300 times at bat?

____

32. A model clipper ship is constructed on the scale of 10 feet to $1\frac{1}{2}$ inches on the model. How high would the mast be on the model if it is actually 75 feet high?

____

33. Hemingway and Gibson are partners who divide their profits in the ratio of 3:5. Find Hemingway's profit if Gibson receives $15,000.

____

34. The owner of a department store has kept the ratio of management-level people to salespeople at 2:9. If the store currently has 180 salespeople, how many management-level employees are there?

____

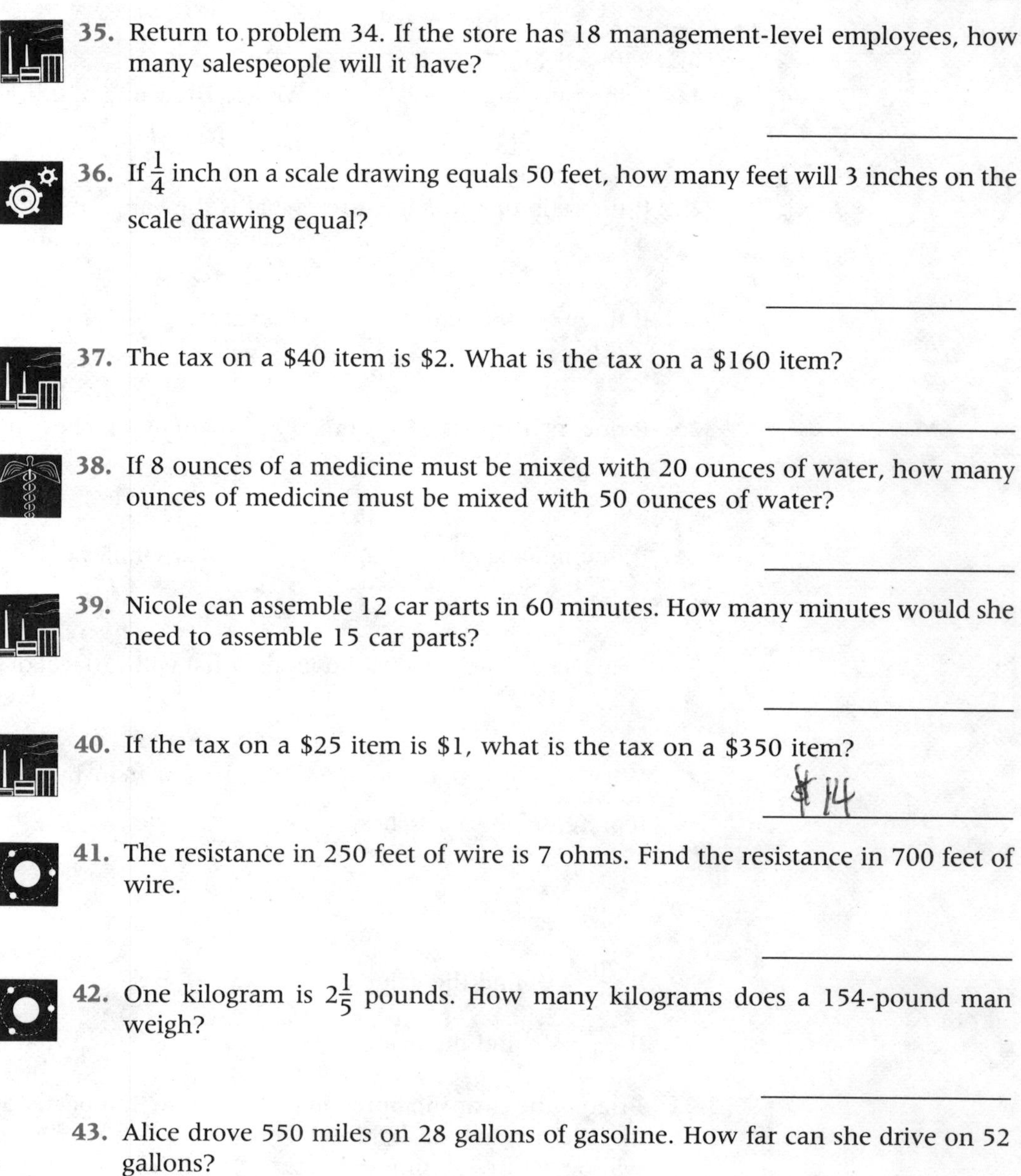

35. Return to problem 34. If the store has 18 management-level employees, how many salespeople will it have?

____

36. If $\frac{1}{4}$ inch on a scale drawing equals 50 feet, how many feet will 3 inches on the scale drawing equal?

____

37. The tax on a $40 item is $2. What is the tax on a $160 item?

____

38. If 8 ounces of a medicine must be mixed with 20 ounces of water, how many ounces of medicine must be mixed with 50 ounces of water?

____

39. Nicole can assemble 12 car parts in 60 minutes. How many minutes would she need to assemble 15 car parts?

____

40. If the tax on a $25 item is $1, what is the tax on a $350 item?

$14
____

41. The resistance in 250 feet of wire is 7 ohms. Find the resistance in 700 feet of wire.

____

42. One kilogram is $2\frac{1}{5}$ pounds. How many kilograms does a 154-pound man weigh?

____

43. Alice drove 550 miles on 28 gallons of gasoline. How far can she drive on 52 gallons?

____

44. If 100 capsules contain $\frac{3}{4}$ of a grain of medication, how many capsules contain $\frac{1}{5}$ of a grain of medication?

____

ANSWERS

1. (a) 1:2 **(b)** 1:3 **(c)** 1:4 **2. (a)** 9:4 **(b)** 4:5 **(c)** 7:17 **3. (a)** 16:1 **(b)** 1:2 **(c)** 9:10 **4. (a)** 3:1 **(b)** 1:3 **5. (a)** 3:10 **(b)** 3:7 **6. (a)** 4, 15 **(b)** 3, 20 **7.** $\frac{3}{2}$; 3:2; $3 \div 2$ **8.** $\frac{6}{5}$; 6:5; $6 \div 5$ **9.** 5:1 **10.** 5:4 **11. (a)** *b, c* **(b)** *a, d* **12.** 1:5 **13.** $x = 35$ **14.** $x = 14$ **15.** $x = 2$ **16.** $n = 15$ **17.** $x = 25$ **18.** $y = 36$ **19.** $x = 63$ **20.** $x = 25$ **21.** $x = 5$ **22.** $a = 72$ **23.** $a = 45$ **24.** 360 cents **25.** 40 inches **26.** 10 miles **27.** 22 pounds **28.** 90 miles **29.** $n = 98$ **30.** $n = 80$ **31.** 205 **32.** $11\frac{1}{4}$ inches **33.** $9,000 **34.** 40 **35.** 81 **36.** 600 feet **37.** $8 **38.** 20 ounces **39.** 75 minutes **40.** $14 **41.** $19\frac{3}{5}$ ohms **42.** 70 kilograms **43.** $1021\frac{3}{7}$ miles **44.** $26\frac{2}{3}$ capsules

FOCUS ON PROBLEM SOLVING

The hectic pace on the New York Stock Exchange floor. *(Courtesy of the New York Stock Exchange)*

The stock market is one of many investment opportunities available to people. Through the purchase of shares of stock in a company a person becomes a part owner in that company. To follow the progress of stock it is only necessary to turn to the business section of a daily newspaper where the daily stock market quotations are carried.

Stocks traded on the previous day will be listed alphabetically and a typical entry with the appropriate column heading for Eastern Athletic, might look like this:

Year to Date		Stock and		Sales					Net
High	Low	Dividend	P/E	in 100's	Open	High	Low	Close	Change
$60\frac{1}{4}$	$40\frac{7}{8}$	Eastern Athletic; 2	26	140	$55\frac{1}{8}$	$57\frac{3}{4}$	$55\frac{1}{8}$	57	$+\ 1\frac{7}{8}$

This year the highest price of the stock was $60\frac{1}{4}$ ($60.25) and the lowest price was $40\frac{7}{8}$ ($40.875). It paid a dividend of 2 ($2.00) and the price-to-earnings ratio, denoted P/E, is 26. Yesterday 140×100 or 14,000 shares of this stock were sold with a low selling price of $55\frac{1}{8}$ ($55.125) and a high selling price of $57\frac{3}{4}$ ($57.75). The net change of $+1\frac{7}{8}$ ($1.875) is the difference between the closing price for one day and the closing price of the preceding day. Here a "plus" sign (+) indicates an increase in the value of a share of stock; a "minus" sign (−) would indicate a decrease in the value of a share of stock.

Example An investment group is holding 120 shares of Eastern Athletic stock on the day of our stock quotation. (a) What is the total value of the 120 shares? (b) How much have the 120 shares increased in value this day? (c) During trading, how much did the stock vary in price this day?

Solution (a) Find the closing price of the stock ($57 per share).
The total value = 120 × $57 = $6,840.

(b) Find the net change for this day $\left(\$1\frac{7}{8}\right)$.

$$\text{The total increase in value} = 1\frac{7}{8} \times 120$$

$$= \frac{15}{8} \times \frac{120}{1} = \$225$$

(c) Find the highest price reached by the stock $\left(\$57\frac{3}{4}\right)$; find the lowest price reached by the stock $\left(\$55\frac{1}{8}\right)$.

The price varied by $\$57\frac{3}{4} - \$55\frac{1}{8} = \$2\frac{5}{8}$ during this day.

BUILDING YOUR MATH VOCABULARY

TERM	DEFINITION	EXAMPLE
Denominator	The number below the fraction line.	$\frac{3}{7}$ ← denominator
Equal or Equivalent Fractions	Fractions having the same value. If $\frac{a}{b} = \frac{c}{d}$ then $a \cdot d = b \cdot c$	$\frac{3}{4} = \frac{6}{8}$ because $3 \cdot 8 = 4 \cdot 6$
Extremes	Given the proportion $\frac{a}{b} = \frac{c}{d}$, a and d are the extremes.	$\frac{2}{5} = \frac{6}{15}$ 2 and 15 are the extremes
Factor (or Divisor)	When two or more numbers are multiplied, each of the numbers is called a factor of the product.	5 and 7 are factors of 35
Fraction	The quotient of two whole numbers, $\frac{a}{b}$ $(b \neq 0)$.	$\frac{1}{2}, \frac{9}{4}, \frac{3}{7}$
Greatest Common Divisor (GCD)	The GCD of a collection of whole numbers is the largest number that will divide into (remainder of 0) each number in the collection.	The GCD of 45 and 60 is 15
Improper Fraction	A fraction in which the numerator is larger than or equal to the denominator.	$\frac{9}{7}$ and $\frac{4}{4}$
Least Common Multiple (LCM)	The LCM of a collection of whole numbers is the smallest whole number that is divisible by each number in the collection.	The LCM of 4, 6, and 8 is 24
Means	Given the proportion $\frac{a}{b} = \frac{c}{d}$, b and c are the means.	$\frac{3}{4} = \frac{9}{12}$; 4 and 9 are the means
Mixed Numbers	The sum of a whole number and a proper fraction.	$5\frac{1}{4}$ and $7\frac{5}{8}$
Numerator	The number above the fraction line.	$\frac{3}{7}$ ← numerator
Prime Number	A whole number greater than 1 that is divisible only by itself and 1.	2, 3, 5, 7, 11, and 13 are prime numbers
Proper Fraction	A fraction in which the numerator is less than the denominator.	$\frac{4}{5}$ and $\frac{1}{7}$
Proportion	A proportion is a statement of two equal ratios.	$\frac{5}{10} = \frac{1}{2}$ or $5:10 = 1:2$

TERM	DEFINITION	EXAMPLE
Ratio	A comparison of two quantities.	24 feet to 36 feet is $\frac{24}{36} = \frac{2}{3}$ or 2:3
Reciprocals	Two numbers whose product is 1.	$\frac{3}{4}$ and $\frac{4}{3}$ are reciprocals
Simplest Form (or lowest terms)	A fraction is expressed in simplest form when its numerator and denominator have no common divisors except 1.	$\frac{7}{8}$ is in simplest form
Terms	Given the proportion $\frac{a}{b} = \frac{c}{d}$, a, b, c, and d are called terms of the proportion.	$\frac{3}{5} = \frac{12}{20}$ 3, 5, 12, and 20 are terms of the proportion
Unequal Fractions	1. If $\frac{a}{b} < \frac{c}{d}$ then $a \cdot d < b \cdot c$	$\frac{2}{7} < \frac{1}{3}$ because $2 \cdot 3 < 1 \cdot 7$; $6 < 7$
	2. If $\frac{a}{b} > \frac{c}{d}$ then $a \cdot d > b \cdot c$	$\frac{4}{5} > \frac{2}{3}$ because $3 \cdot 4 > 5 \cdot 2$; $12 > 10$

Chapter Review

1. Find the reciprocals of (a) $\frac{1}{8}$ ______ (b) $\frac{7}{5}$ ______ (c) $\frac{4}{9}$ ______

2. Find the reciprocals of (a) 6 ______ (b) $\frac{2}{3}$ ______ (c) 10 ______

3. Explain what a proper fraction is. ______

4. Explain what an improper fraction is. ______

In problems 5–8, determine whether the given fractions are equal. If the fractions are not equal, state which one is larger.

5. $\frac{8}{9}, \frac{11}{12}$ ______

6. $\frac{4}{13}, \frac{8}{26}$ ______

7. $\frac{9}{11}, \frac{36}{44}$ ______

8. $\frac{5}{11}, \frac{3}{7}$ ______

9. $\frac{1}{2} + \frac{1}{3} + \frac{1}{4} =$ ______

10. The product of a number and its reciprocal is always ______.

11. Answer true or false.

    (a) When adding, you should always change mixed numbers to improper fractions. ______

    (b) When multiplying you should always change mixed numbers to improper fractions. ______

    (c) 15 is a prime number. ______

    (d) The LCM of 2, 4, 6, and 15 is 60. ______

Reduce the fractions in problems 12–15 to lowest terms.

12. $\frac{25}{40}$ ______

13. $\frac{27}{36}$ ______

14. $\frac{63}{189}$ ______

15. $\frac{605}{660}$ ______

Change the fractions in problems 16–19 to mixed numbers.

16. $\frac{39}{6}$ ______

17. $\frac{98}{8}$ ______

18. $\frac{87}{4}$ ______

19. $\frac{47}{3}$ ______

Change the mixed numbers in problems 20–23 to improper fractions.

20. $5\frac{5}{6}$ ______

21. $7\frac{3}{8}$ ______

22. $9\frac{4}{5}$ ______

23. $33\frac{1}{3}$ ______

In problems 24–27 find the GCD of the given collection of numbers.

24. 24, 54 ______

25. 28, 70 ______

26. 60, 450 ______

27. 231, 294 ______

Raise the fractions in problems 28–31 to higher terms.

28. $\frac{4}{5} = \frac{?}{20}$ ______ **29.** $\frac{6}{8} = \frac{54}{?}$ ______

30. $\frac{3}{4} = \frac{24}{?}$ ______ **31.** $\frac{6}{7} = \frac{?}{35}$ ______

In problems 32–35 find the LCM of the given collection of numbers.

32. 14, 6, 3 ______ **33.** 12, 15, 27 ______

34. 6, 9, 15 ______ **35.** 8, 10, 12 ______

In problems 36–38 find the least common denominator for the given groups of fractions.

36. $\frac{1}{3}, \frac{1}{4}, \frac{1}{5}$ ______ **37.** $\frac{1}{8}, \frac{1}{9}, \frac{1}{6}$ ______

38. $\frac{1}{2}, \frac{1}{5}, \frac{1}{4}$ ______ **39.** $\frac{1}{4}, \frac{1}{6}, \frac{1}{8}$ ______

In problems 40–65 perform the indicated operations.

40. $\frac{2}{3} + \frac{3}{6}$ ______ **41.** $12\frac{1}{5} + 9\frac{3}{4}$ ______

42. $7\frac{3}{10} + \frac{1}{12} + 4\frac{7}{15}$ ______

43.
$$\begin{array}{r} 45\frac{1}{2} \\ 37\frac{2}{3} \\ +\ 16\frac{5}{6} \\ \hline \end{array}$$

44.
$$\begin{array}{r} 21\frac{2}{3} \\ 68\frac{3}{4} \\ +\ 75\frac{3}{8} \\ \hline \end{array}$$

45. $\frac{4}{5} - \frac{1}{2}$ ______ **46.** $4\frac{5}{6} - 3\frac{1}{3}$ ______

47. $8\frac{2}{3} - 2\frac{1}{2}$ ______ **48.** $832\frac{2}{15} - 198\frac{2}{5}$ ______

49. $645\frac{5}{9} - 371\frac{5}{6}$ ______ **50.** $\frac{3}{4} \times \frac{7}{8}$ ______

51. $\frac{10}{9} \times \frac{3}{4}$ ______ **52.** $\frac{12}{7} \times \frac{14}{3}$ ______

53. $\frac{3}{5} \times \frac{2}{4} \times \frac{15}{9}$ ______ **54.** $3\frac{1}{3} \times 2\frac{1}{2}$ ______

55. $4\frac{5}{8} \times 2\frac{2}{3}$ ______ **56.** $\dfrac{6}{\frac{2}{3}}$ ______

57. $\dfrac{\frac{4}{3}}{\frac{8}{5}}$ ______ **58.** $\dfrac{\frac{15}{18}}{\frac{12}{6}}$ ______

59. $\dfrac{4\frac{1}{2}}{\frac{5}{8}}$ ______ **60.** $\frac{9}{16} \div \frac{3}{8}$ ______

61. $\frac{12}{8} \div \frac{3}{4}$ ______

62. $2\frac{3}{5} \div \frac{7}{10}$ ______

63. $2\frac{1}{4} \div 2\frac{1}{2}$ ______

64. $3\frac{3}{4} \div 6\frac{1}{4}$ ______

65. $5\frac{3}{7} \div 2\frac{1}{7}$ ______

Express the ratios given in problems 66–69 in simplest form.

66. 10 feet: 12 feet ______

67. 3 yards: 18 inches ______

68. 50 pounds: 35 ounces ______

69. 2 years: 15 months ______

In problems 70–73 solve for x.

70. $\frac{1}{2} = \frac{x}{9}$ ______

71. $\frac{3}{10} = \frac{5}{x}$ ______

72. $\frac{x}{4} = \frac{7}{8}$ ______

73. $\frac{3}{x} = \frac{6}{11}$ ______

74. Find the total weight of a can and contents if the can weighs $1\frac{6}{8}$ ounces and the contents weigh $23\frac{3}{4}$ ounces.

______

75. A company sends three workers to do a job. The workers require 3 hours, 4 hours, and 5 hours, respectively, to do the job. At the end of one hour how much of the job is done?

______

76. The opening price of a stock was $58\frac{5}{8}$ and the closing price was $61\frac{3}{4}$. Find the increase in the value of the stock.

______

77. A hardware merchant started with a piece of rope that was $220\frac{1}{2}$ feet long. He sold two pieces of rope, $33\frac{5}{16}$ and $52\frac{1}{4}$ feet long, respectively. How much rope was left?

______

78. Compute the total value of 250 shares of stock purchased at $28\frac{5}{8}$ plus 80 shares of stock purchased at $55\frac{1}{4}$.

______

79. How many times is $4\frac{2}{7}$ contained in $34\frac{2}{7}$? __________

80. An office building is under construction in which $\frac{3}{5}$ of the total floor space is to be used for offices. If each office uses $\frac{1}{20}$ of the total floor space of the building, how many offices will the building contain? __________

81. If a distance of 50 miles is represented by $1\frac{3}{4}$ inches on a map, what distance is represented by 7 inches? __________

82. The ratio of faculty to students at a college is 2:35. If the enrollment of the college is 980 students, how many faculty are there? __________

83. If 6 ounces of medicine fill $2\frac{1}{2}$ syringes, how many syringes will hold 4 ounces of medicine? __________

84. If a man earns \$21 for working 4 hours, how much does he earn for working 10 hours? __________

85. A weight of 15 grams stretches a spring 5 centimeters. What weight stretches the spring 12 centimeters? __________

86. Sixteen tons of sulfur are needed to make 49 tons of sulfuric acid. How many tons of sulfur are needed to make 35 tons of acid? __________

ANSWERS

1. (a) 8 **(b)** $\frac{5}{7}$ **(c)** $\frac{9}{4}$ **2. (a)** $\frac{1}{6}$ **(b)** $\frac{3}{2}$ **(c)** $\frac{1}{10}$ **3.** The numerator is less than the denominator. **4.** The numerator is greater than or equal to the denominator. **5.** $\frac{11}{12}$ **6.** equal **7.** equal **8.** $\frac{5}{11}$ **9.** $1\frac{1}{12}$ **10.** 1 **11. (a)** false **(b)** true **(c)** false **(d)** true **12.** $\frac{5}{8}$ **13.** $\frac{3}{4}$ **14.** $\frac{1}{3}$ **15.** $\frac{11}{12}$ **16.** $6\frac{1}{2}$ **17.** $12\frac{1}{4}$ **18.** $21\frac{3}{4}$ **19.** $15\frac{2}{3}$ **20.** $\frac{35}{6}$ **21.** $\frac{59}{8}$ **22.** $\frac{49}{5}$ **23.** $\frac{100}{3}$ **24.** 6 **25.** 14 **26.** 30 **27.** 21 **28.** 16 **29.** 72 **30.** 32 **31.** 30 **32.** 42 **33.** 540 **34.** 90 **35.** 120 **36.** 60 **37.** 72 **38.** 20 **39.** 24 **40.** $1\frac{1}{6}$ **41.** $21\frac{19}{20}$ **42.** $11\frac{17}{20}$ **43.** 100 **44.** $165\frac{19}{24}$ **45.** $\frac{3}{10}$ **46.** $1\frac{1}{2}$ **47.** $6\frac{1}{6}$ **48.** $633\frac{11}{15}$ **49.** $273\frac{13}{18}$ **50.** $\frac{21}{32}$ **51.** $\frac{5}{6}$ **52.** 8 **53.** $\frac{1}{2}$ **54.** $8\frac{1}{3}$ **55.** $12\frac{1}{3}$ **56.** 9 **57.** $\frac{5}{6}$ **58.** $\frac{5}{12}$ **59.** $7\frac{1}{5}$ **60.** $1\frac{1}{2}$ **61.** 2 **62.** $3\frac{5}{7}$ **63.** $\frac{9}{10}$ **64.** $\frac{3}{5}$ **65.** $2\frac{8}{15}$ **66.** 5:6 **67.** 6:1 **68.** 160:7 **69.** 8:5 **70.** $4\frac{1}{2}$ **71.** $16\frac{2}{3}$ **72.** $3\frac{1}{2}$ **73.** $5\frac{1}{2}$ **74.** $25\frac{1}{2}$ ounces **75.** $\frac{47}{60}$ done **76.** $3\frac{1}{8}$ **77.** $134\frac{15}{16}$ feet **78.** $11{,}576\frac{1}{4}$ **79.** 8 **80.** 12 **81.** 200 miles **82.** 56 faculty **83.** $1\frac{2}{3}$ syringes **84.** $\$52\frac{1}{2}$ **85.** 36 grams **86.** $11\frac{3}{7}$ tons

Warmup Test

1. Change the following improper fractions to mixed numbers.

(a) $\frac{10}{7}$ ______ (b) $\frac{9}{4}$ ______ (c) $\frac{23}{8}$ ______

2. Change the following mixed numbers to improper fractions.

(a) $3\frac{3}{5}$ ______ (b) $6\frac{1}{4}$ ______ (c) $10\frac{4}{7}$ ______

3. Raise each fraction to higher terms.

(a) $\frac{3}{4} = \frac{?}{28}$ ______ (b) $\frac{5}{8} = \frac{30}{?}$ ______ (c) $\frac{7}{15} = \frac{?}{75}$ ______

4. Reduce each fraction to lowest terms.

(a) $\frac{48}{60}$ ______ (b) $\frac{35}{56}$ ______ (c) $\frac{18}{54}$ ______

In problems 5–12 do the indicated computation and leave your answer in simplest form.

5. $\frac{1}{8} + \frac{3}{10} + \frac{7}{20}$ ______

6. $5\frac{1}{3} + 4\frac{1}{2} + \frac{5}{6}$ ______

7. $\frac{5}{12} - \frac{7}{18}$ ______

8. $\frac{8}{15} - \frac{11}{25}$ ______

9. $\frac{4}{5} \times \frac{25}{28}$ ______

10. $3\frac{1}{3} \times 5\frac{2}{5} \times \frac{3}{2}$ ______

11. $\frac{25}{24} \div \frac{105}{18}$ ______

12. $\left(3\frac{1}{5} \times \frac{25}{14}\right) \div \frac{10}{21}$ ______

13. An investor watches the progress of a stock during the day. Her stock opened at $\$54\frac{1}{2}$ per share, decreased by $\$1\frac{7}{8}$ per share by noon and then increased by $\$\frac{3}{4}$ per share by the close of the day. What was the value of the stock per share at the close of the day? What would the investor pay if she bought 16 shares of stock at the close of the day?

______

14. Tom's annual salary is $24,300. He pays $\frac{1}{6}$ of his annual salary for mortgage payments and $\frac{1}{5}$ of his annual salary for incomes taxes. How much of his annual salary remains after paying his income taxes and mortgage?

______

15. A dosage of medicine is $\frac{2}{3}$ milliliters. How many dosages can be given from a bottle containing 20 milliliters?

______

16. (a) Express as a ratio: 7 feet to 4 yards. ______

(b) A driver uses 22 gallons of gas while driving 550 miles. What is the ratio of miles driven to gallons of gas?

______

17. Solve for x:

(a) $\frac{5}{x} = \frac{8}{11}$ ______ (b) $\frac{3}{8} = \frac{6}{x}$ ______

18. If a baseball pitcher allows 4 earned runs in 9 innings, how many earned runs will he allow in 198 innings?

______

Challenge Test

1. Change the following improper fractions to mixed numbers.

(a) $\frac{9}{2}$ ______ (b) $\frac{13}{5}$ ______ (c) $\frac{25}{8}$ ______

2. Change the following mixed numbers to improper fractions.

(a) $2\frac{2}{3}$ ______ (b) $6\frac{3}{5}$ ______ (c) $9\frac{7}{8}$ ______

3. Raise each fraction to higher terms.

(a) $\frac{2}{5} = \frac{?}{15}$ ______ (b) $\frac{7}{8} = \frac{14}{?}$ ______ (c) $\frac{6}{13} = \frac{?}{52}$ ______

4. Reduce each fraction to lowest terms.

(a) $\frac{60}{75}$ ______ (b) $\frac{18}{42}$ ______ (c) $\frac{48}{56}$ ______

In problems 5–12, do the indicated computation and leave your answer in simplest form.

5. $\frac{2}{3} + \frac{1}{4} + \frac{5}{12}$ ______

6. $7\frac{1}{4} + 5\frac{3}{5} + \frac{7}{10}$ ______

7. $1\frac{1}{2} - \frac{2}{9}$ ______

8. $\frac{13}{15} - \frac{4}{9}$ ______

9. $\frac{5}{7} \times \frac{21}{15}$ ______

10. $4\frac{2}{5} \times 3\frac{1}{3} \times \frac{5}{11}$ ______

11. $\frac{18}{27} \div \frac{15}{36}$ ______

12. $\left(6\frac{1}{2} \times \frac{8}{15}\right) \div \frac{39}{10}$ ______

13. An advertising campaign hopes to reach $\frac{7}{8}$ of a 160,000-person market. How many people is this?

______

14. A company has $16\frac{1}{4}$ tons of material to ship out. A truckload can transport $1\frac{3}{4}$ tons. How many truckloads will it take to transport the entire load?

______

15. If $\frac{3}{4}$ of an item costs \$36, how much does $\frac{5}{6}$ of this item cost?

______

16. Solve for x:

(a) $\frac{2}{9} = \frac{x}{36}$ ______ (b) $\frac{7}{x} = \frac{49}{56}$ ______

17. (a) Express as a ratio: 6 quarters to 2 dollars. ______

(b) Forty-two students out of 280 students in a freshman course received A's. What is the ratio of students with A's to total number of students?

______

18. A 2-foot stick casts a $\frac{1}{2}$-foot shadow. How tall is a tree that casts an 11-foot shadow?

______

Decimals

CHAPTER OUTLINE

3–0 PRETEST

Solve the following problems. Do not use a calculator.

Problem	Section Reference	Learning Objective
1. Express the following decimals in numerical form. (a) five hundred forty-three and three hundred twenty-four thousandths ______ (b) sixty-five and nine hundred two ten-thousandths ______ (c) three million, eight hundred twelve thousand, six hundred and forty-two ten thousandths ______	3.1	Writing decimal numbers in numerical form
2. Express the following decimals in written form. (a) 502.32 ______ (b) 36.009 ______ (c) 2478.0003 ______ ______	3.1	Expressing decimals in written form
3. Round off the following decimals to the place indicated. (a) 12.57 to the tenths place ______ (b) 240.096 to the hundredths place ______ (c) 25.3415 to the thousandths place ______	3.2	Rounding off decimals
4. Convert the following fractions to decimals and indicate whether they are terminating or nonterminating decimals. (a) $\frac{33}{100}$ ______ (b) $\frac{9}{20}$ ______ (c) $\frac{5}{16}$ ______ (d) $\frac{6}{11}$ ______	3.3	Converting fractions to decimal form
5. Convert the following decimals to fractions. Write your answer in its simplest form. (a) 0.35 ______ (b) 0.0275 ______	3.3	Converting decimals to fraction form

Problem	Section Reference	Learning Objective
(c) 0.125 ______		
(d) 3.15 ______		
6. Perform the following additions.	3.4	Addition of decimals
(a) 94.02 + 312.657 + 28.2004 ______		
(b) 504.002 + 38.14 + 276.582 + 12 ______		
7. Perform the following subtractions.	3.5	Subtraction of decimals
(a) 543.05 − 362.97 ______		
(b) 486.42 − 453.568 ______		
8. Perform the following multiplication.	3.6	Multiplication of decimals
(a) 542.35 × 25.6 ______		
(b) 12.8 × 0.0004 ______		
9. Perform the following divisions.	3.7	Division of decimals
(a) $15.6\overline{)365.82}$ ______		
(b) $0.0040\overline{)21.5}$ ______		
10. A tourist pays $20.75 per night for lodging at a hotel. If he stays for 12 nights, what is the cost of his lodging?	3.6	Solving word problems
11. A carpenter pays $23.04 for lumber. The cost of the lumber is $0.72 per foot. How many feet of lumber did he purchase?	3.7	Solving word problems

3.1 INTRODUCING DECIMAL NUMBERS

Numbers such as 8.3, $52.98, and 176.344 are called **decimal numbers** or simply "decimals." It is the job of the decimal point (.), which appears in each of these numbers, to separate the whole-number portion from the fractional portion of the decimal number.

$$\underbrace{9\ 3\ 1}_{\text{whole number}} \cdot \underbrace{2\ 4\ 5}_{\text{fractional part}}$$

If a number does not have a decimal point, it is understood to be located to the right of the last digit. Thus 37 = 37., 102 = 102., and $308 = $308., for example. You read the digits of a decimal number to the left and right of the decimal point as illustrated in the following table. The table examines the place value of the number

9,235,016.470629. Observe that no commas are used to separate the digits to the right of the decimal point.

Table 3.1

Value	Place	Digit
1,000,000	Millions	9,
100,000	Hundred thousands	2
10,000	Ten thousands	3
1,000	Thousands	5,
100	Hundreds	0
10	Tens	1
1	Units or Ones	6
.	Decimal Point	and
$\frac{1}{10}$	Tenths	4
$\frac{1}{100}$	Hundredths	7
$\frac{1}{1{,}000}$	Thousandths	0
$\frac{1}{10{,}000}$	Ten-thousandths	6
$\frac{1}{100{,}000}$	Hundred-thousandths	2
$\frac{1}{1{,}000{,}000}$	Millionths	9

A number like 78.275 serves as a shortened way of expressing:

$$70 + 8 + \frac{2}{10} + \frac{7}{100} + \frac{5}{1{,}000}$$

Example 1 Use Table 3.1 to express 265.4 as a sum of numbers.

Solution

$$\begin{aligned} 265.4 &= 2(100\text{'s}) + 6(10\text{'s}) + 5(1\text{'s}) \text{ and } 4\left(\tfrac{1}{10}\text{'s}\right) \\ &= 200 + 60 + 5 + \frac{4}{10} \end{aligned}$$

From this it follows that 265.4 is a very convenient way of expressing a sum of four numbers.

Writing Decimals

The decimal 18.32 is read "eighteen and thirty-two hundredths." This is the **written form** of the number. It is important to note that the word "and" is used to represent the decimal point, and it is used exactly once. To illustrate, 3.5 is written "three and five tenths," while 219.097 is written "two hundred nineteen and ninety-seven thousandths."

Example 2 Express the following decimal numbers in written form.
(a) 23.54 (b) 0.46 (c) 5,096.1

Solution
(a) twenty-three and fifty-four hundredths
(b) forty-six hundredths. Because there is no digit to the left of the decimal point, you do not use the word "and."
(c) five thousand, ninety-six and one-tenth

Example 3 Express the following dollar amounts in written form.
(a) $7,526 (b) $0.01 (c) $62.495

Solution
(a) seven thousand, five hundred twenty-six dollars. The decimal point is understood to be to the right of the 6 (that is, $7,526 = $7,526.)
(b) one hundredth of a dollar, or one cent
(c) sixty-two and four hundred ninety-five thousandths dollars

Terminating Decimals

Decimals such as 0.25, 0.5, and 0.63 are called **terminating decimals.** *In each case, it is understood that the digit zero repeats indefinitely after the last nonzero digit.* That is 0.25 = 0.25000 . . ., 0.5 = 0.5000 . . ., and 0.63 = 0.63000 . . . The three dots ". . ." mean that the existing pattern of digits continues on. It is important to note that 0.5 = 0.50 = 0.500 and so on, just as 5 = 5.0 = 5.00 = 5.000 and so on. Decimals that are not terminating decimals are called **nonterminating decimals.** Some nonterminating decimals are: 0.66 . . .; 0.27444 . . .; and 0.383838. . . . A major source of round-off error in mathematics stems from the number of places retained when rounding off a nonterminating decimal.

When a decimal has one or more repeating digits, it is succinctly expressed by placing a bar "—" over the repeating digit or digits. For example, 0.666 . . . may be written $0.\overline{6}$, 0.27444 . . . may be written $0.27\overline{4}$, and 0.383838 . . . may be written $0.\overline{38}$. The decimal 0.5 may be written $0.5\overline{0}$ and $\frac{1}{7}$ = 0.142857 142857 . . . or $0.\overline{142857}$.

Example 4 Classify the following decimals as terminating or nonterminating.

(a) 0.35 35 35 . . . (b) 5.2 (c) 0.375
(d) 0.777 . . . (e) 6.45 (f) 0.893777 . . .

Solution
(a) nonterminating (b) terminating (c) terminating
(d) nonterminating (e) terminating (f) nonterminating

Example 5 Classify the following decimals as terminating or nonterminating.

(a) $0.25\overline{0}$ (b) $0.\overline{3}$ (c) $0.\overline{426}$

Solution (a) terminating (b) nonterminating (c) nonterminating

Equality and Inequality

Two decimals are equal if, and only if, they possess the same digits relative to the decimal point. Decimals that are not equal are said to be unequal, denoted "≠." When two decimals are not equal, you may determine which decimal is larger by inspecting the place value of the digits appearing in each decimal.

In the following examples, determine the missing digit indicated by a ? .

Example 6 If \$7 6 5. 4 9 = \$7 ⓟ 5. 4 9, then the (?) must be 6.

Example 7 If 5, 1 7 3. (?) 4 = 5, 1 7 3. 8 4, then the (?) must be 8.

Just as we have done in previous chapters, we assign a unique point on the number line to each decimal number. When this process is complete, each point on the number line corresponds to a unique number. It is this fact that makes our collection of decimals especially important. We say that the collection of decimals represents the entire number line or the collection of decimals corresponds to the world of **real numbers.** That is, *all real numbers can be expressed as decimals and all decimals correspond to real numbers.* Figure 3.1 shows this unique correspondence between some decimals and their corresponding points on the number line.

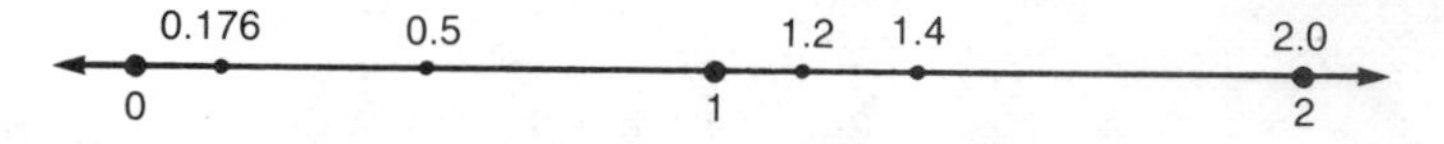

Figure 3.1

Given two decimals (or points) on the number line of Figure 3.1, the decimal farthest to the right is the largest. Observe $1.4 > 1.2$, $2.0 > 0.5$, and $1.2 > 0.176$. You may write either $0.5 > 0.176$ or $0.176 < 0.5$.

Example 8 $7.50 ≠ $7.55. That is, $7.50 does not equal $7.55. The former means seven dollars and fifty cents, while the latter means seven dollars and fifty-five cents. We can write $7.55 > $7.50.

Example 9 $0.09 is less than $0.90, since the former means 9 cents and the latter means 90 cents. Thus may be expressed $0.09 < $0.90.

Example 10 Is 6.275 < 6.471?

Solution The answer can be determined by making a *place by place comparison* (moving from left to right). The number 6 in the units place is the same in both numbers.

6 . 2 7 5 < 6 . 4 7 1

Below you see that the numbers in the tenths place differ.

6 . 2 7 5 < 6 . 4 7 1

Since 2 < 4, 6.275 < 6.471.

Example 11 Is 5.96 > 5.93?

Solution The number 5 in the units place is the same in both numbers.

5 . 9 6 > 5 . 9 3

The number 9 in the tenths place is the same in both numbers.

5 9 6 > 5 . 9 3

The numbers in the hundredths place differ.

5 . 9 6 > 5 . 9 3

Since 6 > 3, 5.96 > 5.93.

Example 12 Arrange the following numbers in order of magnitude from smallest to largest: 0.07, 0.6008, 1.04, 0.056, 0.192. Illustrate the answer on a number line.

Solution 0.056 < 0.07 < 0.192 < 0.6008 < 1.04

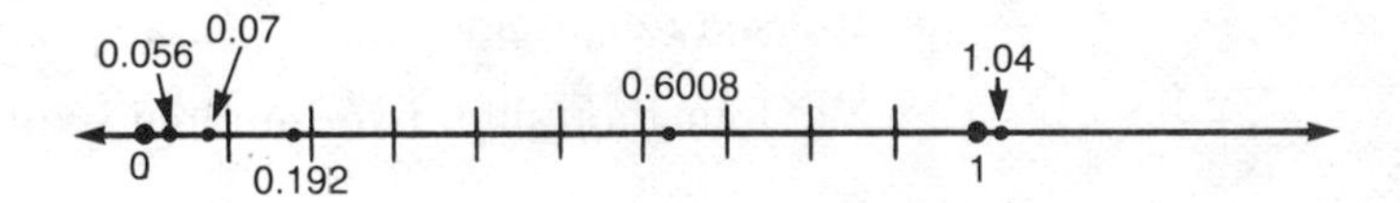

Example 13 Arrange the following numbers in order of magnitude from smallest to largest: 0.66, 0.06, 0.67, 0.666, 0.065.

Solution 0.06, 0.065, 0.66, 0.666, 0.67

3.1 EXERCISES

1. What is the whole number portion of the decimal 29.364? What is the fractional portion?

   ____

2. What is the whole number portion of the decimal 316.45? ______________

   What is the fractional portion? ______________

In problems 3–17 express the decimal in written form.

3. 70.83 ______________
4. 591.12 ______________
5. 6,123 ______________
6. 5,449.18 ______________
7. $607.80 ______________
8. $18.75 ______________
9. 23.07 meters ______________
10. 1.023 ______________
11. 15.1 grams ______________
12. 17.63 ______________
13. 3,940.2 ______________
14. 52,017.376 ______________
15. 19.002 ______________
16. 5.0039 ______________
17. 281.55 ______________

In problems 18–25 you are to express the decimal in numerical form.

18. five and nine tenths ______________
19. sixty-eight and eight thousandths ______________
20. four thousand, two hundred sixty-one and twenty-seven hundredths

    ____

21. thirteen and five hundred forty-six ten thousandths ______________
22. seventy-nine thousand six and twenty-two hundredths______________
23. fifty-one and seven tenths ______________

24. one million, four hundred sixty-five thousand, three hundred thirty-three and four hundred seventy-two ten-thousandths

____

25. nine hundred sixty-four and two hundred five thousandths

____

26. Write $2,359 with a decimal point. ____________

In problems 27–30 classify the following decimals as terminating or nonterminating.

27. 0.33 ____________ **28.** 0.33 . . . ____________

29. 2.5 ____________ **30.** 1.645 ____________

31. Is 4.2 decameters equal to 4.22 decameters? If you say no, which is the larger?

____

In problems 32–39 answer true or false.

32. $0.30 ≠ $0.3 ____________

33. $0.33 > $0.333 ____________

34. 0.44 is a terminating decimal ____________

35. 0.67 < 0.08 ____________

36. 1.8 liters < 0.188 liter ____________

37. 1.80 liters = 1.800 liters ____________

38. 1.88 liters is a nonterminating decimal ____________

39. One and eighty-eight hundredths of a liter is denoted 1.80 ____________

40. Which of the following are correct equalities?

(a) $0.30 = $0.03 ____________ (b) $3.00 = $3.0 ____________

(c) $0.44 = $0.40 ____________ (d) $040 = $40 ____________

In problems 41 and 42 insert the symbol < or >.

41. (a) $23.76 ____ $22.98 (b) 0.0111 ____ 0.01

42. (a) 5.05 liters ____ 5.10 liters (b) 0.255 ____ 0.225

43. Jose claims that $58.632 is approximately equal to $58.63.

(a) Is he correct? ____________

(b) What part of a dollar does he feel that he can neglect? ____________

(c) What part of a cent does he feel that he can neglect? ____________

44. If 0.7838 = 0.78[?]8, then the [?] must equal ____________

45. If 3.[?]61 = 3.461, then the [?] must equal ____________

In problems 46–51 arrange the decimals in order of size from the smallest to the largest.

46. 0.6, 0.07, 0.42 ____________

47. 0.23, 0.071, 0.564 ____________

48. 0.45, 0.63, 0.12, 1.02, 1.23, 0.29 ____________

49. 0.5, 0.16, 0.831, 0.92 ____________

50. 2.05, 0.32, 0.07, 1.98 ____________

51. 1.10, 0.25, 0.01, 1.84, 0.75 ____________

52. 3.25, 3.04, 3.222, 3.65, 3.7 ____________

53. 0.335, 0.33, 0.304, 0.301, 0.3, 0.333 ____________

54. 0.461, 0.416, 0.466, 0.4, 0.415 ____________

ANSWERS

1. 29 is the whole number portion; 364 is the fractional portion. **2.** 316 is the whole number portion; 45 is the fractional portion. **3.** seventy and eighty-three hundredths **4.** five hundred ninety-one and twelve hundredths **5.** six thousand, one hundred twenty-three **6.** five thousand, four hundred forty-nine and eighteen hundredths **7.** six hundred seven dollars and eighty cents **8.** eighteen dollars and seventy-five cents **9.** twenty-three and seven hundredths meters **10.** one and twenty-three thousandths **11.** fifteen and one tenth grams **12.** seventeen and sixty-three hundredths **13.** three thousand, nine hundred forty and two tenths **14.** fifty-two thousand, seventeen and three hundred seventy-six thousandths **15.** nineteen and two thousandths **16.** five and thirty-nine ten thousandths **17.** two hundred eighty-one and fifty-five hundredths **18.** 5.9 **19.** 68.008 **20.** 4,261.27 **21.** 13.0546 **22.** 79,006.22 **23.** 51.7 **24.** 1,465,333.0472 **25.** 964.205 **26.** $2,359.00 **27.** terminating **28.** nonterminating **29.** terminating **30.** terminating **31.** no; 4.22 decameters **32.** false **33.** false **34.** true **35.** false **36.** false **37.** true **38.** false **39.** false **40. (a)** false **(b)** true **(c)** false **(d)** true **41. (a)** $>$ **(b)** $>$ **42. (a)** $<$ **(b)** $>$ **43. (a)** yes **(b)** $\frac{2}{1,000}$ or $\frac{1}{500}$ **(c)** two tenths of a cent **44.** 3 **45.** 4 **46.** 0.07, 0.42, 0.6 **47.** 0.071, 0.23, 0.564 **48.** 0.12, 0.29, 0.45, 0.63, 1.02, 1.23 **49.** 0.16, 0.5, 0.831, 0.92 **50.** 0.07, 0.32, 1.98, 2.05 **51.** 0.01, 0.25, 0.75, 1.10, 1.84 **52.** 3.04, 3.222, 3.25, 3.65, 3.7 **53.** 0.3, 0.301, 0.304, 0.33, 0.333, 0.335 **54.** 0.4, 0.415, 0.416, 0.461, 0.466

3.2 ROUNDING OFF DECIMALS

A motorist is delighted to learn that the next gasoline station is 2.5 kilometers along the highway he is traveling on, even though in fact it is exactly 2.51 kilometers away. If we describe our weight as 170 lb, it may be only approximately true; but for practical purposes this value is quite adequate. Who really cares to know that we weigh 170.36 lb? A bank is being sufficiently accurate when it describes its cash-on-hand as $268,000.

When you round off a decimal number you must be instructed as to the decimal place that you are to round off to. The digit in the place value immediately following the desired place value should be analyzed. If this number is 5 or more, round up (increase the desired place value by 1); otherwise, leave the desired place value as it is and drop the remaining numbers.

Example 1 Round off $63.92 to the nearest dollar.

Solution $ 6 [3]. 9 2 — first digit to the right of the rounding position (9); round off to this position (3)

When we look at the tenths place, we see a 9, which is 5 or larger. Our answer is $64.

Example 2 Round off $185.6621 to the nearest cent (hundredth).

Solution $ 1 8 5.6 [6] 2 1 — first digit to the right of the rounding position (2); round off to this position (6)

Because there is a 2 in the thousandths place, our answer is $185.66.

Example 3 Round off $2,467.49953 to the nearest $\frac{1}{10}$ of a cent.

Solution $\frac{1}{10}$ of a cent is the same as $\frac{1}{1,000}$ of a dollar. Here we wish to round off to the thousandths digit. Because there is a 5 in the ten-thousandths place, our answer is $2,467.500.

Example 4 Round off 456.78 to the nearest tenth.

Solution Because there is an 8 in the hundredths place, our answer is 456.8.

Example 5 Round off 29.3761 to the nearest thousandths.

Solution Because there is a 1 in the ten-thousandths place, our answer is 29.376.

Example 6 Round off 78.975 to the nearest ten.

Solution Because there is an 8 in the units place, our answer is 80.

Example 7 Round off the following number 276.1248:

(a) To the nearest hundred.
(b) To the nearest tenth.
(c) To the nearest hundredth.
(d) To the nearest thousandths.

Solution

(a) Because there is a 7 in the tens place, our answer is 300.
(b) Because there is a 2 in the hundredths place, our answer is 276.1.
(c) Because there is a 4 in the thousandths place, our answer is 276.12.
(d) Because there is an 8 in the ten-thousandths place, our answer is 276.125.

3.2 EXERCISES

Round off the following decimals to the nearest thousandth.

1. 16.9812 ________ 2. 4.5036 ________ 3. 0.7291 ________

Round off the following decimals to the nearest hundredth.

4. 2.761 ________ 5. 55.162 ________ 6. 8.2934 ________

Round off the following decimals to the nearest tenth.

7. $216.42 ________ 8. 0.383 ________ 9. 2.75 ________

10. Round off 23.615 grams to:
    (a) the nearest hundredth ________
    (b) the nearest tenth ________
    (c) the nearest unit ________
    (d) the tens place ________

Round off the following amounts to the nearest cent.

11. $57.445 ________ 12. $229.837 ________ 13. $1.989 ________

Round off each of the following values to the place indicated.

14. 379.91 to units ________
15. 64.02 to tens ________
16. 5.77 to tenths ________
17. 6,532.01 to thousands ________

Round off each of the following numbers to the place indicated.

18. $6,973.258 to the nearest cent ________
19. $999.2567 to the nearest dime ________
20. $5,000.843 to the nearest cent ________
21. $5,000.843 to the nearest dollar ________
22. $878.396 to the tens place ________
23. How much money is needed to purchase 19 units of merchandise at $37 per unit?

    ____

    What is the value of this answer rounded off to the nearest hundred?

    ____

24. A rectangle has a 7-inch length (l) and an 8-inch width (w). Find the area (A) of this rectangle ($A = lw$) and round off your answer to the tens place.

____

25. A bank computes the interest on Mr. Swain's savings account and finds it to be \$37.7034. Round off this interest amount to the nearest cent for the bank.

____

ANSWERS

1. 16.981 **2.** 4.504 **3.** 0.729 **4.** 2.76 **5.** 55.16 **6.** 8.29 **7.** \$216.40 **8.** 0.4 **9.** 2.8 **10.** **(a)** 23.62 **(b)** 23.6 **(c)** 24 **(d)** 20 **11.** \$57.45 **12.** \$229.84 **13.** \$1.99 **14.** 380 **15.** 60 **16.** 5.8 **17.** 7,000 **18.** \$6,973.26 **19.** \$999.30 **20.** \$5,000.84 **21.** \$5,001 **22.** \$880 **23.** \$703; \$700 **24.** 60 **25.** \$37.70

3.3 FRACTION-DECIMAL CONVERSIONS

How do you convert a fraction to an equivalent decimal? How do you convert a terminating decimal to an equivalent fraction? We will answer each of these questions in the order just presented. The problem of converting a nonterminating decimal to an equivalent fraction is more difficult and as such is beyond the scope of this text.

Converting Fractions to Decimals

> *To convert a fraction to an equivalent decimal,* divide the denominator into the numerator.

Example 1 Convert $\frac{7}{8}$ (this may be expressed $7 \div 8$) to an equivalent decimal.

Solution Place a decimal point after the numerator and then add one zero at a time as you perform the indicated division. Thus

$$\begin{array}{r} 0.875 \\ 8\overline{)7.000} \\ \underline{6\,4} \\ 60 \\ \underline{56} \\ 40 \\ \underline{40} \\ 0 \end{array}$$

Therefore $\frac{7}{8} = 0.875$

> *The division of denominator into numerator continues until* (1) the remainder becomes zero, in which case you have a terminating decimal; or (2) the remainder repeats itself, which indicates a nonterminating decimal.

Example 2 Convert $\frac{9}{16}$ to an equivalent decimal.

Solution

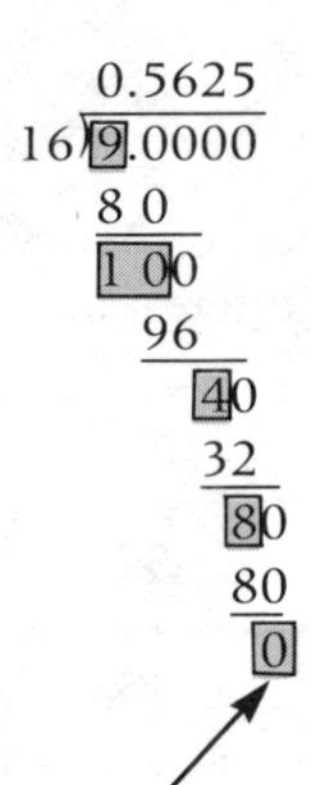

Thus $\frac{9}{16} = 0.5625$

Notice that at each step of the division we have drawn a block around the remainder.

Example 3 Convert $\frac{11}{40}$ to an equivalent decimal.

Solution

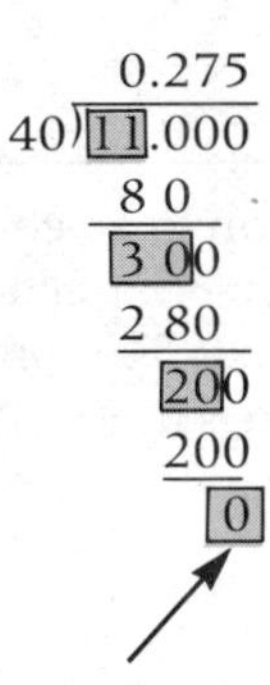

Thus $\frac{11}{40} = 0.275$

Example 4 Convert $\frac{1}{3}$ to an equivalent decimal.

Solution

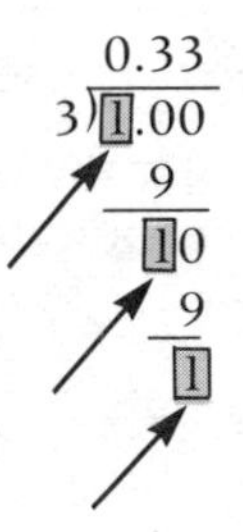

The remainder of 1 repeats itself, indicating a pattern of repeating digits.

Thus $\frac{1}{3} = 0.333\ldots$, or $0.\overline{3}$

Recall, placing a bar over the digit 3 indicates that it repeats indefinitely.

Example 5 Convert $\frac{7}{11}$ to an equivalent decimal.

Solution

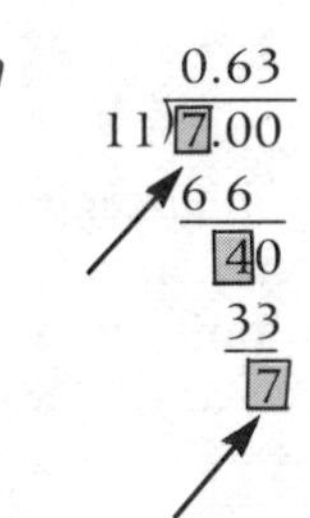

The remainder of 7 repeats itself indicating a pattern of repeating digits.

Thus $\frac{7}{11} = 0.6363\ldots$, or $0.\overline{63}$

Example 6 Convert $\frac{1}{7}$ to an equivalent decimal.

Solution

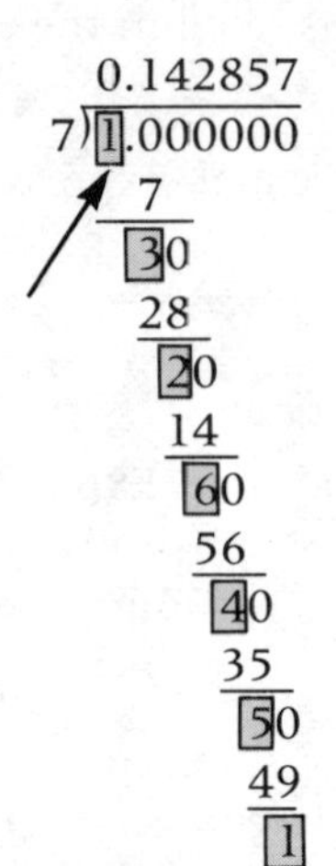

The remainder 1 repeats itself. This indicates that we have a pattern of repeating digits.

$\frac{1}{7} = 0.142857\ 142857\ \ldots$, or

$\frac{1}{7} = 0.\overline{142857}$

Example 7 Convert $\frac{7}{15}$ to an equivalent decimal.

Solution

$$\begin{array}{r} 0.46 \\ 15\overline{)7.00} \\ 6\,0 \\ \hline 100 \\ 90 \\ \hline 10 \end{array}$$

Thus $\frac{7}{15} = 0.4\overline{6}$

Example 8 Convert $\frac{13}{44}$ to an equivalent decimal.

Solution

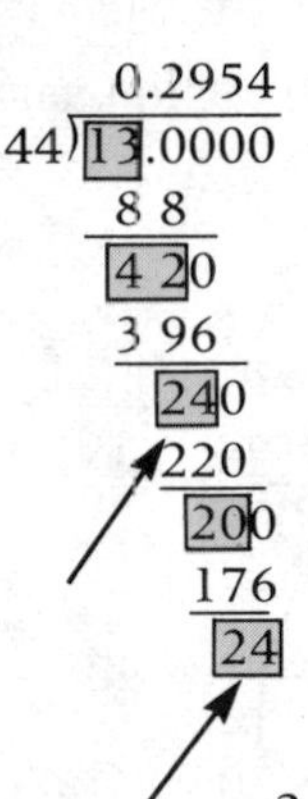

Thus $\frac{13}{44} = 0.29\overline{54}$

Example 9 Convert $3\frac{2}{15}$ to an equivalent decimal.

Solution Recall $3\frac{2}{15}$ means $3 + \frac{2}{15}$ or 3 and $\frac{2}{15}$.

Our answer is the whole number 3 plus the decimal equivalent of $\frac{2}{15}$.

$$\begin{array}{r} 0.13 \\ 15\overline{)2.00} \\ 1\,5 \\ \hline 50 \\ 45 \\ \hline 5 \end{array}$$

Now $\frac{2}{15} = 0.1333\ \ldots$, or $0.1\overline{3}$

Hence $3\frac{2}{15} = 3.1\overline{3}$

Frequently we execute a division to three decimal places and round off to the nearest hundredth. This is illustrated in the next examples.

Example 10 Convert $\frac{3}{7}$ to a decimal. Express the answer to the nearest hundredth.

Solution $\frac{3}{7} = 0.428\ldots$ *Answer:* $\frac{3}{7} = 0.43$

Example 11 Convert $\frac{5}{11}$ to a decimal. Express the answer to the nearest hundredth.

Solution $\frac{5}{11} = 0.454\ldots$ *Answer:* $\frac{5}{11} = 0.45$

Converting Decimals to Fractions

Now that you are able to express a decimal in written form, you are well on your way to converting decimals to fractions. When the decimal 0.2, "two-tenths," is expressed as a fraction you obtain $\frac{2}{10}$. Similarly, 0.37 "thirty-seven hundredths," becomes $\frac{37}{100}$, and 0.245, "two hundred forty-five thousandths," becomes $\frac{245}{1{,}000}$ or $\frac{59}{200}$ in lowest terms. *To convert a decimal to a fraction, place the decimal without the decimal point over the number "1" followed by as many zeros as there are digits after the decimal point.* Whenever possible, reduce this fraction to lowest terms.

Example 12 Convert 0.44 to an equivalent fraction.

Solution There are two digits to the right of the decimal point.

Thus $0.44 = \frac{44}{100} = \frac{11}{25}$

Example 13 Convert 0.125 to an equivalent fraction.

Solution $0.125 = \frac{125}{1{,}000} = \frac{1}{8}$

Example 14 Convert 4.38 to an equivalent fraction.

Solution 1 $4.38 = \frac{438}{100} = \frac{219}{50} = 4\frac{19}{50}$

or

Solution 2 Because 4.38 means 4 and 38 hundredths, we can write:

$4.38 = 4\frac{38}{100} = 4\frac{19}{50}$

We suggest that you memorize the following decimal equivalents of these fractional parts of one: $\frac{1}{2}, \frac{1}{3}, \frac{1}{4}, \frac{1}{5}, \frac{1}{6}, \frac{1}{8}, \frac{1}{10}, \frac{1}{12}$. See Table 3.2. It is also useful to know the decimal equivalent of multiples of these common fractional parts of one. This knowledge will allow you to substitute a fraction for a decimal in an arithmetic computation. The result can be a considerable saving of time.

Example 15 Evaluate $\$88 \times 0.125$.

Solution Begin by noting that $0.125 = \frac{1}{8}$.

$\$88 \times 0.125 = \$88 \times \frac{1}{8} = \$11$

Table 3.2

Fraction	Decimal	Fraction	Decimal	Fraction	Decimal
$\frac{1}{2}$	0.50	$\frac{2}{3}$	$0.66\frac{2}{3}$	$\frac{5}{12}$	$0.41\frac{2}{3}$
$\frac{1}{3}$	$0.33\frac{1}{3}$	$\frac{2}{5}$	0.40	$\frac{7}{8}$	0.875
$\frac{1}{4}$	0.25	$\frac{3}{4}$	0.75	$\frac{7}{12}$	$0.58\frac{1}{3}$
$\frac{1}{5}$	0.20	$\frac{3}{5}$	0.60		
$\frac{1}{6}$	$0.16\frac{2}{3}$	$\frac{3}{8}$	0.375		
$\frac{1}{8}$	0.125	$\frac{4}{5}$	0.80		
$\frac{1}{10}$	0.10	$\frac{5}{6}$	$0.83\frac{1}{3}$		
$\frac{1}{12}$	$0.08\frac{1}{3}$	$\frac{5}{8}$	0.625		

Example 16 Phoenix Manufacturing produces a certain item for \$0.625. How much does it cost the company to produce 96 of these items?

Solution $0.625 = \frac{5}{8}$. So

$$96 \times \$0.625 = 96 \times \$\tfrac{5}{8} = \$60$$

3.3 EXERCISES

In problems 1–20 convert each of the given fractions to decimals.

1. $\frac{7}{10}$ ______ **2.** $\frac{5}{16}$ ______ **3.** $\frac{13}{20}$ ______ **4.** $\frac{8}{9}$ ______

5. $\frac{11}{25}$ ______ **6.** $\frac{11}{4}$ ______ **7.** $2\frac{13}{25}$ ______ **8.** $4\frac{9}{20}$ ______

9. $3\frac{1}{8}$ ______ **10.** $\frac{3}{14}$ ______ **11.** $\frac{5}{9}$ ______ **12.** $2\frac{1}{4}$ ______

13. $\frac{5}{11}$ ______ **14.** $\frac{17}{12}$ ______ **15.** $\frac{7}{22}$ ______ **16.** $5\frac{3}{16}$ ______

17. $\frac{10}{22}$ ______ **18.** $1\frac{3}{11}$ ______ **19.** $\frac{8}{15}$ ______ **20.** $\frac{5}{7}$ ______

In problems 21–30 convert each of the given fractions to decimals; round off to the nearest hundredth.

21. $\frac{11}{14}$ ______ **22.** $\frac{9}{11}$ ______ **23.** $\frac{7}{9}$ ______ **24.** $\frac{4}{15}$ ______

25. $5\frac{1}{9}$ ______ **26.** $8\frac{2}{3}$ ______ **27.** $3\frac{5}{6}$ ______ **18.** $2\frac{4}{11}$ ______

29. $4\frac{5}{14}$ ______ **30.** $1\frac{5}{8}$ ______

In problems 31–42 convert each of the given decimals to a proper fraction or a mixed number.

31. 0.4 ______ **32.** 0.75 ______ **33.** 4.25 ______ **34.** 0.225 ______

35. 3.73 ______ **36.** 0.015 ______ **37.** 2.125 ______ **38.** 0.12 ______

39. 1.95 ______ **40.** 0.894 ______ **41.** 0.204 ______ **42.** 0.45 ______

In problems 43–50 use a fraction equivalent to the given decimal to solve the problem.

43. $663 \times \$0.33\frac{1}{3}$ ________________ **44.** $512 \times \$0.75$ ________________

45. $8232 \times \$0.875$ ________________ **46.** $\$80 \times 4.90$ ________________

47. $132 \times 0.16\frac{2}{3}$ ________________ **48.** $132 \times 0.83\frac{1}{3}$ ________________

49. 75×1.88 ________________ **50.** $945 \times 0.66\frac{2}{3}$ ________________

51. An aluminum scrap dealer gives \$0.375 for each pound of aluminum. How much do you receive for 376 pounds?

____

52. If a woman purchases 250 pads of unlined paper at \$0.90 each, how much does she pay?

____

53. Hardwick Manufacturing estimates that it costs $\$0.66\frac{2}{3}$ to package a certain item. How much does it cost the manufacturer to package 60,000 of these items?

____

ANSWERS

1. 0.7 **2.** 0.3125 **3.** 0.65 **4.** $0.\overline{8}$ **5.** 0.44 **6.** 2.75 **7.** 2.52 **8.** 4.45
9. 3.125 **10.** $0.2\overline{142857}$ **11.** $0.\overline{5}$ **12.** 2.25 **13.** $0.\overline{45}$ **14.** $1.41\overline{6}$ **15.** $0.3\overline{18}$
16. 5.1875 **17.** $0.\overline{45}$ **18.** $1.\overline{27}$ **19.** 0.53 **20.** $0.\overline{714285}$ **21.** 0.79 **22.** 0.82
23. 0.78 **24.** 0.27 **25.** 5.11 **26.** 8.67 **27.** 3.83 **28.** 2.36 **29.** 4.36
30. 1.63 **31.** $\frac{2}{5}$ **32.** $\frac{3}{4}$ **33.** $4\frac{1}{4}$ **34.** $\frac{9}{40}$ **35.** $3\frac{73}{100}$ **36.** $\frac{3}{200}$ **37.** $2\frac{1}{8}$
38. $\frac{3}{25}$ **39.** $1\frac{19}{20}$ **40.** $\frac{447}{500}$ **41.** $\frac{51}{250}$ **42.** $\frac{9}{20}$ **43.** \$221 **44.** \$384
45. \$7,203 **46.** \$392 **47.** 22 **48.** 110 **49.** 141 **50.** 630 **51.** \$141
52. \$225 **53.** \$40,000

3.4 ADDITION (+)

Myrna checked her gasoline credit slips after completing a five-day trip and found that she used the following amounts of gasoline: 10.75 gallons, 11.2 gallons, 9.6 gallons, 14.84 gallons, and 6.7 gallons. How many gallons of gasoline did she use for the trip? We will answer this question in this section.

When we add decimals we always want to add digits that have the same place value. That is, we must add units to units, tens to tens, tenths to tenths,and so forth. Here is the method used to add decimals.

> *To add decimals,* line up the numbers with all of the decimal points in a column and then add, beginning with the column on the right.

Example 1 Add 27.65 grams + 691.88 grams + 674.53 grams.

Solution

$$\begin{array}{rl} 27.65 & \text{grams} \\ 691.88 & \text{grams} \\ +\ 674.53 & \text{grams} \\ \hline 1{,}394.06 & \text{grams} \end{array}$$

In addition problems, the numbers being added are called **addends,** and the answer is called the **sum** or the **total.** This vocabulary, which is identical to that of whole numbers, is illustrated in the next example. *To check an addition,* repeat the addition by adding each column in a manner that is different from the original addition. Once you have lined up your decimal points in a vertical line, you will observe that adding decimals is similar to adding whole numbers.

When you are adding numbers such as 7.645 and 5.3 you may place two zeros after the 3 if you wish. Thus

$$\begin{array}{r} 7.645 \\ +\ 5.300 \\ \hline 12.945 \end{array}$$

Example 2 Add 7.68 + 97 + 18.4.

Solution If it is helpful, add place-keeping zeros after the decimal point. Remember 97 = 97. = 97.00

$$\begin{array}{rl} 7.68 & \text{addend} \\ 97.00 & \text{addend} \\ +\ 18.40 & \text{addend} \\ \hline 123.08 & \text{sum} \end{array}$$

Example 3 Myrna checked her gasoline credit slips after completing a five-day trip and found that she used the following amounts of gasoline: 10.75 gallons, 11.2 gallons, 9.6 gallons, 14.84 gallons, and 6.7 gallons. How many gallons of gasoline did she use on this trip?

Solution

$$\begin{array}{rl} 10.75 & \\ 11.20 & \\ 9.60 & \\ 14.84 & \\ +\ 6.70 & \\ \hline 53.09 & \text{gallons} \end{array}$$

Example 4 Theresa recently had her automobile serviced. The parts description and cost section of the work order sheet is illustrated below.

Quantity	Parts Description	Cost
1	Oil Filter	4.75
4	Quarts Oil	7.00
2	Windshield Wipers	4.10
4	Spark Plugs	5.60
1	Gas Filter	3.25
1	Air Filter	6.53
1	PCV Valve	3.11
1	Can—Parts Cleaner	4.95
	Total Parts Sale	39.29

Example 5 Many people use credit cards to make purchases. A credit card sales slip is shown below.

4502 731 251 414 5 9 9 4 3 3 1

02/85 01/86 VISA

Tom Barry

Date 6/10/85			
Qty	Description	Price	Amount
1	Pants	26.95 ea	26.95
1	Shirt	15.95 ea	15.95
1	Tie	6.75 ea	6.75
1 pr	Shoes	40.49	40.49
Customer Signature	Tom Barry	Sub Total	90.14
		Tax	—
Sales Slip		Total	90.14

Retain this copy for your records

3.4 EXERCISES

In problems 1–15 perform the indicated additions.

1. $23.47 + 86.25$

2. $8.043 + 15.021$

3. $904.63 + 687.56$

4. $97.0015 + 19.9985$

5. $12.8 + 6.9 + 32.5$

6. $25.6 + 43.9 + 31.5$

7. $7.95 + 24.26 + 39.68$

8. $57.08 + 68.97 + 37.47$

9. $102.005 + 79.826 + 212.293$

10. $52.058 + 63.17 + 89.54$

11. $5.237 + 476.81 + 1.023 + 16.47$

12. $188.66 + 54.78 + 18.779 + 8.008$

13. $6.0402 + 7.15 + 28.6 + 9.827 =$ ______________

14. $357.63 + 63.4 + 592.685 + 86.245 =$ ______________

15. $0.456 + 8.9 + 0.81 + 10.7 =$ ______________

Check your answers to problems 1–15. If you did not get at least 12 problems correct turn to Practice Problem Set 9A in the Appendix.

In problems 16–31 perform the indicated additions.

16.
$$\begin{array}{r} 7.5 \\ +\ 8.6 \\ \hline \end{array}$$

17.
$$\begin{array}{r} 15.63 \\ +\ 35.058 \\ \hline \end{array}$$

18.
$$\begin{array}{r} 8.04 \\ 15.3 \\ +\ 27.012 \\ \hline \end{array}$$

19.
$$\begin{array}{r} 56.05 \\ 0.003 \\ +\ 2.17 \\ \hline \end{array}$$

20.
$$\begin{array}{r} 78.9 \\ 45.04 \\ +\ 112.063 \\ \hline \end{array}$$

21.
$$\begin{array}{r} 262.53 \\ 87.4095 \\ +\ 3.204 \\ \hline \end{array}$$

22.
$$\begin{array}{r} 12.002 \\ 0.908 \\ 5.6 \\ +\ 4.82 \\ \hline \end{array}$$

23.
$$\begin{array}{r} 10.52 \\ 6.31 \\ 0.3582 \\ +\ 4 \\ \hline \end{array}$$

24.
$$\begin{array}{r} 205.6 \\ 42.03 \\ 2.7 \\ +\ 3.004 \\ \hline \end{array}$$

25.
$$\begin{array}{r} 34.2 \\ 3 \\ 12.05 \\ +\ 40 \\ \hline \end{array}$$

26.
$$\begin{array}{r} 15.8 \\ 64.32 \\ 5.79 \\ +\ 2.006 \\ \hline \end{array}$$

27.
$$\begin{array}{r} 20.02 \\ 13.18 \\ 9.842 \\ 13 \\ +\ 8.0049 \\ \hline \end{array}$$

28.
$$\begin{array}{r} 6.78 \\ 5.923 \\ 14.853 \\ +\ 0.0685 \\ \hline \end{array}$$

29.
$$\begin{array}{r} 92.66 \\ 104.503 \\ 58.673 \\ 0.098 \\ +\ 16.23 \\ \hline \end{array}$$

30.
$$\begin{array}{r} 13.402 \\ 12 \\ 0.0632 \\ 37.92 \\ 14 \\ +\ .05 \\ \hline \end{array}$$

31. $206 + 0.003 + 42.56 + 3.0102 + 159.5 + 3.0008 =$ _______________

32. For the first three months of last year the Lacostes' electric bills were $43.71, $39.68, and $41.55. Find the total.

____

33. Mrs. Eldridge hopes to save money by canning vegetables. She figures that gas heat for canning will cost $11.25; sugar, salt, vinegar, and spices, $14.57; and jar lids and rings, $8.20. Find her total cost.

____

34. Alice purchased a dress for herself two years ago and paid $59.00 for it. To restyle the dress she recently spent $6.50 for a new collar, $4.75 for a new belt, and $3.65 for new buttons. What is the total amount that Alice has spent on this dress?

____

35. A salesman drove the following distances last week: 242.6 miles, 312 miles, 98.75 miles, 128.3 miles, and 74.55 miles. What is the total distance he drove last week?

____

36. The Carle family had a new hot water heater installed. The heater cost $205.78; the plumber charged $52.98 to install it. What was the cost of the heater including installation?

____

37. An athlete competes in a triathalon. He swims 2.25 miles in 30.04 minutes, bicycles 45.5 miles in 122.301 minutes, and runs 9.3 miles in 55.8 minutes. What is the total distance of the three events? How long does it take the athlete to finish the triathalon?

____

38. Mrs. Martin is a computer programmer for the First National Bank. Yesterday she ran five programs through the bank's computer and noted the computer time of each program. They were: 24.045 seconds, 58.6 seconds, 107.94 seconds, 30 seconds, and 64.738 seconds. What was the total computer time needed to run her programs?

____

39. Part of an automobile service work order sheet is shown. Find the total cost.

____

Quantity	Parts Description	Cost
1	Head Lamp	5.50
1	Muffler	49.95
1	Radial Tire	69.95
1	Oil Filter	4.75
4	Quarts Oil	7.00
	Total: Parts Sale	

40. Part of a credit card sales is shown. Find the subtotal and the total amount.

____

Date 5/12/85			
Qty	Description	Price	Amount
1	Garden Hoe	11.95 ea	11.95
1	Garden Rake	12.95 ea	12.95
1	Bag Peat Moss	10.59 ea	10.59
1	Bag Fertilizer	8.99 ea	8.99
		Subtotal	
		Tax	2.22
	Sales Slip	Total	

41. The perimeter of a polygon is found by finding the sum of the lengths of its sides. Find the perimeter of the given polygon.

____

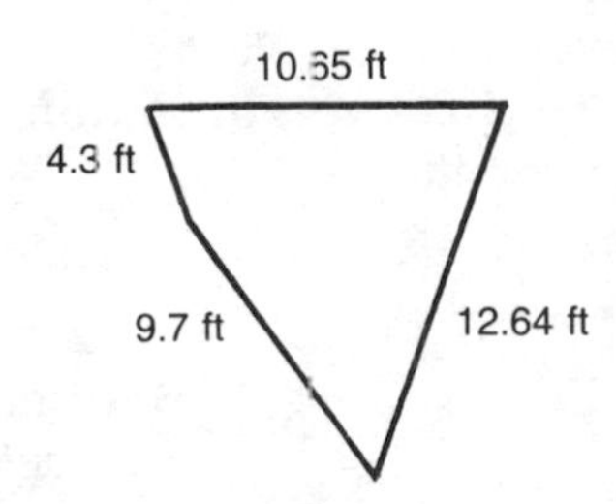

42. Find the perimeter of the given polygon.

____

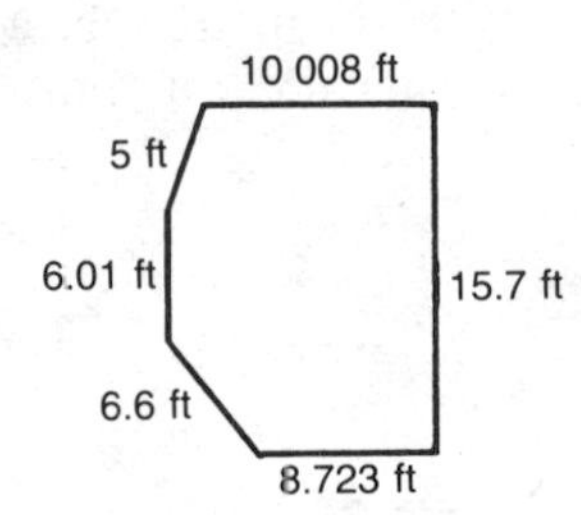

ANSWERS

1. 109.72 **2.** 23.064 **3.** 1592.19 **4.** 117 **5.** 52.2 **6.** 101 **7.** 71.89 **8.** 163.52 **9.** 394.124 **10.** 204.768 **11.** 499.54 **12.** 270.227 **13.** 51.6172 **14.** 1099.96 **15.** 20.866 **16.** 16.1 **17.** 50.688 **18.** 50.352 **19.** 58.223 **20.** 236.003 **21.** 353.1435 **22.** 23.33 **23.** 21.1882 **24.** 253.334 **25.** 89.25 **26.** 87.916 **27.** 64.0469 **28.** 27.6245 **29.** 272.164 **30.** 77.4352 **31.** 414.074 **32.** $124.94 **33.** $34.02 **34.** $73.90 **35.** 856.2 miles **36.** $258.76 **37.** 57.05 miles; 208.141 minutes **38.** 285.323 seconds **39.** $137.15 **40.** $44.48; $46.70 **41.** 37.29 feet **42.** 52.041 feet

3.5 SUBTRACTION (−)

How much is $435.66 minus $398.76? Mohammed purchased three items costing $19.50, $39.95, and $52.25 at Bulton's Inc. In addition to paying for these items, the salesperson informed him that he must pay a sales tax of $5.59. Assuming he has $200.00 in his pocket, how much money does he have left after purchasing these items? We will answer these questions in this section.

When we subtract decimals, we always want to subtract digits that have the same place value. That is, we must subtract units from units, tens from tens, tenths from tenths and so forth. Here is the method used to subtract decimals.

> To *subtract* decimals, line up the numbers with all of the decimal points in a column and then subtract beginning with the column on the right.

Example 1 How much is $435.67 minus $398.76?

Solution

```
   $435.67 ← minuend
 − $398.76 ← subtrahend
 ---------
   $ 36.91 ← difference
```

In a subtraction problem the number that we start with is called the **minuend,** the number being subtracted is called the **subtrahend,** and the answer is called the **difference.** This vocabulary, which is identical to that of whole numbers, is illustrated in the next example. Once you have lined up your decimal points vertically, you will observe that subtracting decimals is similar to subtracting whole numbers. **To check a subtraction,** add the difference to the subtrahend; the sum should be the minuend.

Example 2

$$\begin{array}{rl} 75.35 & \text{minuend} \\ -\ 27.62 & \text{subtrahend} \\ \hline 47.73 & \text{difference} \end{array} \qquad \textit{Check:} \qquad \begin{array}{rl} 27.62 & \text{subtrahend} \\ +\ 47.73 & \text{difference} \\ \hline 75.35 & \text{minuend} \end{array}$$

When you are subtracting numbers, such as 8.2 less 5.369, you may place two zeros after the 2 if you wish. Thus,

$$\begin{array}{r} 7\ 11\ 9 \\ \not{8}.\,{}^{1}\not{2}\,{}^{1}\not{0}\,{}^{1}0 \\ -\ 5.\ 3\ 6\ 9 \\ \hline 2.\ 8\ 3\ 1 \end{array}$$

Example 3 How much is left if you begin with \$19 and deduct \$6.46?

Solution If it is helpful, add "place-keeping" zeros after the decimal point in \$19.

$$\begin{array}{rl} \$19.00 & \\ -\quad 6.46 & \\ \hline \$12.54 & \text{is left} \end{array}$$

Example 4 If a gallon of milk equals 3.785 liters and during a given day Gary consumes 0.757 liter of milk, how many liters from this gallon remain?

Solution

$$\begin{array}{rl} 3.785 & \\ -\ 0.757 & \\ \hline 3.028 & \text{liters remain} \end{array}$$

Example 5 Subtract: 2.0076 less 1.5008.

Solution

$$\begin{array}{r} 2.0076 \\ -\ 1.5008 \\ \hline 0.5068 \end{array}$$

Example 6 Mohammond purchased three items costing \$19.50, \$39.95, and \$52.25 at Bulton's Inc. In addition to paying for these items, the salesperson informed him that he must pay a sales tax of \$5.59. Assuming that he has \$200.00 in his pocket, how much money does he have left after purchasing these items?

Solution The items cost \$19.50 + \$39.95 + \$52.25 = \$111.70.
\$111.70 + \$5.59 (sales tax) = \$117.29.
He has \$200.00 − \$117.29 = \$82.71 left.

Example 7 A checkbook transaction register is shown below. It is an application of addition and subtraction of decimals.

Date	Number	Transaction Description	(+) Amount of Deposit	(−) Amount of Withdrawal	Balance Forward 1113.51
11–27	501	Eastern Electric		60.10	−60.10
					1053.41
11–27	502	Plymouth Savings (mortgage)		288.49	−288.49
					764.92
12–1		Deposit	368.75		+368.75
					1133.67
12–5	503	Dr. James Maloney, DDS		60.00	−60.00
					1073.67

3.5 EXERCISES

In problems 1–15 perform the indicated subtractions

1. $35.9 - 17.5$

2. $212.82 - 57.61$

3. $50.089 - 28.408$

4. $27.5 - 19.9$

5. $273.04 - 185.06$

6. $23.642 - 19.587$

7. $49.44 - 38.76$

8. $359.81 - 217.39$

9. $683.74 - 509.98$

10. $0.007143 - 0.000968$

11. $0.00828 - 0.00739$

12. $5.01 - 0.000046$

13. $16.52 - 4.9000038 =$ ______

14. $982.47 - 905.68 =$ ______

15. $68 - 12.4 =$ ______

Check your answers to problems 1–15. If you did not get at least 12 problems correct, turn to Practice Problem Set 10A in the Appendix.

In problems 16–28 perform the indicated subtractions.

16. $105. - 83.56$

17. $603.854 - 259.93$

18. $852.351 - 268.56$

19. $94.6 - 39.82$

20. $\begin{array}{r} 56.5 \\ -\ 48.64 \\ \hline \end{array}$

21. $\begin{array}{r} 204.08 \\ -\ 174.9 \\ \hline \end{array}$

22. $\begin{array}{r} 325.04 \\ -\ 234.285 \\ \hline \end{array}$

23. $\begin{array}{r} 65 \\ -\ 52.607 \\ \hline \end{array}$

24. $\begin{array}{r} 48.77 \\ -\ 40.892 \\ \hline \end{array}$

25. $\begin{array}{r} 0.00002174 \\ -\ 0.00001368 \\ \hline \end{array}$

26. $\begin{array}{r} 0.0000794 \\ -\ 0.0000086 \\ \hline \end{array}$

27. $\begin{array}{r} 4 \\ -\ 0.0012 \\ \hline \end{array}$

28. $\begin{array}{r} 9 \\ -\ 0.0675 \\ \hline \end{array}$

29. Subtract: 2502.5 less 304.25. ______________

30. Subtract: 72 less 65.34. ______________

31. Subtract: 372.64 less 285.763. ______________

32. Subtract: 584.263 less 302.405. ______________

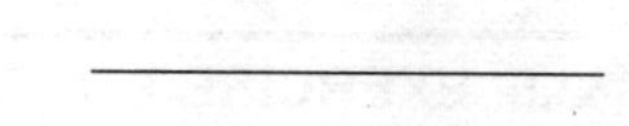

33. Subtract: 401.001 less 259.1235. ______________

34. Last year the Ulz family paid $2,660 in property taxes on their home. This year their tax bill is $2,995.50. How much more must they pay this year?

____

35. A student paid $27.50 for a textbook, $1.95 for a notebook, and $0.89 for a pen. He pays for his materials with a $50 bill. How much change does he receive?

____

36. Infrared light waves approximately range in length from 0.03 centimeter to 0.000076 centimeter. What is the difference between the shortest and longest wavelength?

____

37. Mary's checking account had a balance of $408.82. She made deposits of $276.65 and $154.92 and wrote checks for $75.68, $93.71, $104.20, $38.57 and $12.46. What is the present balance in her checking account?

____

38. The Dow-Jones stock market average opened at 1116.29, reached a high of 1129.56, reached a low of 1106.71, and closed at 1117.65.

(a) What is the difference between the opening average and the low average?

____

(b) What is the difference between the high average and the opening average?

____

(c) What is the difference between the closing average and the opening average?

____

(d) What is the difference between the high average and the low average?

____

39. Fill in the missing balance items in the following checkbook transaction register.

Date	Number	Transaction Description	(+) Amount of Deposit	(−) Amount of Withdrawal	Balance Forward 425.62
9–4	415	N.E. Telephone		35.10	
9–6		Deposit	384.56		
9–7	416	First Card Visa		135.99	
9–7	417	Sun-House Solarium		112.65	

40. Is the balance in the following checkbook transaction correct? If not, what is the correct balance?

Date	Number	Transaction Description	(+) Amount of Deposit	(−) Amount of Withdrawal	Balance Forward 369.23
1–5		Deposit	158.00		
1–8	604	Smith Pharmacy		89.00	
1–10	605	Standard Oil		125.60	
1–12	606	Andersonville Savings Bank		278.00	36.63

ANSWERS

1. 18.4 **2.** 155.21 **3.** 21.681 **4.** 7.6 **5.** 87.98 **6.** 4.055 **7.** 10.68
8. 142.42 **9.** 173.76 **10.** 0.006175 **11.** 0.00089 **12.** 5.009954
13. 11.6199962 **14.** 76.79 **15.** 55.6 **16.** 21.44 **17.** 343.924 **18.** 583.791
19. 54.78 **20.** 7.86 **21.** 29.18 **22.** 90.755 **23.** 12.393 **24.** 7.878
25. 0.00000806 **26.** 0.0000708 **27.** 3.9988 **28.** 8.9325 **29.** 2198.25 **30.** 6.66
31. 86.877 **32.** 281.858 **33.** 141.8775 **34.** $335.50 **35.** $19.66
36. 0.029924 cm **37.** $515.77 **38.** (a) 9.58 **(b)** 13.27 **(c)** 1.36 **(d)** 22.85
39. 390.52; 775.08; 639.09; 526.44 **40.** no; 34.63

3.6 MULTIPLICATION (×)

How much is 4.3 × 2.1? If one meter equals 3.281 feet, how many feet are contained in 75 meters? What is Jose's gross pay if he is paid $6.40 per hour and he works 39.75 hours? These questions will be answered in this section.

Below is the method used to multiply decimal numbers.

> To *multiply* two decimal numbers, ignore the decimal point for a moment and multiply the numbers. Point off, right to left, a number of decimal places equal to the sum of the decimal places in the given numbers.

Example 1 Multiply: 43 × 2.1

Solution

4.3	1 decimal place
× 2.1	1 decimal place
4 3	partial product
8 6	partial product
9.0 3	1 + 1 = 2 decimal places to be pointed off

In multiplication problems the number being multiplied (the top one) is called the **multiplicand,** the number doing the multiplying is called the **multiplier,** and the answer is called the **product.** This vocabulary, which is identical to that of whole numbers, is illustrated in the next example. Except for inserting a decimal point in the product, multiplying decimals is similar to multiplying whole numbers. Indeed the rationale, for "indenting" each successive partial product one place to the left is the same for decimals as it is for whole numbers. *To check a multiplication,* reverse the order of multiplicand and multiplier and compute the product again.

Example 2 Multiply: 5.43 × 7.9

Solution

5.4 3	multiplicand
× 7.9	multiplier
4 8 8 7	
38 0 1	
42.8 9 7	product

Check:

7.9	multiplicand
× 5.4 3	multiplier
2 3 7	
3 1 6	
39 5	
42.8 9 7	product

Example 3 Multiply: 36.1 × 0.00028

Solution

36.1	1 decimal place
× 0.0002 8	5 decimal places
288 8	
722	
0.01010 8	1 + 5 = 6 decimal places to be pointed off

Example 4 Evaluate (37.53 − 21.47) × ($5.60 + $2.20).

Solution First evaluate what is inside each pair of parentheses. This yields (16.06) × ($7.80).

```
     1 6.0 6        2 decimal places
×    $  7.8         1 decimal place
   ---------
    12 8 4 8
   112 4 2
   ---------
  $125.2 6 8        2 + 1 = 3 decimal places to be pointed off
```

Example 5 If an employee is paid by the hour, the employee's *gross pay* is computed by multiplying the number of hours worked by the hourly rate of pay. Compute the gross pay for Jose, who worked 39.75 hours at $6.40 per hour.

Solution

```
       39.75
×      $6.40
   ----------
     15 90 00
    238 50
   ----------
  $254.40 00 or just $254.40
```

Example 6 If 1 meter = 3.281 feet, how many feet are contained in 75 meters?

Solution

```
     3.281   feet
×       75
   --------
    16 405
   229 67
   --------
   246.075   feet
```

Observe what happens when a decimal number is multiplied by 10, 100, 1000, and so on:

8.35 × 1[0] = 83.5	Decimal point moves 1 place to the right.
8.35 × 1[00] = 835	Decimal point moves 2 places to the right.
8.35 × 1[000] = 8,350	Decimal point moves 3 places to the right.

The decimal point moves the same number of places to the right as there are zeros in the multiplier.

What happens when a decimal number is multiplied by .1, .01, .001 and so on?

8.35 × .[1] = 0.835	Decimal point moves 1 place to the left.
8.35 × .[01] = 0.0835	Decimal point moves 2 places to the left.
8.35 × .[001] = 0.00835	Decimal point moves 3 places to the left.

The decimal point moves the same number of places to the left as there are digits in the multiplier.

3.6 EXERCISES

1. Explain the movement of the decimal point when 7.61 is multiplied by 100 and give the resulting answer.

____

2. Explain the movement of the decimal point when 38.4 is multiplied by 0.001 and give the resulting answer.

____

In problems 3–15 perform the indicated multiplications.

3. **(a)** $5.43 \times 10^3 =$ ______________ **(b)** $5.43 \times 10^2 =$ ______________

4. $\begin{array}{r} 3.5 \\ \times \quad 7 \\ \hline \end{array}$

5. $\begin{array}{r} 3 \\ \times\ 2.1 \\ \hline \end{array}$

6. $\begin{array}{r} 20.6 \\ \times\ \ 6.8 \\ \hline \end{array}$

7. $\begin{array}{r} 10.45 \\ \times \quad 3.2 \\ \hline \end{array}$

8. $\begin{array}{r} 35.76 \\ \times \quad 100 \\ \hline \end{array}$

9. $\begin{array}{r} 18.065 \\ \times \qquad 43 \\ \hline \end{array}$

10. $\begin{array}{r} 49.73 \\ \times\ \ 2.74 \\ \hline \end{array}$

11. $\begin{array}{r} 39.27 \\ \times\ 16.77 \\ \hline \end{array}$

12. $\begin{array}{r} 0.47 \\ \times\ 0.01 \\ \hline \end{array}$

13. $876.05 \times 2.34 =$ ______________

14. $3.2 \times 0.00068 =$ ______________

15. $65.4 \times 0.045 =$ ______________

Check your answers to problems 1–15. If you did not get at least 12 problems correct turn to Practice Problem Set 11A in the Appendix.

In problems 16–25 perform the indicated multiplications.

16. $\begin{array}{r} 0.85 \\ \times\ 0.009 \\ \hline \end{array}$

17. $\begin{array}{r} 9.864 \\ \times\ 0.002 \\ \hline \end{array}$

18. $\begin{array}{r} 82.7 \\ \times\ 0.004 \\ \hline \end{array}$

19. $\begin{array}{r} 7.5 \\ \times\ 0.0001 \\ \hline \end{array}$

20. $\begin{array}{r} 8.82 \\ \times\ 1000 \\ \hline \end{array}$

21. $\begin{array}{r} 0.0488 \\ \times\ \quad 100 \\ \hline \end{array}$

22. $\begin{array}{r} 51.82 \\ \times\ 0.305 \\ \hline \end{array}$

23. $\begin{array}{r} 6.02 \\ \times\ 0.0205 \\ \hline \end{array}$

24. $\begin{array}{r} 45.2 \\ \times\ 0.1003 \\ \hline \end{array}$

25. $\begin{array}{r} 0.058 \\ \times\ 0.00077 \\ \hline \end{array}$

In problems 26–28 complete the calculations and round off your answer to the nearest hundredth.

26. $(231.62 + 88.88) \times (43.55 - 19.63) =$ ____________________

27. $(540.82 - 364.78) \times (15.17 + 18.33) =$ ____________________

28. $(25.032 + 6.024) \times (8.444 - 5.708) =$ ____________________

29. Find the area of a square whose side dimension is 5.4 inches. (*Hint:* $A = s \times s$ or s^2.)

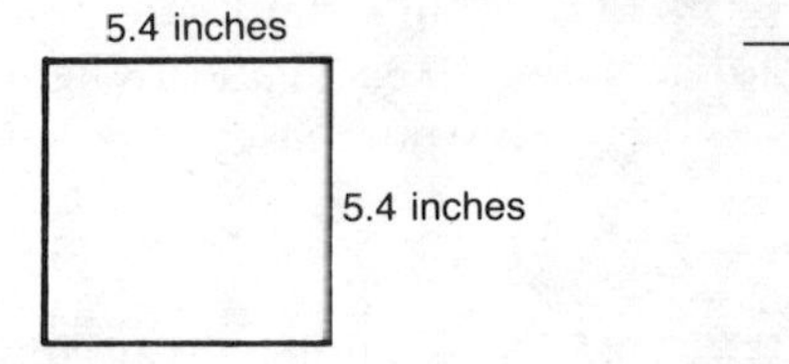

____

30. If 1 kilometer = 0.621 mile, how many miles are in 10 kilometers?

____

31. If a metric ton = 2,204.62 lb, how many pounds are in 8 metric tons?

____

32. If one liter = 0.264 liquid gallons, how many gallons are in 57 liters?

____

33. Find the area of a rectangle whose width is 3.61 inches and whose length is 2.74 inches. (*Hint:* $A = l \times w$.)

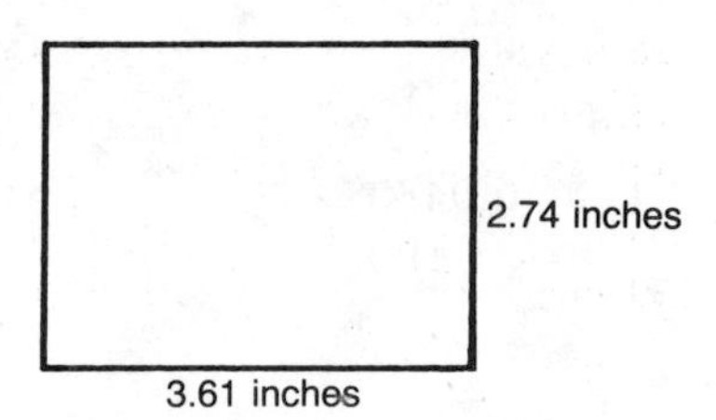

34. Find the area of a triangle whose base is 5.88 inches and whose height is 4.07 inches. (*Hint:* $A = \frac{1}{2}bh$.)

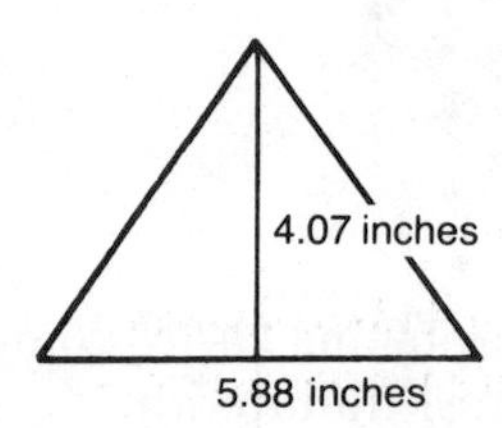

35. Mr. Nelson used 998 gallons of fuel oil last year. How much did this cost him at $1.22 per gallon?

36. Allison deposits $37.50 in her savings account every month. What is the total amount deposited by Allison at the end of 10 full years?

37. Brian's Hardware purchased two dozen rakes at $1.85 each, a dozen hoes at 1.89 each, and 18 light wheelbarrows at $15.00 each. Find the total amount of his purchase.

38. Mr. Sandler always buys fire insurance on his house at the 5-year rate, which is four times the rate for a single year. How much does he pay on his $60,000 policy if the 1-year rate is $5.80 per $1,000?

39. A manuscript of 415 pages is typed on paper that is 0.003 inch thick. Will the manuscript fit in a box 1-inch deep?

40. In a certain city the telephone rate is $15.80 per month plus 12.5 cents for each call above 40 during the month. What is the cost of 68 calls during the month?

ANSWERS

1. Decimal point is moved 2 places to the right; 761 **2.** Decimal point is moved 3 places to the left; 0.0384 **3. (a)** 5,430 **(b)** 543 **4.** 24.5 **5.** 6.3 **6.** 140.08 **7.** 33.44 **8.** 3,576 **9.** 776.795 **10.** 136.2602 **11.** 658.5579 **12.** 0.0047 **13.** 2049.957 **14.** 0.002176 **15.** 2.943 **16.** 0.00765 **17.** 0.019728 **18.** 0.3308 **19.** 0.00075 **20.** 8,820 **21.** 4.88 **22.** 15.8051 **23.** 0 12341 **24.** 4.53356 **25.** 0.00004466 **26.** 7666.36 **27.** 5897.34 **28.** 84.97 **29.** 29.16 square inches **30.** 6.21 miles **31.** 17,636.96 lb **32.** 15.048 gallons **33.** 9.8914 square inches **34.** 11.9658 square inches **35.** $1,217.56 **36.** $4,500 **37.** $337.08 **38.** $1,392 **39.** no; 1.245 > 1 **40.** $19.30

3.7 Division (÷)

What is the value of 38.25 divided by 17? How many items, each costing $0.98, can you purchase for $18? If 1 kilogram equals 2.205 pounds, what is the value of 1 pound in kilograms? We will answer these questions in this section.

Not only must you be able to add, subtract, and multiply decimals, you must also be able to divide decimals. There are two cases to be considered: (1) dividing a decimal by a whole number and (2) dividing a decimal by a decimal. The former case is the easiest and the one that we will discuss first.

> *To divide a decimal by a whole number,* write the division as though both numbers are whole numbers. Perform the division and place your decimal point in the quotient directly above the decimal point in your dividend.

Example 1a Evaluate $\frac{38.25}{17}$, which means 38.25 ÷ 17.

Solution Begin by writing $17\overline{)38.25}$.
Next, divide just as you would any two whole numbers.
Finally, place your decimal point in the quotient directly above the decimal point in the dividend, 38.25. Thus

```
                2.25  ← quotient
divisor →  17)38.25  ← dividend
              34
               4 2
               3 4
                 85
                 85
                  0  ← remainder
```

Example 1b Evaluate $\frac{0.3825}{17}$, which means 0.3825 ÷ 17.

Solution

```
   0.0225
17)0.3825
     34
     42
     34
      85
      85
       0
```

It is important to notice that we had to insert a "place-keeping" zero, in the tenths place, in our answer.

In a division problem the number doing the dividing is called the **divisor,** the number being divided is called the **dividend,** and the answer is called the **quotient.** Of course, just as in a division of whole numbers, there may be a nonzero remainder. This vocabulary, which is identical to that of whole numbers, is illustrated in the preceding example. **To check a division,** multiply the divisor by the quotient and, to this product, add the remainder; you should obtain the dividend. We encourage you to return to Example 1b and verify that (17) (0.0225) = 0.3825.

> *To divide a decimal by a decimal,* convert the division into an equivalent division with a whole-number divisor. This is accomplished by multiplying both dividend (numerator) and divisor (denominator) by 10, 100, 1,000, and so on that will move the decimal point in the divisor all the way to the right. When this is done, proceed as before.

Example 2 Evaluate $\frac{7.68}{1.2}$, which means $7.68 \div 1.2$.

Solution $\frac{7.68}{1.2} = \frac{7.68 \times 10}{1.2 \times 10} = \frac{76.8}{12} = 6.4$ or

```
     6.4
12)76.8
   72
    4 8
    4 8
      0
```

Alternatively, you may find it convenient to set up the division of decimals and "point off" the same number of decimal places in divisor and dividend as you work. See the illustration below:

```
      6.4
1.2)7.68
    7 2
     48
     48
      0
```

Observe, when you move the decimal point the same number of places in both the divisor and the dividend, you do not change the value of the problem. The next examples will illustrate division of decimals.

Example 3 How many items, costing \$0.98 each, can you purchase for \$18?

Solution To begin, \$0.98 is nearly \$1.00 which implies that the answer must be close to 18.

```
        18.36
0.98)18.00 00
     9 8
     8 20
     7 84
       36 0
       29 4
       06 60
       05 88
       00 72   remainder
```

Because you can only purchase a whole item, the answer is 18 items.

The remainder is read 0.0072. Position of the digits in the remainder is done with respect to the "original position" of the decimal point.

We generally round off a decimal quotient to the nearest hundredth. If this were done in Example 3, the answer would have been 18.37. Finally, to check this last example multiply the quotient, 18.36, by the divisor, 0.98 and add the remainder 0.0072 to this product. The answer should be 18. We encourage you to complete the check to this example.

Example 4 Evaluate 25.536 ÷ 4.8.

Solution Because the divisor has one decimal place, begin by moving the decimal point one place in the divisor, 4.8, and in the dividend, 25.536. The decimal point in your quotient lies directly above its new location in your dividend. Thus

```
                   5.32   quotient
divisor     4.8)25.5 36   dividend
                24 0
                 1 5 3
                 1 4 4
                    96
                    96
                     0    remainder
```

Example 5 Evaluate 4.016 ÷ 3.2

Solution

```
        1.255
3.2)4.0 160
    3 2
      8 1
      6 4
      1 76
      1 60
        160
        160
          0
```

Example 6 If one kilogram equals 2.205 pounds, express one pound in kilograms, rounded off to three decimal places.

Solution 1 pound = 1 ÷ 2.205 kilograms = 0.454 kilograms

```
          0.4535       R = 0.0000325
2.205)1.000 0000
        882 0
        118 00
        110 25
          7 750
          6 615
          1 1350
          1 1025
             325
```

Example 7 Evaluate $(15.47 + 33.98) \div (51.6 - 38.85)$. Carry this division to three decimal places and round off to two decimal places.

Solution First, complete the work inside each pair of parentheses. This yields 49.45 and 12.75. Next, divide the decimals adding additional zeros to the dividend as necessary.

```
          3.878      R = 0.00550
12.75)49.45 000
      38 25
      11 20 0
      10 20 0
       1 00 00
         89 25
         10 750
         10 200
         00 550
```

Answer 3.88

Example 8 Tom O'Hara used 38.5 gallons of gasoline in driving the 858-mile roundtrip from Boston to Washington, D.C. How many miles did Tom get per gallon? (Round off to the nearest tenth.)

Solution

```
        2 2.28
38.5)858.0 00
     770
      88 0
      77 0
      11 0 0
       7 7 0
       3 3 00
       3 0 80
         2 20
```

22.28 miles per gallon rounded off to the nearest tenth is 22.3 miles per gallon.

Round-Off Error The following formula is important in business mathematics: $I = PRT$ (Simple Interest = Principal × Rate × Time measured in years). Let's use this formula to illustrate round-off error.

Assume you are to find the simple interest, I, on a principal, P, of \$5,000 at a rate, R, of 6%, for $\frac{2}{3}$ of a year (*Note:* $6\% = \frac{6}{100} = 0.06$.) The value of $I = PRT = \$5{,}000 \times \frac{6}{100} \times \frac{2}{3} = \$200$. Now consider the following solutions of this same problem using a decimal approximation for $\frac{2}{3}$.

Principal	Rate (Decimal Approximate)	Time	Interest	Round-off Error
\$5,000	0.06	0.67	\$201.00	\$1.00
\$5,000	0.06	0.667	\$200.10	\$0.10
\$5,000	0.06	0.6667	\$200.01	\$0.01
\$5,000	0.06	0.66667	\$200	none

The use of a nonterminating decimal, such as 0.666 . . ., in an arithmetic calculation can cause round-off error, and frequently this error is magnified by further mathematical calculations. *How can we avoid round-off error?*

> You can avoid round-off error by using fractions instead of nonterminating decimals whenever it is reasonable to do so. In a situation where you must use a nonterminating decimal, be aware that many decimal places may be needed to obtain the desired degree of accuracy.

This is why many business mathematics tables will be written to ten decimal places. The number of decimal places that you retain in a calculation will vary according to the size of the number you are dealing with. For example, you will need more decimal places when dealing with a principal of $500,000 than with a principal of $5,000.

Example 9 You are to determine $\frac{1}{6}$ of $87,203.40.

Solution The *best solution* is obtained by multiplying $\frac{1}{6}$ by $87,203.40 to realize $14,533.90. It is clear that multiplying a number by $\frac{1}{6}$ is the same as dividing it by 6. A *poor solution* is obtained by rounding off $0.16\frac{2}{3}$ to 0.17 and evaluating (0.17) ($87,203.40) = $14,824.58. The round-off error is $14,824.58 − $14.533.90 = $290.68. A better but *still poor solution* is obtained by rounding off $\frac{1}{6}$ to 0.1667 and evaluating
(0.1667) ($87,203.40) = $14,536.81.
Now the round-off error is
$14,536.81 − $14,533.90 = $2.91.

Example 10 Find $\frac{1}{8}$ of $887.92.

Solution

(a) $\frac{1}{8}$ of $\$887.92 = \frac{1}{8} \times \$887.92 =$

$$\frac{\$887.92}{8} = \$110.99$$

(b) Observe that $\frac{1}{8}$ is a terminating decimal

$$\frac{1}{8} = 0.125$$

$$\frac{1}{8} \text{ of } \$887.92 = (0.125)(\$887.92) = \$110.99$$

Both solutions are equal and neither produces any round-off error. This is true because we used all of the digits in our terminating decimal.

Note: As in the preceding example, when you are working with terminating decimals, retain all of the digits to avoid round-off error.

3.7 EXERCISES

In problems 1–15 perform the indicated division

1. $\frac{7.65}{9}$ ______
2. $\frac{40.64}{8}$ ______
3. $\frac{27.6}{12}$ ______
4. $\frac{49.78}{19}$ ______
5. $4.5\overline{)409.5}$ ______
6. $2.6\overline{)993.2}$ ______
7. $1.6\overline{)3.15}$ ______
8. $8.5\overline{)1751}$ ______
9. $0.7\overline{)0.231}$ ______
10. $3.06\overline{)76.5}$ ______
11. $0.704\overline{)2.5344}$ ______
12. $18.06\overline{)581.532}$ ______
13. $2123.1377 \div 23.81 =$ ______
14. $46 \div 9.2 =$ ______
15. $1.274 \div 45.5 =$ ______

Check your answers to problems 1–15. If you did not get at least 12 problems correct, turn to Practice Problem Set 12A in the Appendix.

In problems 16–26 perform the indicated division.

16. $9.002\overline{)0.684152}$ ______
17. $0.13\overline{)1144}$ ______
18. $0.044\overline{)27.2844}$ ______
19. $7.3\overline{)36.5146}$ ______
20. $0.12\overline{)0.000684}$ ______
21. $0.00125\overline{)0.00075}$ ______
22. $0.0042\overline{)0.00105}$ ______
23. $0.0019\overline{)0.0152}$ ______
24. $0.0463\overline{)0.3241}$ ______
25. $0.0078\overline{)0.003432}$ ______
26. $3.6783 \div 6.1 =$ ______

In problems 27–31 round off your answer to the indicated place.

27. $29 \div 7.5$ (hundredths) $=$ ______
28. $0.3065 \div 0.42$ (hundredths) $=$ ______
29. $84.76 \div 2.1$ (tenths) $=$ ______
30. $80.5 \div 6.4$ (thousandths) $=$ ______
31. $68 \div 3.2$ (tenths) $=$ ______

In problems 32– 37 complete the calculations and round off your answer to two decimal places (hundredths).

32. $(477.02 - 263.58) \div (24.61 - 15.39) =$ ______
33. $(505.56 + 175.34) \div (63.03 - 17.48) =$ ______
34. $(86.314 + 98.507) \div (57.921 - 46.017) =$ ______
35. If one foot $=$ 0.3048 meter, express one meter in feet.

______

36. If 1 mile = 1.609 kilometers, express 1 kilometer in miles.

37. If one yard = 0.9144 meter, express one meter in yards.

38. Joe figures that it costs him $1,883.60 to drive his car last year. If he drove 9,450 miles, how much did it cost him per mile? (Round to nearest cent.)

39. In a bicycle race Harold rode 2,462.8 miles in 123.56 hours. What was his average speed in miles per hour correct to the nearest tenth?

40. If you divide 6.35 by 2.5 and multiply the quotient by 3.14, what is the result?

41. (a) The carat is a measure of weight used in weighing precious stones and equals 3.086 grains. How many grains does a 3.4-carat diamond weigh?

multiplyA to find the Answer

(b) If a gram weighs 15.432 grains, how many grams does this diamond weigh? (Round off to the nearest thousandth.)

42. Felipe finds that the gasoline tax in his home state is 15.9 cents per gallon. He estimates that he drives 18,300 miles each year in his home state and that he averages 25.6 miles per gallon. To the nearest cent, how much gasoline tax does he pay annually?

43. How many shares of Electron Corporation stock may be purchased with $30,870 if the stock is selling at $85.75 per share?

44. A taxi charges $1.00 for the first 0.2 mile and 25 cents for each additional 0.2 mile or fraction thereof. Frieda took a taxi from her home to the airport, a distance of 15.7 miles. She gave the driver a tip of $1.25. How much did the trip cost her including the tip?

45. The odometer of Susan's car read 6353.2 miles at the start of a trip. When Susan returned, it read 8247.6 Susan used 72.5 gallons of gasoline. How many miles did she get per gallon? (Round off to the nearest tenth.)

46. (a) Find $\frac{1}{7}$ of $3,898.23 using fractions.

(b) Find $\frac{1}{7}$ of \$3,898,23 using the approximation $\frac{1}{7} = 0.14$.

(c) Determine the round-off error, if any, that results from using the decimal approximation for $\frac{1}{7}$.

47. (a) Find $\frac{3}{11}$ of \$9,002.51 using fractions.

(b) Find $\frac{3}{11}$ of \$9,002.51 using the approximation $\frac{3}{11} = 0.27$.

(c) Determine the round-off error, if any, that results from using the decimal approximation for $\frac{3}{11}$.

48. (a) Find $\frac{3}{8}$ of \$20,267.28 using fractions.

(b) Find $\frac{3}{8}$ of \$20,267.28 using $\frac{3}{8} = 0.375$.

(c) Determine the round-off error, if any, that results from using this decimal for $\frac{3}{8}$.

49. (a) Find $\frac{4}{9}$ of \$2,210.58.

(b) Find $\frac{4}{9}$ of 2,210.58 using $\frac{4}{9} = 0.444$.

(c) Determine the round-off error, if any, that results from using this decimal for $\frac{4}{9}$.

ANSWERS

1. 0.85 **2.** 5.08 **3.** 2.3 **4.** 2.62 **5.** 91 **6.** 382 **7.** 1.96875 **8.** 206 **9.** 0.33 **10.** 25 **11.** 3.6 **12.** 32.2 **13.** 89.17 **14.** 5 **15.** 0.028 **16.** 0.076 **17.** 8800 **18.** 620.1 **19.** 5.002 **20.** 0.0057 **21.** 0.6 **22.** 0.25 **23.** 8 **24.** 7 **25.** 0.44 **26.** 0.603 **27.** 3.87 **28.** 0.73 **29.** 40.4 **30.** 12.578 **31.** 21.3 **32.** 23.15 **33.** 14.95 **34.** 15.53 **35.** 3.28 feet **36.** 0.62 mile **37.** 1.09 yards **38.** \$0.20 **39.** 19.9 miles per hour **40.** 7.9756 **41. (a)** 10.4924 **(b)** 0.680 **42.** \$113.66 **43.** 360 **44.** \$21.75 **45.** 26.1 **46. (a)** \$556.89 **(b)** \$545.75 **(c)** \$11.13 **47. (a)** \$2,455.23 **(b)** \$2,430.68 **(c)** \$24.55 **48. (a)** \$7,600.23 **(b)** \$7,600.23 **(c)** none **49. (a)** 982.48 **(b)** 981.50 **(c)** \$0.98

3.8 THE HAND-HELD CALCULATOR (OPTIONAL)

You wish to multiply \$39,367.84 by 88,349; you wish to divide 79.86 into 98,174.06; or you wish to evaluate the product \$25,000 × 0.0625 × 24.75. Each of these calculations represents a chore when done with paper and pencil. *When you have mastered arithmetic, a hand-held calculator becomes a valuable mathematics tool.* There are many brands of calculators on the market. *Consequently, it is important that you read the instructions that accompany your calculator because processes may vary.*

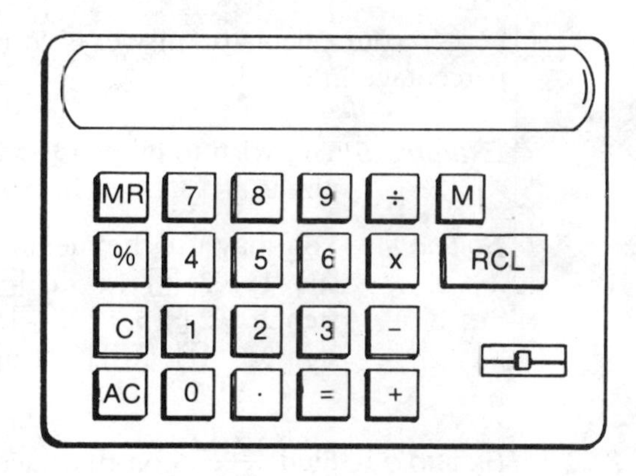

A hand-held calculator that performs the four basic operations of addition, subtraction, multiplication, and division is sufficient for most arithmetic exercises. Hand-held calculators capable of replacing cumbersome desk calculators, which were both expensive to buy and repair, are now within the price range of all students. This remarkable technological achievement has come about within the past ten years. The pattern to execute each of the four basic operations is:

enter number, hit operation key, enter number; depress [=] key

By hitting the = key you instruct the calculator to complete the entered calculation. If you enter a wrong number and complete the calculation, you receive a wrong answer.

Example 1 (a) 73 × 89 on the calculator becomes
7 3 [×] 8 9 [=] 6497
In this case, 6497 appears as the answer on the display.
(b) 1,201 − 789 on the calculator becomes:
1 2 0 1 [−] 7 8 9 [=] 412
Here, 412 appears as the answer on the display. Note that you do not enter a comma as part of your calculation.

Your calculator has a key for entering a decimal point (.). When entering a number such as 26.95, you must use this key. That is, to enter 26.95 you hit five keys: 2 6 . 9 5. If you wish to enter 26. (or 26). you only enter 26, for the calculator will automatically insert the decimal point after the 6 for you. Your calculator has a *floating decimal point,* which simply means that it will locate the decimal point in the answer for you.

Three keys that appear on most simple calculators are the clear key [C], the percentage key [%], and the memory key [M]. The following examples will illustrate how to use these helpful keys. In conjunction with the memory key there is the recall key, RCL.

Example 2 439.65 ÷ 25 on the calculator becomes:
4 3 9 . 6 5 [÷] 2 5 [=] 17.586.

When you hit the clear key [C], it will erase the last number entered.

Example 3 You wish to compute 9×8 and you accidentally entered 9×9.
9 [×] 9 [C] 8 [=] 72
erases the preceding 9

Example 4 Find 25% of 36.
2 5 [%] [×] 3 6 [=] 9

Note: To find 25% of a number means to find $\frac{25}{100}$ or $\frac{1}{4}$ of the number because $\frac{1}{4} \times 36 = 9$, our calculator answer is correct. In Chapter 4 we will discuss the topic of percentage in detail.

Example 5 You wish to purchase 3 small cheese pizzas at \$3.50 each and 2 large pizzas at \$5.25 each.

Solution You may use the memory key [M] as follows:
step 1: 3 [×] 3 . 5 0 [=] [M] (at this point 10.5 is stored)
step 2: 2 [×] 5 . 2 5 [=]; 10.5 appears (large pizzas)
step 3: [+] [RCL]; 21 appears as your total purchase

Example 6 Find $\frac{17 \times 45}{2 \times 7}$ on the calculator.

Solution 1 7 [×] 4 5 [=] [÷] 2 [÷] 7 [=]; 54.64 appears.
Explanation [÷] 2 [÷] 7: Calculator will multiply $17 \times 45 = 765$, then divide 765 by 2 ($765 \div 2 = 382.5$). If we had pressed [×] 7, it would have multiplied 382.5 by 7. Thus the problem solved would have been

$$\frac{17 \times 45}{2} \times 7$$

Example 7a Find $\frac{57+64}{33 \div 3}$ on the calculator without using memory keys.
step 1: For the numerator, 5 7 [+] 6 4 [=]; 121 appears.
step 2: For the denominator 3 3 [÷] 3 [=]; 11 appears.
Now you enter the values 121 and 11 again:
step 3: 1 2 1 [÷] 1 1 [=]; 11 appears

Example 7b Repeat Example 7a using memory keys.
step 1: For the numerator 57 [+] 64 =; 121 appears. Press [M] 1; 1 is a memory location.
step 2: For the denominator 33 [÷] 3 =; 11 appears. Press [M] 2; 2 is a memory location.
step 3: [RCL] 1 =; 121 appears ÷ [RCL] 2 =; 11 appears [=]; 11 appears

3.8 Exercises

Do each of the following problems on a hand-held calculator and round off your answer to the nearest hundredth.

1.
```
   15.54
   72.86
 + 77.05
```

2.
```
  571.47
   19.235
 + 208.3
```

3.
```
   88.56
    8.753
  905.7
 +  63.821
```

4.
```
  652.3215
   36.0307
 + 185.4398
```

5. $\begin{array}{r} 16 \\ -\ 7.91 \\ \hline \end{array}$

6. $\begin{array}{r} 32.7 \\ -\ 18.14 \\ \hline \end{array}$

7. $\begin{array}{r} 439.762 \\ -\ 188.865 \\ \hline \end{array}$

8. $\begin{array}{r} 2,602.395 \\ -1,855.297 \\ \hline \end{array}$

9. $\begin{array}{r} 87.9 \\ \times\ \ \ 6 \\ \hline \end{array}$

10. $\begin{array}{r} 54.237 \\ \times\ \ 7.8 \\ \hline \end{array}$

11. $\begin{array}{r} 19.83 \\ \times\ 47.56 \\ \hline \end{array}$

12. $\begin{array}{r} 3,409 \\ \times\ 0.0081 \\ \hline \end{array}$

13. $\frac{3,875}{25} =$ ______

14. $\frac{563.2}{17} =$ ______

15. $\frac{37.81}{47.65} =$ ______

16. $\frac{6,504.5}{413.9} =$ ______

17. $\frac{5 \times 4}{3 \times 6} =$ ______

18. $\frac{47 \times 81}{2.5 \times 7} =$ ______

19. $\frac{18 \times 8}{17 \times 4} =$ ______

20. $\frac{9.3 \times 5.2}{1.8 \times 6.4} =$ ______

21. $\frac{17+63+91}{48-5.6} =$ ______

22. $\frac{75+32-23}{1.7+2.1} =$ ______

23. $\frac{54+18}{23 \times 0.5} =$ ______

24. $\frac{0.0028}{0.00015} =$ ______

25. $(153,000) \times (0.000273) =$ ______

26. $\frac{0.0005}{0.0045} =$ ______

27. $(4,200)(0.000036) =$ ______

28. Barbara drove 331.1 miles on 14 gallons of gasoline; how many miles is she averaging per gallon?

______

29. If $(1 + 0.025)^3$ means $(1 + 0.025)\ (1 + 0.025)\ (1 + 0.025)$ and $A = \$2,975 \times (1 + 0.025)^3$, find the value of A.

______

30. Find the property tax on a home with a market value of $75,800, if the tax is computed by multiplying the market value by the property tax rate of 0.04632.

______

31. Use trial and error to estimate the square root of 19 (a number times itself whose product is 19). Correct to two decimal places.

______

32. A woman purchases a new car for $7,950. She must pay a sales tax which is 0.05 of this figure and an excise tax of $25.00 per $1,000 of this price. What is the total of these two tax bills?

______

33. If one meter is approximately 39.37 inches, how many inches are contained in 562 meters?

______

34. Evaluate $I = \$98,000 \times 0.16125 \times 4.5$.

______

35. A publisher manufactures a certain book for $18.00. The publisher sells the book to the college book store for $18.00 plus 0.15 of $18.00. The college book store takes this total and adds on 0.20 of the total. This final figure is what the student pays. How much does the student pay?

____

ANSWERS

1. 165.45 **2.** 799.01 **3.** 1,066.83 **4.** 873.79 **5.** 8.09 **6.** 14.56 **7.** 250.90 **8.** 747.10 **9.** 527.4 **10.** 423.05 **11.** 943.11 **12.** 27.61 **13.** 155 **14.** 33.13 **15.** 0.79 **16.** 15.72 **17.** 1.11 **18.** 217.54 **19.** 2.12 **20.** 4.20 **21.** 4.03 **22.** 22.11 **23.** 6.26 **24.** 18.67 **25.** 41.77 **26.** 0.11 **27.** 0.15 **28.** 23.65 **29.** $3,203.75 **30.** $3,511.06 **31.** 4.36 **32.** $596.25 **33.** 22,125.94 inches **34.** $71,111.25 **35.** $24.84

FOCUS ON PROBLEM SOLVING

Independent Bank
114 Elm Street
Topeka, Kansas

period ending
10–08–8X

Mr Earl Nichols
Mrs Margaret Nichols
263 Walnut Street
Topeka, Kansas

CHECKING ACCOUNT SUMMARY

ACCOUNT NUMBER	BEGINNING BALANCE	CHECKS PAID AMOUNT	NO.	TOTAL DEPOSITS AMOUNT	NO.	SERVICE CHARGE	ENDING BALANCE
749515	241.73	1,388.35	14	1,485.00	3	5.00	333.38

* * * * * * CHECKING ACTIVITY * * * * *

DATE	CHECKS	CHECKS	DEPOSITS	BALANCE
09-08	150.00			91.73
09-09	34.63			57.10
09-10	6.00	28.00	460.00	483.10
09-21			125.00	608.10
09-23	60.00	62.45	900.00	1,385.65
09-27	61.55	80.11		
	203.05			1,040.94
09-28	45.00	45.98		
	86.58			863.38
09-30	225.00			638.38
10-07	300.00			338.38
10-08	5.00SC			333.38
-20-				

LINE OF CREDIT
The **ANNUAL PERCENTAGE RATE** AND METHOD OF DETERMINING **FINANCE CHARGE**
The **FINANCE CHARGE** is equal to the product obtained by multiplying the Average Daily Balance by a Daily Periodic Rate of .04931% (**ANNUAL PERCENTAGE RATE OF 18%**) times the Cycle Days.
NOTICE—SEE REVERSE SIDE FOR IMPORTANT INFORMATION

One of the everyday problems faced by many people is the problem of "balancing" their checkbook against their bank statement. Here one endeavors to make the ending balance on the bank statement ($333.38 in our illustration) agree with the current balance recorded in the checkbook.

In this balancing process we must remember any late deposits into our checking account that do not appear on the bank statement, we must be aware of all outstanding checks (checks that remain uncashed), we must consider any bank service charges ($5.00 and labeled SC in our illustration), and interest given to us by the bank.

Example Consider Earl Nichols, whose bank statement is illustrated. His current checkbook balance is $443.63; he made one late deposit of $500.00 not recorded on the bank statement; and he has outstanding checks in the following amounts: $56.25, $240.00, and $98.50. As you can see, the bank statement indicates a service charge of $5.00 and no interest. Balance Earl's bank statement against his checkbook.

Solution

Statement balance		$333.38
To this add all late deposits	+	500.00
		$833.38
From this total subtract all outstanding checks		
no. 143 for $ 56.25		
no. 148 for 240.00		
no. 155 for 98.50		
	−	$394.75
		$438.63
Corrected statement balance		$438.63

Checkbook balance		$443.63
To this add all interest	+	0.00
		$443.63
From this subtract all service charges	−	5.00
Corrected checkbook balance		$438.63

BUILDING YOUR MATH VOCABULARY

TERM	DEFINITION	EXAMPLE
Addend	A number being added in an addition problem.	8.45 ← addend + 6.38 14.83
Decimal Numbers	A number with a decimal point.	71.96 34 or 34.; the decimal point is understood to follow the four
Difference	The answer to a subtraction problem.	9.043 − 3.625 5.418 ← difference
Dividend	The number being divided by another number in a division problem.	$5.3\overline{)22.26}$ = 4.2 ← dividend
Divisor	The number doing the dividing in a division problem.	divisor → $5.3\overline{)22.26}$ = 4.2
Minuend	A number from which another number is subtracted.	9.043 ← minuend − 3.625 5.418
Multiplicand	A number being multiplied.	5.84 ← multiplicand × 6.3 1 752 35 04 36.792
Multiplier	A number doing the multiplying.	5.84 × 6.3 ← multiplier 1 752 35 04 36.792
Nonterminating Decimal	All decimals that are not terminating decimals.	$0.5454\ldots = 0.\overline{54}$
Product	The answer to a multiplication problem.	5.84 × 6.3 1752 3504 36.792 ← product
Quotient	The answer to a division problem.	$5.3\overline{)22.26}$ = 4.2 ← quotient
Real Numbers	The collection of numbers associated with all of the points on a line.	$0.65, \frac{1}{2}, 0.\overline{32}$ and 6 are real numbers
Remainder	The number left after you determine the maximum number of times the divisor is contained in the dividend.	$3.1\overline{)7.26}$ = 2.3 6 2 1 0 6 9 3 1 3 ← remainder R = 0.13

TERM	DEFINITION	EXAMPLE
Sum	The answer to an addition problem.	$\begin{array}{r} 8.45 \\ +\ 6.38 \\ \hline 14.83 \end{array} \leftarrow$ sum
Terminating Decimal	A decimal in which the digit zero repeats indefinitely.	0.4, 0.75, 1.92

Chapter Review

In problems 1 and 2 express the decimal in numerical form.

1. (a) forty-seven and thirty-six thousandths

____

(b) five million, five hundred one thousand, two hundred fifty and ninety-nine hundredths

____

2. (a) six hundred eighteen and twenty-eight ten-thousandths

____

(b) thirty-four thousand, one hundred seventy-six and two hundred twenty-three thousandths

____

In problems 3–5 express the given decimals in words.

3. 52.71 ______________

4. 63.045 ______________

5. 869.31 ______________

Round off the decimals given in problems 6–9 to the place indicated.

6. 6,832.61 to the units place ______________

7. 5.503 to the tenths place ______________

8. 46.09 to the tenths place ______________

9. 8.765 to the hundredths place ______________

In problems 10–13 convert each of the given fractions to an equivalent decimal. If the decimal does not terminate, round it off to the hundredths place.

10. $\frac{5}{6}$ ______________ **11.** $\frac{7}{8}$ ______________

12. $2\frac{1}{3}$ ______________ **13.** $5\frac{4}{13}$ ______________

In problems 14–17 convert the given decimals to equivalent fractions.

14. 0.88 ______________ **15.** 1.44 ______________

16. 0.355 ______________ **17.** 2.02 ______________

Perform the operation indicated in problems 18–30.

18.
$$\begin{array}{r} 32.71 \\ 17.82 \\ +\ \ 6.8 \\ \hline \end{array}$$

19. 6.5 + 18.29 + 0.763 = ______

20. 5,896.27 + 2,063.77 + 898.84 =

____

21.
$$\begin{array}{r} 51.98 \\ -\ 27.89 \\ \hline \end{array}$$

22. $61.7 - 18.52 =$ ________

23. $79{,}632.15 - 53{,}998.08$

24. 5.8×6.4

25. 18.47×0.0009

26. 95.25×4.85

27. $16.56 \times .379$

28. $5.6\overline{)87.64}$ ________

29. $1.98\overline{)7.024}$ ________

30. $(19.632 \times 44.68) \div (55{,}639 - 54{,}709.63) + 875.047 =$ ________
(*Hint:* Use a hand-held calculator. Round off to hundredths.)

31. Brian bought a new car having a base price of $7,595. He opted for the following extras: automatic transmission, $250; radio, $85.50; power steering, $225.25; and whitewall tires, $48.65. What was the total price of the car?

____

32. A pieceworker for the Ace Zipper Company received the following pay daily for a week: $28.50, $32.60, $37.50, $29.65, $37.70. What was the total gross pay?

____

33. If the worker in problem 32 had $13.50 deducted for federal income taxes, $4.20 deducted for state taxes, $3.29 deducted for social security, and $1.54 for medical coverage; find:

(a) the total amount of payroll deductions for this worker. ________

(b) the take-home pay for this worker. ________

34. Mike Cardinal drives 27.6 miles per day to work and another 27.6 miles home. What is the total number of miles that he travels in a 5-day work week?

____

35. Each month Mary must pay 0.015 times her unpaid charge account balance as interest. How much is her interest charge for December if her unpaid balance is $560?

____

36. If a woman is paid $3.95 per hour and she works 32 hours in a given week, what is her gross pay?

____

37. If a woman is paid $6.40 per hour and time-and-a-half after 40 hours of work, what is her gross pay for a 48-hour week?

____

38. A man receives three checks each for $15.63, two checks each for $48.92, and five checks each for $18.71. What is the total sum received?

____

39. A store receives a shipment of 800 radios and an invoice bill of $44,208. What is the cost of each radio?

____

40. If Don received $150 (gross) for a 40-hour week, what is his hourly rate of pay?

____

ANSWERS

1. (a) 47.036 **(b)** 5,501,250.99 **2. (a)** 618.0028 **(b)** 34,176.223 **3.** fifty-two and seventy-one hundredths **4.** sixty-three and forty-five thousandths **5.** eight hundred sixty-nine and thirty-one hundredths **6.** 6,833 **7.** 5.5 **8.** 46.1 **9.** 8.77 **10.** 0.83 **11.** 0.875 **12.** 2.33 **13.** 5.31 **14.** $\frac{22}{25}$ **15.** $1\frac{11}{25}$ **16.** $\frac{71}{200}$ **17.** $2\frac{1}{50}$ **18.** 57.33 **19.** 25.553 **20.** 8,858.88 **21.** 24.09 **22.** 43.18 **23.** 25,634.07 **24.** 37.12 **25.** 0.016623 **26.** 461.9625 **27.** 6.27624 **28.** 15.65 **29.** 3.5, $R = 0.094$ **30.** 875.99 **31.** $8,204.40 **32.** $165.95 **33. (a)** $22.53 **(b)** $143.42 **34.** 276 miles **35.** $8.40 **36.** $126.40 **37.** $332.80 **38.** $238.28 **39.** $55.26 **40.** $3.75

3

Warmup Test

1. Express the following decimals in numerical form.

(a) seven hundred nine and forty-seven hundredths ______

(b) three million ninety-two and seven hundred forty-three thousandths ______

(c) two thousand, three hundred forty-two and sixty-two thousandths ______

2. Express the following decimals in written form.

(a) 98.204

______

(b) 50,386.04

______

(c) 7,000.0002

______

3. Convert the following decimals to fractions. Reduce to lowest terms.

(a) 0.125 ______ (b) 3.32 ______ (c) 2.48 ______

4. Convert the following fractions to decimals.

(a) $\frac{9}{20}$ ______ (b) $2\frac{3}{8}$ ______ (c) $\frac{9}{11}$ ______

In problems 5–8 perform the indicated computations.

5. (a) 879.004 + 35.68 + 205.77 + 75.6 ______

(b)
$$\begin{array}{r} 17.96 \\ 28.205 \\ 33 \\ +\ 40.7 \\ \hline \end{array}$$

6. (a) 305.062 − 288.794 ______

(b)
$$\begin{array}{r} 1205.32 \\ -\ 987.68 \\ \hline \end{array}$$

7. (a) 28.65 × 5.4 ______

(b)
$$\begin{array}{r} 647.4 \\ \times\ 0.0034 \\ \hline \end{array}$$

(c) 2.856 × 100 ______

8. (a) $4.032 \div 0.72$ ________________

(b) $0.1215 \div 0.089$ (round off to hundredths) ________________

9. A family's home is assessed at $65,100. The property taxes on the home are $19.25 for every $1,000 of assessed value. How much does the family pay in property taxes?

____

10. A seamstress pays $34.97 for some woolens. The cost of the woolens was $5.45 per yard. How many yards, to the nearest hundredth, of material did she buy?

____

11. In 1980 George Brett had the highest batting average in the American League since 1941. He had 175 hits in 449 at bats. What was his batting average? (Round off to thousandths.)

____

12. A person's monthly salary is $2050.75. He pays $410.15 in taxes, $395.92 for his mortgage, $30.35 life insurance premium, $58.65 automobile costs, $116.54 for heat and electricity, and $30.26 for the telephone. How much of his monthly salary remains after he pays his bills?

____

3

Challenge Test

1. Express the following decimals in numerical form:

   (a) twenty-five and sixty-three hundredths ________

   (b) nine hundred seventy-one and two tenths ________

   (c) five thousand, six hundred ninety-eight and four hundred thirty-six thousandths ________

2. Express the following in written form.

   (a) 37.056 ________

   (b) 8,860.22 ________

   (c) 62,500.447 ________

3. Convert the following decimals to fractions. Reduce to lowest terms.

   (a) 0.75 ________ (b) 8.40 ________ (c) 3.62 ________

4. Convert the following fractions to decimals.

   (a) $\frac{7}{8}$ ________ (b) $4\frac{9}{10}$ ________ (c) $\frac{8}{25}$ ________

In problems 5–8 perform the indicated computations.

5. (a) 48.92 + 11.81 + 506.003 + 87.9 ________

   (b)
   $$\begin{array}{r} 9.37 \\ 82.05 \\ 163.6 \\ +\ 22.005 \\ \hline \end{array}$$

6. (a) 798.63 − 581.74 ________

   (b)
   $$\begin{array}{r} 526.84 \\ -\ 369.76 \\ \hline \end{array}$$

7. (a) 32.18 × 6.6 ________

   (b) 0.087 × 1,000 ________

   (c)
   $$\begin{array}{r} 158.75 \\ \times\ 46.02 \\ \hline \end{array}$$

8. (a) 4.725 ÷ 0.63 ________

   (b) 1.481 ÷ 0.35 (round off to hundredths) ________

9. In six jobs, a painting contractor used the following gallon amounts of a certain type of paint: 7.20, 5.45, 8.50, 6.75, 4.25, and 6.85. What is the total number of gallons of paint used?

   ____

10. A patient is to receive 0.06 grain of atropine sulfate. Thus far, she has received 0.045 grain of atropine sulfate. How much more of the medication does she need?

____

11. Because of inflation, food now costs 1.16 times what it did last year. If the Trivone family's food bill was $126.90 per week last year, what is their weekly food bill this year to the nearest cent?

____

12. A company hopes to ship 65 drums, each weighing 85.42 pounds, in a truck that can carry 6,000 pounds. Can they ship the drums? How much is the truck under (or over) capacity?

____

Percentage

CHAPTER OUTLINE

4–0 PRETEST

Solve the following problems. Do not use a calculator.

Problem	Section Reference	Learning Objective
1. Convert the following percents to decimals: (a) 35% _____ (b) 0.42% _____ (c) $\frac{1}{4}$% _____ (d) 10.3% _____	4.1	Converting percents to decimals
2. Convert the following percents to fractions: (a) 12% _____ (b) 9.5% _____ (c) $\frac{1}{2}$% _____ (d) 0.3% _____	4.1	Converting percents to fractions
3. Convert the following numbers to percents (%): (a) $\frac{3}{5}$ _____ (b) $\frac{5}{8}$ _____ (c) 0.35 _____ (d) 2 _____	4.1	Converting fractions and decimal numbers to percents
4. A worker earns $24,500 annual salary. He pays 24% of his annual salary in taxes. How much does he pay in taxes? _____	4.2	Solving percent problems
5. A professional basketball team won 55% of its games. They won 44 games. How many games did the team play? _____	4.3	Solving percent problems
6. A student takes a 72-question final exam. He answers 60 of the questions correctly. What percent of the questions does he answer correctly? _____	4.4	Solving percent problems
7. Tom's salary has been increased from $420 to $480. What is the percent of increase? _____	4.5	Solving percent problems

Problem	Section Reference	Learning Objective
8. Maria borrows $2500 at 12% simple interest for 2 years to buy a home computer. How much interest does she pay on her loan? ________	4.6	Solving simple interest problems
9. An investment group invests $12,000 at 16.5% interest for 5 years. (a) How much simple interest is accrued? ________ (b) How much money has the group accumulated after 5 years? ________	4.6	Solving simple interest problems
10. Find the bank discount and proceeds on a simple discount note of $12,000 at 11% for 180 days. ________	4.7	Solving bank discount problems

4.1 CONVERTING FROM ONE PERCENT FORM TO ANOTHER

Discounts on clothing and on other goods are stated in percent form. Bank discounts on loans are given in percent form. The interest on a savings account is given as a percent of the amount of money on deposit. The daily standings of major league baseball teams are given in percent form.

A particular garment is 15% wool, 10% silk, and 75% cotton. A coat is advertised as 100% waterproof. Today, 2% of our class is absent. The chance of rain is 80%. The property tax on a home is a percent of its current market value. Our federal income tax and our social security tax are both computed as a percent of our gross earnings. These examples represent a few cases where percent enters into our daily lives. It is a virtual certainty that each of us will use percents during our lifetime.

We have some quantity—perhaps a cake, a piece of cloth, a sum of money, or maybe just a number; and we divide it into 100 equal parts. Each of these 100 equal parts represents 1 percent of the whole quantity.

One percent of a quantity is a hundredth $\left(\frac{1}{100} \text{ or } 0.01\right)$ of that quantity. One percent is commonly denoted "1%."

Each square represents 1% of the whole

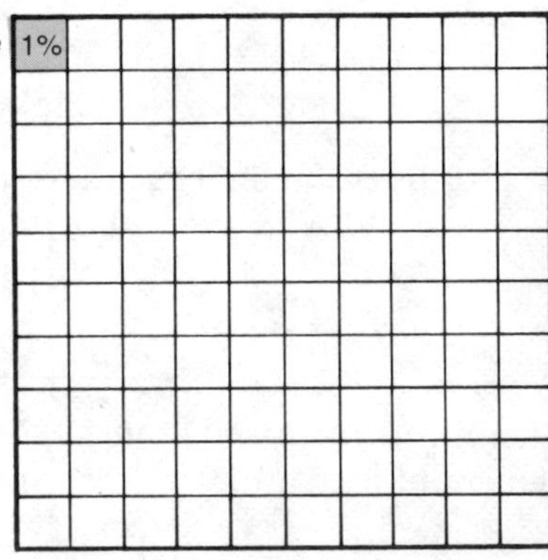

Figure 4.1

Three percent of a quantity is three hundredths and may be expressed as 3%, $\frac{3}{100}$, or 0.03. Eighty-six percent of a quantity may be expressed as 86%, $\frac{86}{100}$, or 0.86.

One hundred percent of any number *N* is the number *N* itself. This makes good sense, for it simply means that if a number is divided into 100 equal parts that you are including all these hundred parts.

There are, in addition to words, three ways of expressing percent:

1. With a percent sign (%).
2. As a fraction whose denominator is 100.
3. As a decimal.

Example 1 Seven percent may be expressed as (a) 7%, (b) $\frac{7}{100}$, or (c) 0.07.

Since there are three ways of expressing a percent, it is important that we be able to convert from one form to another with ease. In addition, we may be given a fraction, which we wish to express as a percent.

Changing a Percent to a Decimal or Fraction

> To change a percent (with a percent sign, %) to an equivalent decimal, drop the percent sign and move the decimal point two places to the left.

Thus 72% = 72.% = 0.72.

Example 2 Change from a percent to a decimal.

(a) 18% = 0.18

(b) $44\frac{3}{4}\% = 44.75\% = 0.4475$

If the percent includes a fraction, change that fraction to a decimal first, then convert the percent to a decimal.

(c) $\frac{1}{4}\% = 0.25\% = 0.0025$

(d) 125% = 1.25

(e) 8.4% = 0.084

(f) 0.5% = 0.005

> To change a percent (with a percent sign, %) to an equivalent fraction, drop the percent sign and place the number over 100.

Thus $55\% = \frac{55}{100} = \frac{11}{20}$. Because calculations with small numbers are usually easier to do, we always want the fraction in simplest form.

Example 3 Change from a percent to a fraction.

(a) $40\% = \frac{40}{100} = \frac{2}{5}$

(b) $12.5\% = \frac{12.5}{100} = \frac{12.5 \times 10}{100 \times 10} = \frac{125}{1{,}000} = \frac{1}{8}$ Multiply by $\frac{10}{10}$ to get a whole numb in the numerator

(c) $250\% = \frac{250}{100} = \frac{5}{2}$

(d) $46\% = \frac{46}{100} = \frac{23}{50}$

(e) $0.6\% = \frac{0.6}{100} = \frac{0.6 \times 10}{100 \times 10} = \frac{6}{1{,}000} = \frac{3}{500}$

Example 4 Change from a percent to a fraction.

(a) $\frac{3}{8}\% = \frac{\frac{3}{8}}{100} = \frac{\frac{3}{8} \times 8}{100 \times 8} = \frac{3}{800}$ Multiply by $\frac{8}{8}$ to get a whole number in the numerator

(b) $3\frac{1}{2}\% = \frac{3\frac{1}{2}}{100} = \frac{\frac{7}{2}}{100} = \frac{\frac{7}{2} \times 2}{100 \times 2} = \frac{7}{200}$

Computational Shortcut In converting percents to either fractions or decimals, the following computations may be used. Since percent means hundredths we can:

1. Replace the percent symbol (%) by its fractional equivalent, $\frac{1}{100}$, or its decimal equivalent, 0.01.
2. Insert a multiplication sign between the number to be converted and the fractional or decimal equivalent of the percent symbol.
3. Perform the indicated multiplication.

Example 5 Convert the following percents to fractions.

(a) 15% (b) 32%

Solution (a) $15\% = 15 \times \frac{1}{100} = \frac{15}{100} = \frac{3}{20}$

(b) $32\% = 32 \times \frac{1}{100} = \frac{15}{100} = \frac{8}{25}$

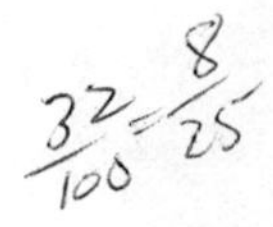

Example 6 Convert the following percents to fractions.

(a) $12\frac{1}{2}\%$ (b) $3\frac{1}{8}\%$

Solution (a) $12\frac{1}{2}\% = 12\frac{1}{2} \times \frac{1}{100} = \frac{\overset{1}{\cancel{25}}}{2} \times \frac{1}{\underset{4}{\cancel{100}}} = \frac{1}{8}$

(b) $3\frac{1}{8}\% = 3\frac{1}{8} \times \frac{1}{100} = \frac{\overset{1}{\cancel{25}}}{8} \times \frac{1}{\underset{4}{\cancel{100}}} = \frac{1}{32}$

Example 7 Convert the following percents to decimals.

(a) 6% (b) 23% (c) 14.5% (d) $5\frac{3}{4}\%$

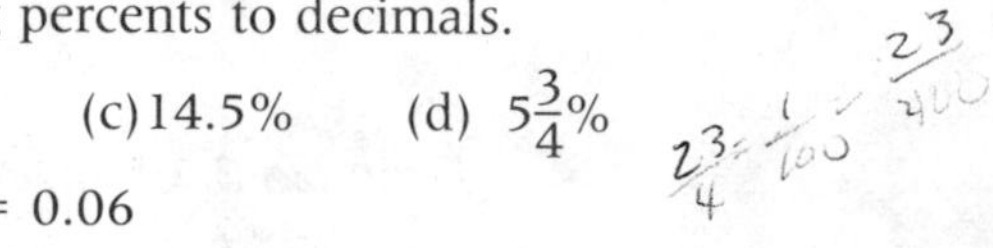

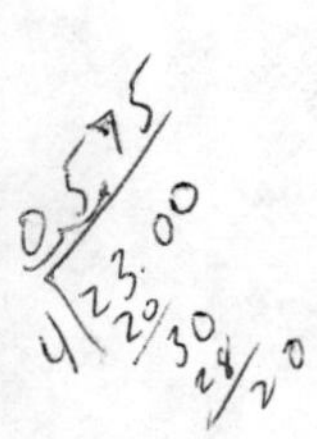

Solution (a) $6\% = 6 \times 0.01 = 0.06$

(b) $23\% = 23 \times 0.01 = 0.23$

(c) $14.5\% = 14.5 \times 0.01 = 0.145$

(d) $5\frac{3}{4}\% = 5\frac{3}{4} \times 0.01 = 5.75 \times 0.01 = 0.0575$

Changing Decimals and Fractions to Percents

> To change a decimal to an equivalent percent (with a percent sign, %) move the decimal point two places to the right and add the percent sign.

Thus, $0.54 = 0.54.\% = 54\%$.

Example 8 Change the following decimals to percents.

(a) $0.80 = 80\%$

(b) $0.09 = 9\%$

(c) $0.526 = 52.6\%$

(d) $7.85 = 785\%$

(e) $0.005 = 0.5\%$

Example 9 Change the following repeating decimals to percents (to the nearest hundredths).

(a) $0.\overline{3} = 0.3333\ldots = 33.33\%$

(b) $0.\overline{17} = 0.1717\ldots = 17.17\%$

> To change a fraction to a percent (with a percent sign, %), change the fraction to a decimal and proceed as before.

Thus, $\frac{4}{5} = 0.80 = 80\%$

Example 10 Change the following fractions to percents.

(a) $\frac{3}{8} = 0.375 = 37.5\%$

$$\begin{array}{r} 0.375 \\ 8\overline{)3.000} \\ \underline{2\,4} \\ 60 \\ \underline{56} \\ 40 \\ \underline{40} \\ 0 \end{array}$$

(b) $2\frac{1}{4} = 2.25 = 225\%$

(c) $\frac{17}{20} = 0.85 = 85\%$

Example 11 Change the following fractions to percents (to the nearest hundredth).

(a) $\frac{7}{11} = 0.\overline{63} = 63.64\%$

$$\begin{array}{r} 0.63 \\ 11\overline{)\boxed{7}.00} \\ \underline{6\,6} \\ 40 \\ \underline{33} \\ \boxed{7} \end{array}$$

(b) $1\frac{2}{3} = 1.6666\ldots$

$= 166.67\%$

4.1 EXERCISES

In problems 1–20 convert the given percent values to decimals and to fractions.

	Percents	*Decimals*	*Fractions*
1.	5%		
2.	18%		
3.	175%		
4.	300%		
5.	8.5%		
6.	45%		
7.	10.3%		
8.	0.6%		
9.	0.25%		
10.	0.26%		
11.	$\frac{3}{8}\%$		
12.	112.5%		
13.	$62\frac{1}{5}\%$		
14.	15.25%		
15.	0.08%		
16.	$\frac{2}{5}\%$		
17.	250%		
18.	0.04%		

19. 32.5% ____________ ____________

20. 100% ____________ ____________

In problems 21–50 convert the given numbers to percents with a percent sign (%).

21. 0.37 ______ **22.** 0.042 ______ **23.** $\frac{3}{10}$ ______ **24.** $\frac{2}{3}$ ______

25. $\frac{7}{4}$ ______ **26.** 1.035 ______ **27.** 3.03 ______ **28.** $\frac{1}{8}$ ______

29. $\frac{4}{15}$ ______ **30.** 0.006 ______ **31.** 0.09 ______ **32.** 0.62 ______

33. $\frac{3}{4}$ ______ **34.** $\frac{1}{5}$ ______ **35.** $\frac{11}{12}$ ______ **36.** 1.74 ______

37. 0.002 ______ **38.** $\frac{7}{8}$ ______ **39.** $\frac{18}{25}$ ______ **40.** $\frac{11}{6}$ ______

41. $3\frac{1}{8}$ ______ **42.** $\frac{5}{6}$ ______ **43.** 0.07 ______ **44.** 2.25 ______

45. 0.705 ______ **46.** 0.89 ______ **47.** $\frac{7}{9}$ ______ **48.** $\frac{2}{7}$ ______

49. 0.005 ______ **50.** 2.04 ______

51. A student read this statement: "Milk is 87% water by volume." Explain what this means. ____________

52. The label on a winter coat describes the material that the coat is made of as 80% wool and 20% polyester. Explain what this means. ____________

53. The content of a 1950 U.S. half dollar is 90% silver and 10% nickel. Explain what this means. ____________

In problems 54–57 list the correct answers.

54. 20 percent is: 20%, 0.20, $\frac{20}{100}$, 20. ____________

55. 16 percent is: 0.16, $\frac{16}{100}$, 16, 16%. ____________

56. 1% is: $\frac{1}{100}$, 0.1, 0.01, 0.001. ____________

57. $12\frac{1}{2}$ percent is: 12.5, 0.125, $12\frac{1}{2}$%, $\frac{1}{8}$. ____________

ANSWERS

1. 0.05; $\frac{1}{20}$ **2.** 0.18; $\frac{9}{50}$ **3.** 1.75; $1\frac{3}{4}$ **4.** 3; 3 **5.** 0.085; $\frac{17}{200}$ **6.** 0.45; $\frac{9}{20}$ **7.** 0.103; $\frac{103}{1{,}000}$ **8.** 0.006; $\frac{3}{500}$ **9.** 0.0025; $\frac{1}{400}$ **10.** 0.0026; $\frac{13}{5{,}000}$ **11.** 0.00375; $\frac{3}{800}$ **12.** 1.125; $1\frac{1}{8}$ **13.** 0.622; $\frac{311}{500}$ **14.** 0.1525; $\frac{61}{400}$ **15.** 0.0008; $\frac{1}{1{,}250}$ **16.** 0.004; $\frac{1}{250}$ **17.** 2.5; $2\frac{1}{2}$ **18.** 0.0004; $\frac{1}{2{,}500}$ **19.** 0.325; $\frac{13}{40}$ **20.** 1; 1 **21.** 37% **22.** 4.2%

23. 30% **24.** $66\frac{2}{3}\%$ **25.** 175% **26.** 103.5% **27.** 303% **28.** 12.5% **29.** $26\frac{2}{3}\%$

30. 0.6% **31.** 9% **32.** 62% **33.** 75% **34.** 20% **35.** $91\frac{2}{3}\%$

36. 174% **37.** 0.2% **38.** 87.5% **39.** 72% **40.** $183\frac{1}{3}\%$ **41.** 312.5%

42. $83\frac{1}{3}\%$ **43.** 7% **44.** 225% **45.** 70.5% **46.** 89% **47.** $77\frac{7}{9}\%$

48. $28\frac{4}{7}\%$ **49.** 0.5% **50.** 204% **51.** 87 parts out of 100 parts are water.

52. 80 parts out of 100 parts are wool; 20 parts out of 100 parts are polyester.

53. 90 parts out of 100 parts are silver; 10 parts out of 100 parts are nickel.

54. 20%, 0.20, $\frac{20}{100}$ **55.** 0.16, $\frac{16}{100}$, 16% **56.** $\frac{1}{100}$, 0.01 **57.** 0.125, $12\frac{1}{2}\%$, $\frac{1}{8}$

4.2 DETERMINING THE PERCENTAGE, *P*

To obtain 1% of a number, we multiply it by 0.01 or $\frac{1}{100}$. For example, 1% of 253 is $0.01 \times 253 = 2.53$. One percent of 6.20 is $0.01 \times 6.20 = 0.062$. Fifteen percent of 6.20 is then $15 \times (1\% \text{ of } 6.20) = 15 \times (0.062) = 0.93$. Of course, this is equivalent to $(15 \times 1\%) \times 6.20 = (15 \times 0.01) \times 6.20 = 0.15 \times 6.20 = 0.93$. To find 15 percent of any number means that when this number is divided into 100 equal parts (percents), we desire 15 of these parts.

For example, to compute 32% of 400 we multiply $0.32 \times 400 = 128$.

Example 1 Find 28% of 456.

Solution $0.28 \times 456 = 127.68$

456

127.68

28%

The majority of percentage problems may be solved by using the following formula:

$$P = RB; \text{ Percentage} = \text{Rate} \times \text{Base}$$

Example 2 Find 78% of \$600.

Solution $R = 78\%$ or 0.78, $B = \$600$, we must find P.
$P = (0.78)(\$600) = \$468$

Figure 4.2 serves as a standard memory device. Because the letter P occupies an area equal to the combined areas of R and B we remember that $P = RB$.

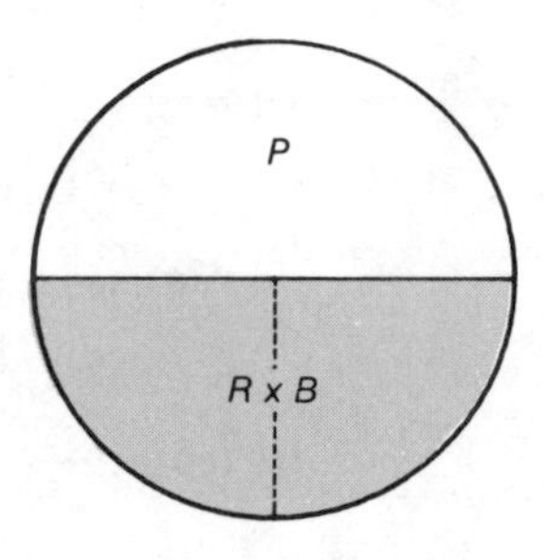

Figure 4.2

1. **Base** (B): This is the number that represents the whole and to which the rate (R) refers.
2. **Rate** (R): This refers to the number of hundredths of the base (what part of the base) we desire. A rate is a percent. For example, $R = 8\%$. For computations involving R, convert R to a fraction or decimal.
3. **Percentage** (P): The percentage is the result of multiplying the base and the rate. The percentage is part of the base. It is a number.

Have you noticed that the words **percent** and **percentage** have different meanings? The word **percent** refers to the rate (the number of hundredths, %). On the other hand, **percentage** is a number. It is part of the base number. For example, if $R = 30\%$ and $B = \$8.40$, then $P = (0.30)(\$8.40) = \$2.52$. This distinction is an important aid in establishing quick recognition of given data.

Example 3 The mayor of Pleasant Hills estimates that the population of his town has decreased by $9\frac{7}{8}\%$ since the last census. If the population was 28,624 at that time, what is the present population?

Solution $9\frac{7}{8}\% = 9.875\% = 0.09875$

$P = 0.09875 \times 28{,}624 = 2{,}826.62$ (decrease)

The present population is approximately $28{,}624 - 2{,}827 = 25{,}797$.

Example 4 The Gibsons' water bill was $75.80 last month. If this bill is paid within 10 days, a 5% discount is given.

(a) Assuming that the bill was paid within 10 days, how much money was saved?

(b) What amount was sent to the electric company?

Solution (a) $P = 0.05 \times \$75.80 = \$3.79$

(b) They paid $\$75.80 - \$3.79 = \$72.01$.

Example 5 In 1984 workers paid 6.7% of the first $35,700 of their earned income to social security. If Sue Lee earned $18,500 in 1984, how much of her salary went to social security?

Solution $6.7\% \times \$18{,}500 = 0.067 \times \$18{,}500 = \$1{,}239.50$

In a percentage problem we are always given two of the three values in the formula $P = RB$ and asked to solve for the third value Your method of solution depends on your recognizing which two of the three quantities P, R, or B are given. To help you determine which value is P and which value is B use the following: the number following the phrase "percent of" (% of) is the value of B.

Can you find more than 100% of a number? Yes. Here we are asking for a number greater than the original base. For example, 200% of 25 means that we desire twice (two 100%s of) 25 or 50. Assume you want to find 101% of 500. This means that should 500 be divided into 100 equal parts, each of size 5, you want all 100 equal parts plus one additional part. Thus, 101% of 500 is $500 + 5 = 505$, which is greater than the base value 500. Because 101% equals 1.01 when expressed as a decimal, 101% of 500 may be determined by computing $1.01 \times 500 = 505$. Three hundred percent of $8 = 3.00 \times 8 = 24$, and 350% of $8 = 3.50 \times 8 = 28$.

Example 6 Find 120% of 66.

Solution $R = 120\% = 1.20,\ B = 66$
$P = 1.20 \times 66 = 79.20 > 66$
(which is our original figure)

79.20

66

120%

Example 7 Find $33\frac{1}{3}\%$ of 420.

Solution If the percent involves a fraction that is a nonterminating decimal, convert the percent to a fraction.

$$R = 33\tfrac{1}{3}\% = 33\tfrac{1}{3} \times \frac{1}{100}$$

$$= \frac{\cancel{100}^{\,1}}{3} \times \frac{1}{\cancel{100}_{\,1}} = \frac{1}{3}$$

$$P = \frac{1}{\cancel{3}_{\,1}} \times \cancel{420}^{\,140} = 140$$

4.2 EXERCISES

In problems 1–10 determine the value of *P*.

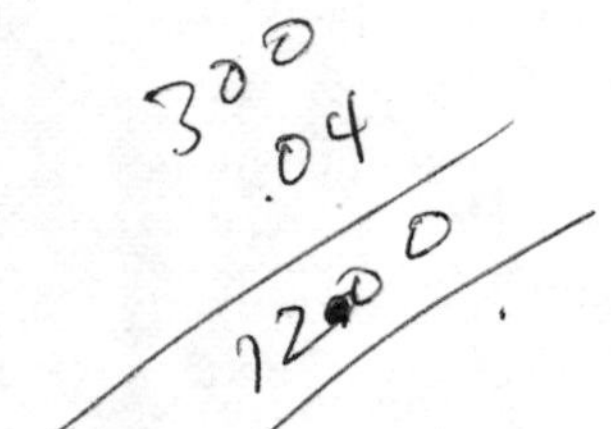

1. $R = 4\%,\ B = \$300$ ________
2. $R = 35\%,\ B = \$8{,}760$ ________
3. $R = 10\%,\ B = 47.64$ ________
4. $R = 8.5\%,\ B = 505$ ________
5. $R = 105\%,\ B = 36$ ________
6. $R = 15\%,\ B = 80$ ________
7. $R = 65\%,\ B = 400$ ________
8. $R = 12\frac{1}{2}\%,\ B = 40.8$ ________
9. $R = 5\frac{1}{4}\%,\ B = 128$ ________
10. $R = 10.2\%,\ B = 84$ ________
11. 6% of 120 is what number? ________
12. 12% of 300 is what number? ________
13. $5\frac{1}{2}\%$ of 64 is what number? ________
14. 3.5% of 40 is what number? ________
15. 75% of 196 is what number? ________
16. 0.15% of 30 is what number? ________
17. 0.6% of 35 is what number? ________
18. 24% of 75 is what number? ________
19. $2\frac{1}{4}\%$ of 800 is what number? ________

20. 225% of 60 is what number? ________

21. Mrs. Brown sold her home through a real estate agent to whom she paid a commission of $7\frac{1}{2}$% of the sales price. If the sale price was $70,000, how much commission did Mrs. Brown pay?

____

22. A farmer sold 1,000 quarts of milk in one week. The milk contained 4% butterfat, for which he was paid $.65 per quart.

(a) How many quarts of butterfat did he sell? ________
(b) How much money did he make from the sale of this butterfat? ________

23. A solution of a storage battery consists of 16% sulfuric acid. If we have 8 gallons of the solution, how much of it is sulfuric acid?

____

24. A rule of thumb is that a one-year-old baby weighs 400% of his birth weight. Using this rule of thumb, what should a one-year-old weigh if he weighed 7 lb at birth?

____

25. A worker paid a tax of 20.6% of his income, which was $10,250.00. How much tax did he pay?

____

26. Ordinary brass contains 61.6% copper, 2.9% lead, 0.2% tin, and 35.3% zinc. How many pounds of each metal would there be in 180 lb of brass?

____

27. In making a steel rod, 50 inches in length, at the Alcot Cast Iron Works, the allowable error is $\frac{1}{4}$% of an inch. What are the minimum and maximum allowable lengths?

____

28. The monthly finance charge on a credit card is 1.5% of the previous month's outstanding balance. Amanda had an outstanding balance of $250 last month. What was the monthly finance charge?

____

29. A state worker has to pay 18% of his salary for federal and state taxes, 2.2% for health insurance, and 5% for retirement. He earns $1,400 per month. Determine how much he pays each month for (a) federal and state taxes ________, (b) health insurance, ________ and (c) retirement ________

(d) how much of his monthly salary does he receive after deductions ________.

30. A final exam contains 120 questions, of which 75% are multiple choice, $20\frac{5}{6}$% are true-false, and $4\frac{1}{6}$% are essays. How many questions of each type are on the exam?

____

ANSWERS

1. \$12 **2.** \$3,066 **3.** 4.764 **4.** 42.925 **5.** 37.8 **6.** 12 **7.** 260 **8.** 5.1 **9.** 6.72 **10.** 8.568 **11.** 7.2 **12.** 36 **13.** 3.52 **14.** 1.4 **15.** 147 **16.** 0.045 **17.** 0.21 **18.** 18 **19.** 18 **20.** 135 **21.** \$5,250 **22.** **(a)** 40 quarts **(b)** \$26 **23.** 1.28 gallons **24.** 28 lb **25.** \$2,111.50 **26.** copper; 110.88 lb; lead; 5.22 lb; tin, 0.36 lb; zinc, 63.54 lb **27.** minimum, 49.0075 inches; maximum, 50.0025 inches **28.** \$3.75 **29.** **(a)** \$252 **(b)** \$30.80 **(c)** \$70 **(d)** \$1047.20 **30.** 90; 25; 5

4.3 DETERMINING THE BASE, *B*

How do we find a number when a percentage of it is given? If \$300 represents 75% of a number, what is the number? To begin with, the number we are seeking must be greater than \$300. Do you see why? Our problem is one of determining the **base value,** *B*, given: (1) a part of *B* (the percentage, *P*), and (2) the rate, *R*. All such problems may be solved by direct substitution of known values into the formula below:

$$B = \frac{P}{R}$$

This formula may always be remembered by returning to Figure 4.2. Cover the *B* and you are left with $\frac{P}{R}$. Try it!

Example 1 If \$300 represents 75% of a number, what is the number?

Solution We are given that $P = \$300$ and $R = 75\%$ or 0.75. Thus,

$$B = \frac{P}{R} = \frac{\$300}{0.75} = \$400$$

Example 2 If \$407 represents 74% of a number, what is the number?

Solution Solve by direct substitution into the formula for *B*.

$P = \$407$ and $R = 0.74$ Thus

$$B = \frac{\$407}{0.74} = \$550$$

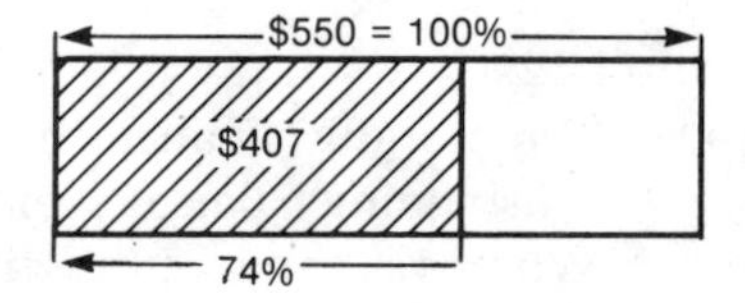

Example 3 Lisa's French test paper was marked 80% correct. She answered 16 questions correctly. How many questions must have been on the French test?

Solution Here $P = 16$ and $R = 80\%$ or 0.80.

$$B = \frac{P}{R} = \frac{16}{0.80} = 20 \quad \text{questions}$$

Example 4 If 120 represents 160% of a number, what is the number?

Solution Our formula is perfectly valid when *R* is more than 100%. Thus

$$B = \frac{120}{1.60} = 75$$

Example 5 The total amount of money deducted from Marie's gross pay last week amounted to \$90 and represented $37\frac{1}{2}\%$ of her gross pay. What was her gross pay?

Solution $37\frac{1}{2}\% = 0.375$. Hence

$$B = \frac{\$90}{0.375} = \$240, \text{ her gross pay}$$

Example 6 Fred received a 6% pay raise. He now earns \$12,720.00. What was his salary prior to the salary increase?

Solution His present salary is 106% of his old salary.

$R = 106\% = 1.06$
$P = \$12{,}720$
Hence

$$B = \frac{\$12{,}720}{1.06} = \$12{,}000, \text{ his old salary}$$

4.3 EXERCISES

In problems 1–10 determine the value of B.

1. $R = 40\%,\ P = 60$ ________

2. $R = 60\%,\ P = 120$ ________

3. $R = 18\%,\ P = 54$ ________

4. $R = 16\%,\ P = 128$ ________

5. $R = 5\frac{1}{2}\%,\ P = 385$ ________

6. $R = 4\frac{1}{4}\%,\ P = 11.05$ ________

7. $R = 6.2\%,\ P = 4.65$ ________

8. $R = 12.5\%,\ P = 510$ ________

9. $R = 0.04\%,\ P = 0.0268$ ________

10. $R = 37\frac{1}{2}\%,\ P = 24$ ________

11. 45% of what number is 90? ________

12. 30% of what number is 45? ________

13. 12% of what number is 36? ________

14. 75% of what number is 60? ________

15. 150% of what number is 90? ________

16. 105% of what number is 1050? ________

17. 2.5% of what number is 48? ________

18. 10.25% of what number is 6.355? ________

19. 37.5% of what number is 16.875? ________

20. $22\frac{1}{2}\%$ of what number is 23.4? ________

21. $6\frac{1}{4}\%$ of what number is 0.575? ______________

22. Lynite is an alloy used in the manufacture of piston rods for car engines. It contains 11% copper. How many pounds of Lynite can be made from 66 pounds of copper added to the right amount of other materials?

____

23. Carl purchased a suit in Jason's for $115.50. He purchased the suit at a "30% off" sale. What was the original selling price of the suit?

____

24. Mr. Jones works on a 15% commission. If he wishes to earn $300 next week, how much must he sell?

____

25. This semester there are 600 married students on campus, which represents 15% of the total enrollment. What is the total enrollment?

15 · X = 600

____

26. A building is insured for 80% of its value. If the insurance coverage is for $52,000, what is the value of the building?

____

27. A bank gives Max McGee $51.64 interest at 8% on his savings account balance. What is the balance?

____

28. Mr. Smith has 23% of his weekly salary withheld. This amounts to $98.90; what is his weekly salary?

____

29. It is projected that 84% of the incoming freshman class at Metropolitan University will graduate in 4 years. According to school officials this will mean a graduating class of 1,743 students. How many freshman entered Metropolitan University?

____

30. A team won 90 games last season. This was $56\frac{1}{4}\%$ of the games played. How many games were played?

____

31. Meghan received an 8% pay raise. She now earns $16,740. What was her salary prior to the salary increase?

____

32. A professional athlete received a 40% pay raise over last year's salary. He now earns $63,000. What was the athlete's salary prior to the increase?

____

ANSWERS

1. 150 **2.** 200 **3.** 300 **4.** 800 **5.** 7,000 **6.** 260 **7.** 75 **8.** 4,080 **9.** 67 **10.** 64 **11.** 200 **12.** 150 **13.** 300 **14.** 80 **15.** 60 **16.** 1,000 **17.** 1,920 **18.** 62 **19.** 45 **20.** 104 **21.** 9.2 **22.** 600 pounds **23.** $165 **24.** $2,000 **25.** 4,000 **26.** $65,000 **27.** $645.50 **28.** $430 **29.** 2,075 **30.** 160 **31.** $15,500 **32.** $45,000

4.4 DETERMINING THE RATE, *R*

How do we find what percent one number is of another? The number 3 is what percent of 40? To begin with, 3 is certainly less than 50% of 40 because 50% of 40 is 20. Three is less than 25% of 40. Do you see why? Our problem is one of determining the **rate,** *R,* given (1) the base, *B,* and (2) the percentage, *P.* All such problems may be solved by direct substitution into the formula below:

$$R = \frac{P}{B}$$

This formula may always be remembered by returning to Figure 4.2. Cover the *R* and you are left with $\frac{P}{B}$. Try it!

Example 1 The number 3 is what percent of 40? Solve this problem by direct substitution into the formula.

Solution Recall that the number following "percent of" is the value of *B*.

$P = 3,\ B = 40$. Thus

$$R = \frac{P}{B} = \frac{3}{40} = 0.075 = 7.5\%$$

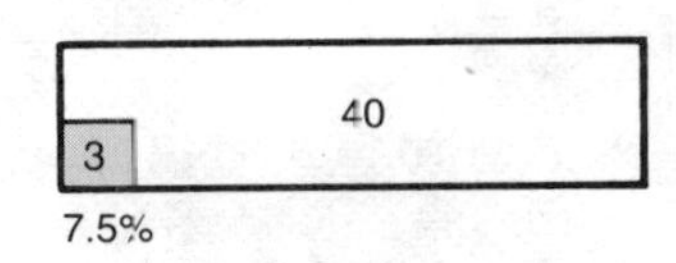

Example 2 Determine what percent 14 is of 70.

Solution $P = 14,\ B = 70$. Thus

$$R = \frac{14}{70} = \frac{1}{5} = 0.20, \text{ or } 20\%$$

Example 3 Five is what percent of 40?

Solution $P = 5,\ B = 40,\ R = ?$

$$R = \frac{5}{40} = 0.125, \text{ or } 12.5\%$$

Example 4 Fifteen is what percent of 5?

Solution $P = 15,\ B = 5,\ R = ?$

$$R = \frac{15}{5} = 3, \text{ or } 300\%$$

Example 5 Twelve is what percent of 28?

Solution $P = 12,\ B = 28,\ R = ?$

$$R = \frac{12}{28} = 0.429, \text{ or } 42.9\%$$

Example 6 Susan had a sales goal of 140 electric typewriters this month. She actually sold 175. What percent of her sales goal was achieved?

Solution $P = 175,\ B = 140,\ R = ?$

$$R = \frac{175}{140} = 1.25, \text{ or } 125\%$$

Example 7 The Green Growth Company found that 765 of 848 grass seeds germinated in a test. What percent of the grass seeds germinated? Round off to nearest hundredth of a percent.

Solution $P = 848,\ B = 765,\ R = ?$

$$R = \frac{765}{848} = 0.90212, \text{ or } 90.21\%$$

4.4 EXERCISES

In problems 1–11 determine the value of R.

1. $P = 7,\ B = 50$ ____________

2. $P = 9,\ B = 20$ ____________

3. $P = 14,\ B = 70$ ____________

4. $P = 35,\ B = 105$ ____________

5. $P = 4\frac{1}{2},\ B = 5$ ____________

6. $P = 12\frac{1}{2},\ B = 40$ ____________

7. $P = 4,\ B = 2$ ____________

8. $P = 18,\ B = 5$ ____________

9. $P = 10.5,\ B = 42$ ____________

10. $P = 5.2,\ B = 31.2$ ____________

11. $P = 6.8,\ B = 54.4$ ____________

12. 48 is what percent of 60? ____________

13. 15 is what percent of 75? ____________

14. 24 is what percent of 60? ____________

15. 30 is what percent of 80? ____________

16. 28 is what percent of 8? ____________

17. 34 is what percent of 20? ____________

18. 7.8 is what percent of 78? ____________

19. 0.023 is what percent of 4.6? ____________

20. 3.4 is what percent of 0.68? ____________

21. 6.4 is what percent of 76.8? ______

22. If we read 55 pages of a book that has 200 pages, we have read what percent of the book?

______

23. Joe correctly answered 12 questions of 15 questions on his French exam. What percent of the questions did he have correct?

______

24. A secretary who earns $250 per week saves $25 per week. What percent of her salary does she save each week?

______

25. Assume that the United Fund has a present goal of $3,000,000. Further assume that, as of today, it has received $1,650,000 in pledges. What percent of its goal is presently pledged?

______

26. An 80-degree angle is what percent of a complete rotation (360°)?

______

27. At South High the girls' basketball team has a record of 6 wins and 9 losses, while the boys' basketball team has a record of 8 wins and 11 losses.

(a) Find the percent of wins for each team. ______

(b) Which team has the better winning percent? ______

28. Sam's monthly salary is $1,425. His deductions amount to $313.50. What percent of Sam's monthly salary is take-home pay?

______

29. A real estate broker sold a parcel of land for $125,500. Her commission was $9,412.50. What percent of the selling price was her commission?

______

30. A gallon of unleaded gasoline sells for $1.20 per gallon. A new federal tax of $0.05 per gallon will be levied on gasoline. What percent of the selling price of gasoline is the new tax?

______

31. A car salesman had a sales goal of 16 new cars for the month of February. He sold 24 new cars. What percent of his sales goal was reached?

______

32. A baseball player set a personal goal of hitting 25 home runs for the season. He hit 32 home runs. What percent of his personal goal did he reach?

______

ANSWERS

1. 14% **2.** 45% **3.** 20% **4.** $33\frac{1}{3}\%$ **5.** 90% **6.** 31.25% **7.** 200% **8.** 360% **9.** 25% **10.** $16\frac{2}{3}\%$ **11.** 12.5% **12.** 80% **13.** 20% **14.** 40% **15.** 37.5% **16.** 350% **17.** 170% **18.** 10% **19.** 0.5% **20.** 500% **21.** $8\frac{1}{3}\%$ **22.** 27.5% **23.** 80% **24.** 10% **25.** 55% **26.** $22\frac{2}{9}\%$ **27.** **(a)** 40%; $42\frac{2}{19}\%$ **(b)** boys **28.** 78% **29.** 7.5% **30.** $4\frac{1}{6}\%$ **31.** 150% **32.** 128%

> Each of the preceding sections deals with a specific type of percentage problem. It is important that you be able to distinguish which type of percentage problem you are asked to solve. The following set of exercises will include all of the types of percentage problems that you have previously encountered.

MISCELLANEOUS REVIEW EXERCISES

1. In a class of 32 students, 18 students are female. What percent of the students are female?

____

2. An Asian country set a production goal of 40 tons of rice. Through improved agriculture they produced 88 tons of rice. What percent of their goal did they reach?

____

3. A lawyer's fee was 15% of the settlement in a personal damages suit. If the settlement was $20,750, what was the lawyer's fee?

____

4. There are 45 handicapped students attending City Community College. This is $1\frac{1}{2}\%$ of the total student enrollment. How many students attend City Community College?

____

5. The tax on meals in Massachusetts is 8%. A family of four spends $44.50 for dinner at a restaurant. What will the tax be?

____

6. A final exam of 125 questions consists of two parts: 60 true-false questions and 65 multiple-choice questions. What percent of the test consists of multiple-choice questions?

____

7. Of the 128 players who qualified for the American League batting championship last season, 16 of the players hit above 300. What percent of the qualifying players hit above 300?

____

8. Of the graduating class at Hillside Community College, 546 members continued their education at four-year colleges. This was 65% of the graduating class. How many students graduated?

____

9. A house that was bought for $46,000 five years ago sold for 145% of its original cost. What was the selling price of the home?

____

10. The Environmental Protection Agency conducted a study of the concentration of ozone in the air over a large city for 50 days; 18 days had an ozone concentration between 80 and 120 parts per billion. What percent of days had an ozone concentration between 80 and 120?

____

11. In a poll conducted by a travel bureau to determine people's favorite vacation spots, 420 favored the mountains. This represented 35% of the people polled. How many people were polled?

____

12. Susan paid $7,800 for a new car. The car depreciated by 32% after one year. How much did the car depreciate and what is its present value?

____

13. Mary spends $437 a month for mortgage payments and property taxes. This amount is 23% of her monthly salary. What is Mary's monthly salary?

____

14. A National Hockey League goaltender plays in 54 games of his team's 80 game schedule. What percent of the games did he play?

____

15. At a clearance sale, a Pendleton 100% wool sweater sold for $55.30. This was 70% of the original price. What was the original price?

____

16. Condensed milk is evaporated whole milk that contains 45% sugar. How many pounds of sugar are there in 34 pounds of condensed milk?

____

17. In a newborn infant, 12% of the weight is composed of blood. If the blood of a newborn infant weighs 1.02 lb, how much does the newborn weigh?

____

18. There were 1982 births at South Shore Hospital last year, and $2\frac{1}{2}\%$ of the newborns had insignificant heart murmurs. How many infants had insignificant heart murmurs? (Round off to units.) __________

ANSWERS

1. 56.25% **2.** 220% **3.** $3,112.50 **4.** 3,000 **5.** $3.56 **6.** 52% **7.** 12.5% **8.** 840 **9.** $66,700 **10.** 36% **11.** 1,200 **12.** $2,496; $5,304 **13.** $1,900 **14.** 67.5% **15.** $79 **16.** 15.3 lb **17.** 8.5 lbs **18.** 50

4.5 INCREASING AND DECREASING PERCENT PROBLEMS

By using a percent (%) we can often clarify the increase or decrease taking place in an everyday situation. A school enrollment has increased from 400 students to 425 students. Presenting this increase as $6\frac{1}{4}\%$ $\left(\frac{25}{400}\right)$ helps us visualize the increase in enrollment. You know that a half-gallon of milk will cost you $0.98 at your supermarket, but because it is closed you must purchase this half-gallon for $1.07 at a local store. Now this is only an increase of $0.09, and we all know that $0.09 doesn't purchase much in today's economy; however, when we look at $0.09 as an increase in our supermarket price we see that $\frac{\$0.09}{\$0.98} = 0.092$ or 9.2%. As a percent, it represents a substantial increase.

Due to a weak economy, Eaton Industries asks a worker to reduce his $500 salary voluntarily by 10%. This represents a decrease of $(.10)(\$500) = \$50$ in current salary and results in a new salary of $450. Due to a lack of rainfall, a town reservoir currently has 1,600,000 gallons of water. The town manager tells the town meeting that it must appropriate funds to drill wells that will add 25% more gallons to the present water supply in order to satisfy the coming summer demand. What increase, in gallons, is necessary? What is the total number of gallons that the town needs for the summer? The town must add $(0.25)(1{,}600{,}000) = 400{,}000$ gallons. If this is done, the total will be 2,000,000 gallons.

Example 1 In February last year 40 inches of snow fell in the White Mountains of New Hampshire. This year 48 inches fell during February. What was the percent increase in snowfall?

Solution The actual increase in snowfall is 48 inches − 40 inches = 8 inches. The percent increase is $\frac{8}{40} = 0.20 = 20\%$.

Example 2 Nancy's landlord has just told her that her rent is to be raised from $360 to $400 per month.

(a) How much is this in dollars?
(b) What percent of her present rent does this increase represent? (Round off to nearest tenth of a percent.)

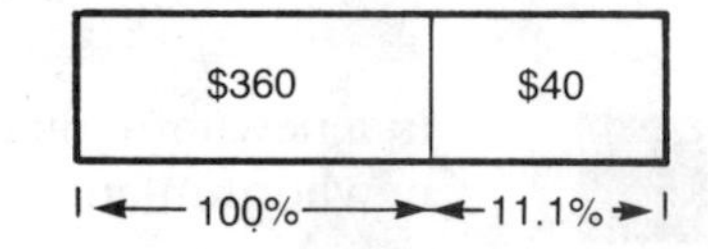

Solution (a) Amount of increase = $400 − $360 = $40

(b) The percent increase is $\frac{\$40}{\$360} = 0.111 = 11.1\%$.

Example 3 If you increase \$500 by 12%, how much do you have?

Solution 1 Amount of increase = (0.12)(\$500) = \$60
New amount = \$500 + \$60 = \$560

Often when finding a percent increase we first find the amount of increase and then add it to the original amount. As an alternative, consider adding the decimal representation of the percent to 1, and then multiplying your original amount by this value. Using this method, we could have solved the preceding example as follows:

Solution 2 New amount = (1 + 0.12)(\$500) = (1.12)(\$500) = \$560

For the case of a percent decrease, the decimal representation can be subtracted from 1 before multiplying the original amount to find the answer.

Example 4 Last year the city of Maddon had 900 automobile accidents, but through the efforts of a safety campaign the number of automobile accidents has been reduced by 30% this year. How many automobile accidents occurred this year?

Solution 1 30% of 900 = (0.30)(900) = 270; subtract this from 900. The answer is 900 − 270 = 630.

Solution 2 Since 100% − 30% = 70%, find 70% of 900. The answer is

(0.70)(900) = 630.

Example 5 This year the annual sales for Fitzgibbons, Inc., is \$150,000. Last year, the annual sales were \$190,000. This decrease is what percent of last year's annual sales?

Solution The annual sales decreased by \$40,000.

$$r = \frac{\$40{,}000}{\$190{,}000} = \frac{4}{19} = 0.21, \text{ or } 21\%$$

Example 6 A store item listing for \$632 is sold for \$551.42. What is the percent decrease in list price?

Solution Amount of decrease = \$632 − \$551.42 = \$80.58

$$\text{Percent of decrease} = \frac{\$80.58}{\$632} = 0.1275 = 12.75\%, \text{ or } 12\tfrac{3}{4}\%$$

4.5 EXERCISES

In problems 1–6 find the total after the percent increase or decrease.

1. 300; 42% increase ____________
2. 260; 36% decrease ____________
3. \$840; 7% increase ____________
4. \$1,272; 5% decrease ____________
5. \$945; 7.1% decrease ____________
6. \$684; 53% increase ____________

7. In a local election year 54,750 people voted. This election the number of people voting has increased by 4% of the previous election. How many people voted this time?

8. A collection agency collected a doctor's bill for $2,698 and deducted $7\frac{1}{2}$% of this amount for services rendered.

   (a) How much did the agency receive?

   (b) How much did the doctor receive?

9. Last year Paula paid $1,568.40 in federal taxes. Due to a pay raise she estimates that she will pay 10% more in federal taxes this year. How much will Paula pay this year?

10. You know that you can drive 32,000 miles on brand A tires and, while brand B tires are more expensive, you believe that they will last 15% longer than brand A. How many miles do you believe that brand B tire will provide?

11. Find the percent increase from 98 to 148.

12. Find the percent increase from 39.6 to 49.5.

13. Find the percent increase from 40 to 55.

14. Find the percent increase from 12.5 to 20.

15. Find the percent increase from 4.8 to 9.6.

16. Find the percent decrease from 250 to 200.

17. Find the percent decrease from 20.4 to 15.3.

18. Find the percent decrease from 36.3 to 24.2.

19. Find the percent decrease from 110 to 40.

20. Find the percent decrease from $1,200,000 to $84,000.

21. Last year your charge account balance was $320, and exactly one year later it has increased to $480. What is the percent of increase?

22. A share of Economy Oil stock priced at $95 yesterday is worth $100 today. What is the percent increase in the value of this stock? Round off to hundredths of a percent.

23. During the previous month Garvey Shoe sold 408 pairs of shoes. This month they only sold 370 pairs of shoes. What is the percent decrease in sales?

24. Steven saved $88 last month and $56 this month. By what percent did his savings decrease this month?

25. The second semester Harold improved his class average from 72 to 84. What is the percent of increase in his average?

26. During a warm spell the snow depth at a New England ski resort dropped from 56 inches to 12 inches. Express the drop in snow depth as a percent of the original depth.

27. In 1900 only 39 wild bison survived in the United States—all in Yellowstone National Park. The herd in Yellowstone today numbers 801 animals. What is the percent of increase in wild bison?

28. In 1964, the population of Valdez, Alaska, was 1,200. Today, the population has increased by 500%. What is the new population of Valdez?

29. Two years ago the price of a home computer was $700. Today the price of the same model has decreased by 35.6%. What is the price today?

30. At California's sole breeding colony on Anacapa Island for the brown pelican, nesting attempts dropped from 1,272 in 1969 to 247 in 1973. What was the percent of decrease in nesting attempts? (Round off to tenths of a percent.)

31. Tom and Joanne have taken a variable rate mortgage out on their home. Their mortgage payments increased from $280 per month to $283 per month. What is the percent of increase? (Round off to tenths of a percent.)

32. The Trauma Center at City Hospital noted that victims of automobile accidents were now being treated within 45 minutes of the accident. When they were brought to the Emergency Department they were treated within 75 minutes. What is the percent of decrease in the time the victims are treated at the Trauma Center?

33. Jim Rice hit 24 home runs in 1982 and 39 home runs in 1983. What was the rate of increase in Jim Rice's home run production?

34. Maria bought a new car that had a list price of \$8,500. If she paid cash, the price of the car would be discounted to \$7,800. What is the percent of decrease in the car's price? (Round off to tenths of a percent.)

____

35. Janine bought a stereo system for \$480. She sold it for \$570. What was the percent of increase in price?

____

ANSWERS

1. 426 **2.** 166.4 **3.** \$898.80 **4.** \$1,208.40 **5.** \$877.91 **6.** \$1,046.52 **7.** 56,940 **8. (a)** \$202.35 **(b)** \$2495.65 **9.** \$1725.24 **10.** 36,800 miles **11.** 51.02% **12.** 25% **13.** 37.5% **14.** 60% **15.** 100% **16.** 20% **17.** 25% **18.** $33\frac{1}{3}\%$ **19.** $63\frac{7}{11}\%$ **20.** 93% **21.** 50% **22.** 5.26% **23.** 9.31% **24.** $36.\overline{36}\%$ **25.** $16.\overline{6}\%$ **26.** 78.57% **27.** 1953.85% **28.** 7,200 **29.** \$450.80 **30.** 80.6% **31.** 1.1% **32.** 40% **33.** 62.5% **34.** 8.2% **35.** 18.75%

4.6 SIMPLE INTEREST ($I = PRT$)

Ellen needed a short-term loan to carry her through Christmas. She needed \$850 from December 5 to May 4, and her local bank agreed to lend her the money at 12% simple interest. How much interest will she pay? We will solve Ellen's problem in this section.

An amount of money that is borrowed is called the **principal,** *P*. It may be that someone, such as Ellen, is borrowing from a bank; or it may be that the bank is borrowing from depositors who have placed money in savings accounts. The money paid by the **borrower** for the privilege of using the principal is called **interest,** *I*. The **rate of interest,** *R*, will tend to vary from one lending institution to another depending upon the supply of money. The prevailing rate of interest on any given day is called the *worth* of money. As of January 1, 1983, banks were free to pay whatever they feel is a competitive interest rate to savings account depositors whose balance exceeds \$2,500. If a bank tells you that the rate of interest to be charged on a given day is 16.5%, you are being told that the worth of money on that day is 16.5%. Never assume that the worth of money is exactly the same in all banks on any given day. This is just not true.

In addition to the principal, *P*, and the rate of interest, *R*, there is one other important factor to be considered when evaluating simple interest. This is time, *T*, measured in years, or fractions thereof, for which the borrower needs the principal. The longer you wish to use the principal, the more you should expect to pay.

In summation, **simple interest,** *I*, depends upon three factors: principal, *P*; rate of yearly interest, *R*; and length of borrowing time, *T*. Here is the formula for computing simple interest.

$$\text{Interest} = \text{Principal} \times \text{Rate} \times \text{Time}$$
$$I = P \times R \times T \text{ or } I = PRT$$

Example 1 Determine the simple interest on $P = \$2{,}000$ at $R = 15\%$ for one year. That is, $T = 1$.

Solution $I = PRT = (\$2{,}000)(0.15)(1) = \$300$

Figure 4.3 is a memory device to help you remember this formula. Because the letter I occupies an area equal to the combined areas of P, R, and T, we remember that $I = PRT$.

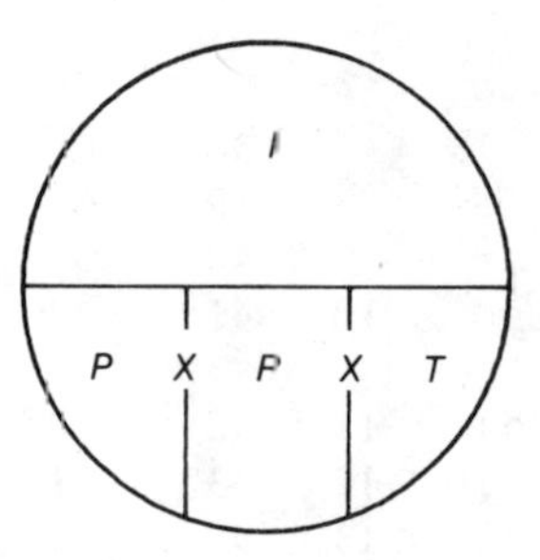

Figure 4.3

Example 2 Determine the simple interest on $P = \$10{,}000$ at $R = 8\%$ for 3 years.

Solution $I = PRT = 10{,}000(0.08)(3) = \$2{,}400$

Example 3 Determine the simple interest on $P = \$18{,}400$ at $R = 16\%$ for $2\frac{1}{2}$ years.

Solution $I = PRT = \$18{,}400(0.16)(2.5) = \$7{,}360$

The most common time factor in use today is a fraction whose numerator is the exact number of days in the length of the loan, called the "term" of the loan, and whose denominator is based upon a 360-day year. This time factor is called the **Bankers' Rule.** You may assume from now on that when we speak of time we mean Bankers' Rule time.

$$\text{Bankers' Rule Time, } T = \frac{\text{Exact Number of Days in the Term}}{\text{360 days}}$$

Example 4 Find the exact number of days from March 17 to August 2. Use Table 4.1.

Solution In Table 4.1, months are recorded across the top of the table and days of the month are recorded in the first column. August 2 is the 214th day of the year and March 17 is the 76th day of the year. Subtracting these two table values tells us that there are exactly $214 - 76 = 138$ days from March 17 to August 2.

Table 4.1 The Number of Each of the Days of the Year (Add 1 to each date after February 29 for a leap year)

Day of Month	Jan.	Feb.	March	April	May	June	July	Aug.	Sept.	Oct.	Nov.	Dec.
1	1	32	60	91	121	152	182	213	244	274	305	335
2	2	33	61	92	122	153	183	214	245	275	306	336
3	3	34	62	93	123	154	184	215	246	276	307	337
4	4	35	63	94	124	155	185	216	247	277	308	338
5	5	36	64	95	125	156	186	217	248	278	309	339
6	6	37	65	96	126	157	187	218	249	279	310	340
7	7	38	66	97	127	158	188	219	250	280	311	341
8	8	39	67	98	128	159	189	220	251	281	312	342
9	9	40	68	99	129	160	190	221	252	282	313	343
10	10	41	69	100	130	161	191	222	253	283	314	344
11	11	42	70	101	131	162	192	223	254	284	315	345
12	12	43	71	102	132	163	193	224	255	285	316	346
13	13	44	72	103	133	164	194	225	256	286	317	347
14	14	45	73	104	134	165	195	226	257	287	318	348
15	15	46	74	105	135	166	196	227	258	288	319	349
16	16	47	75	106	136	167	197	228	259	289	320	350
17	17	48	76	107	137	168	198	229	260	290	321	351
18	18	49	77	108	138	169	199	230	261	291	322	352
19	19	50	78	109	139	170	200	231	262	292	323	353
20	20	51	79	110	140	171	201	232	263	293	324	354
21	21	52	80	111	141	172	202	233	264	294	325	355
22	22	53	81	112	142	173	203	234	265	295	326	356
23	23	54	82	113	143	174	204	235	266	296	327	357
24	24	55	83	114	144	175	205	236	267	297	328	358
25	25	56	84	115	145	176	206	237	268	298	329	359
26	26	57	85	116	146	177	207	238	269	299	330	360
27	27	58	86	117	147	178	208	239	270	300	331	361
28	28	59	87	118	148	179	209	240	271	301	332	362
29	29		88	119	149	180	210	241	272	302	333	363
30	30		89	120	150	181	211	242	273	303	334	364
31	31		90		151		212	243		304		365

Example 5 Ellen needed a short-term loan to carry her through Christmas. She borrowed \$850 from her local bank, at 12% simple interest, from December 5 to May 4 of the next year. How much simple interest will she pay?

Solution $P = \$850$, $R = 12\%$ or $\frac{12}{100}$. Table 4.1 tells us that December 5 is the 339th day of the year and May 4 is the 124th day of the year. There are then $26 + 124 = 150$ days in the term of this loan, so $T = \frac{150}{360} = \frac{5}{12}$.

$$I = PRT$$

$$= (\$850)\left(\frac{12}{100}\right)\left(\frac{5}{12}\right)$$

$$I = \$42.50$$

The solution to the preceding example could have been obtained using decimals to compute $I = PRT$. In this case, $P = \$850$, $R = 0.12$, and $T = \frac{5}{12} = 0.42$ correct to two decimal places. Thus, $I = (\$850)(0.12)(0.42) = \$42.84$. The fact that we have rounded off our time factor, T, to the nearest hundredth resulted in a difference of \$0.34 in interest. We encourage the reader to verify both calculations. Round-off error, as we have already said, is important.

Example 6 (a) Find the simple interest, I, on \$1,400 at $R = 9\%$ for 75 days.
(b) Add the interest, I, to the principal, P, to determine the total amount due after the term has expired.

Solution (a) $I = \$1{,}400 \times \frac{9}{100} \times \frac{75}{360} = \$26.25$
(b) The total amount is $\$1{,}400 + \$26.25 = \$1{,}426.25$.

The **amount,** A, sometimes called the **"maturity value,"** is the sum of money accumulated at the end of the term. You determine A by adding the principal, P, to the simple interest, I.

$$A = P + I, \text{ or } A = P + PRT$$

In Example 5, the amount, A, or maturity value of Ellen's loan is $A = P + I = \$850.00 + \$42.50 = \$892.50$.

Example 7 Alex Kesselheim invested \$3,500 at 18% simple interest for 3 years.
(a) Determine the simple interest, I.
(b) Determine the amount, A.

Solution $P = \$3{,}500$, $T = 3$ (years is understood), and $R = 0.18$

(a) $I = (\$3{,}500)(0.18)(3) = \$1{,}890$
(b) $A = P + I = \$3{,}500 + \$1{,}890 = \$5{,}390$

4.6 EXERCISES

A calculator may be used in solving these problems. Round off all answers to the nearest hundredth. If the problem involves a fraction whose decimal equivalent is nonterminating, use fractions rather than nonterminating decimals in solving the problem.

In problems 1–5 you are given *P*, *R*, and *T*. Find the values of *I* and *A*.

	I	A
1. $P = \$400$, $R = 8\%$, $T = \frac{1}{2}$	______	______
2. $P = \$5{,}000$, $R = 6\%$, $T = 2$	______	______
3. $P = \$1{,}280$, $R = 18.5\%$, $T = 0.25$	______	______
4. $P = \$90{,}000$, $R = 11.2\%$, $T = 3$	______	______

		I	A
5.	$P = \$150{,}000$, $R = 15\%$, $T = 4\frac{1}{2}$	________	________

6. Find the interest on $50 at 8% for 6 months. ________

7. Find the interest on $10,000 at 9.5% for 4 months. ________

8. Find the interest on $1,575 at 11.5% for 9 months. ________

9. Find the interest on $600 at 7.5% for 18 months. ________

10. Find the interest on $700 at 10% for 20 months. ________

11. Find the interest on $200 at 8% for 50 days. ________

12. Find the interest on $500 at 10.5% for 90 days. ________

13. Find the interest on $1,050 at 7% for 120 days. ________

14. Find the interest on $1,500 at 12.5% for 180 days. ________

15. Find the interest on $800 at $6\frac{1}{4}$% for 270 days. ________

16. Find the interest on $1,000 at $7\frac{1}{2}$% for 30 days. ________

17. Find the interest on $840 at 6% from September 16 to December 5.

____

18. Find the interest on $2,500 from May 13 to November 29 at 12%.

____

19. Find the interest on $1386.25 at $R = 0.075$ from April 14 to October 1.

____

20. Acme Wallpaper Company borrowed $5,000 at $R = 0.145$ from March 15 until September 10. Compute A.

____

21. A man borrows $1,000 for 9 months at 13% per year. Find the interest *(I)* and the amount *(A)*.

I = 97.5

A = 1097.5

____

22. Camille borrows $650 for 10 months at 15.5% per year. Find the interest and the amount.

____

23. John deposited $1,500 in the bank at $8\frac{1}{4}$% interest for two years. Find the interest John's money earned and the amount he has in the bank after two years.

____

24. Anita bought a new car for $7,550. She paid $1,550 in cash and financed the remaining amount for 3 years at $12\frac{1}{2}$%. How much interest does she pay?

____

25. A woman borrowed $18,000 to buy inventory for her new business at a rate of 12% from May 16 to August 14. How much interest did she pay and what amount did she pay back?

____

26. As a device to attract new customers, New World Bank is offering 10% simple interest on deposits of $10,000 or more for 180 days. Meghan O'Hara deposits $11,500 in New World Bank. How much interest will she earn and what amount will she have after 180 days?

____

27. A department store charges 13.2% interest per year on charge accounts. Dan had an unpaid balance of $120 last month. How much interest did Dan pay on his unpaid balance?

____

28. A student borrows $750 for tuition at 8% from January 23 to May 23. How much interest does the student pay and what amount does the student pay back?

____

ANSWERS

1. $16; $416 **2.** $600; $5,600 **3.** $59.20; $1,339.20 **4.** $30,240; $120,240 **5.** $101,250; $251,250 **6.** $2 **7.** $316.67 **8.** $135.84 **9.** $67.50 **10.** $116.67 **11.** $2.22 **12.** $13.13 **13.** $24.50 **14.** $93.75 **15.** $37.50 **16.** $6.25 **17.** $11.20 **18.** $166.67 **19.** $49.10 **20.** $5,360.49 **21.** interest, $97.50; amount, $1097.50 **22.** interest, $83.96; amount, $733.96 **23.** interest, $247.50; amount, $1,747.50 **24.** $2,250 **25.** $540; $18,540 **26.** $575; $12,075 **27.** $1.32 **28.** interest, $20; amount, $770

4.7 BANK DISCOUNT

Louise needs $5,000 to purchase stock for the upcoming summer season at her beach gift shop, Louise's Home of Gifts. Her local banker agreed to lend her the $5,000 she needs for a term of 60 days at 12% interest. She signed a simple discount note, at her local bank, payable in 60 days at 12%. When she returned to the bank to pick up her money, she received a check for $4,900. The bank has computed the interest as follows: $\$5{,}000 \times \frac{12}{100} \times \frac{1}{6} = \$100$, and proceeded to keep this interest. She received the **proceeds,** S, given by the **maturity value,** M, less the interest. That is, she received $\$5{,}000 - \$100 = \$4{,}900$.

When you sign a simple discount note, you agree to pay the maturity value stated on the note at the end of the specified time period. The interest on a simple discount note is called **bank discount,** D, or simply discount. You receive the **proceeds,** represented by the face value of the note less the bank discount. Hence, you receive a sum of money that is less than the sum you have agreed to pay back.

Bank discount, D, is computed exactly like simple interest. D is a product of maturity value, M, times rate, R, times time, T, measured in years.

Bank Discount = Maturity Value × Rate × Time
$D = M \times R \times T$, or $D = MRT$

Proceeds = Maturity Value − Bank Discount
$S = M - D$, or $S = M - MRT$

Example 1 Barbara Swain took out a simple discount note for $2,000 at 15% for 270 days.

(a) Find the bank discount, D.
(b) Find the proceeds, S.

Solution $M = \$2{,}000,\ R = 15\% = 0.15,\ T = \frac{270}{360} = \frac{3}{4} = 0.75$

(a) $D = (\$2{,}000)(0.15)(0.75) = \$225$
(b) $S = M - D = \$2{,}000 - \$225 = \$1{,}775$

a2

Example 2 Victor Riese took out a simple discount note for $2,800 at 18% for 225 days.

(a) Find the bank discount, D.
(b) Find the proceeds, S.

Solution $M = \$2{,}800,\ R = 18\%$ or $0.18,\ T = \frac{225}{360} = 0.625$

(a) $D = (\$2{,}800)(0.18)(0.625) = \$315$
(b) $S = M - D = \$2{,}800 - \$315 = \$2{,}485$

Lending institutions write simple discount notes rather than simple interest notes for two reasons: (1) because they obtain a higher effective rate of interest, and (2) because the maturity value is more convenient for record-keeping purposes.

Let's review: In the last section we discussed simple interest; **simple interest** is a percentage of the principal, P. When you have a simple interest note, you add the interest, $I = PRT$, to the principal P, to find the amount, A. Thus, $A = P + PRT$. In this section we are discussing **simple discount** or bank discount; bank discount is a percentage of the maturity value, M. When you have a simple discount note, you subtract the bank discount, D, from the maturity value, M, to determine the proceeds, S. Thus, $S = M - MRT$. Simple interest, I, and bank discount, D, are applications of percentage.

In our next example, we will compare a simple interest note and a simple discount note.

Example 3 Consider two notes, a simple interest note and a simple discount note, each with a face value of $1,000, an interest rate of 9% and a term of 120 days. Compare these two notes.

Solution 1 For a *simple interest note,* $P = \$1{,}000,\ R = \frac{9}{100},\ T = \frac{1}{3}$

$$A = P + PRT$$
$$A = \$1{,}000 + \left(1000 \times \frac{9}{100} \times \frac{1}{3}\right)$$
$$= \$1{,}030$$

Solution 2 For a *simple discount note,* $M = \$1{,}000$, $R = \frac{9}{100}$, $T = \frac{1}{3}$

$$S = M - MRT$$
$$S = \$1{,}000 - \left(1000 \times \frac{9}{100} \times \frac{1}{3}\right)$$
$$= \$970$$

In the first case the borrower pays $30 for the use of $1,000. In the second case the borrower pays $30 for the use of $970.

4.7 EXERCISES

A calculator may be used in solving these problems. Round off all answers to the nearest hundredth. If the problem involves a fraction whose decimal equivalent is non-terminating, use fractions rather than non-terminating decimals in solving the problem.

Find the bank discount, *D*, and proceeds, *S*, in problems 1–7.

	M	Rate, *R*	Time, *T*	*D*	*S*
1.	$250	8%	1 year	______	______
2.	$395	10%	$\frac{1}{4}$ year	______	______
3.	$6,750	7%	$\frac{3}{4}$ year	______	______
4.	$8,120	18%	$\frac{2}{3}$ year	______	______
5.	$9,700	9%	$\frac{5}{8}$ year	______	______
6.	$75,000	10%	90 days	______	______
7.	$150,000	16%	120 days	______	______

8. Find the bank discount and the proceeds on a simple discount note taken out at the Hancock Bank by Joe Maggiore for 180 days at 15% interest for $6,000.

______

9. Find the bank discount and the proceeds on a simple discount note for $5,000 at 12% interest for 120 days.

______

10. Find the bank discount and the proceeds on a simple discount rate for $2,500 at 10.5% interest for 270 days.

______

11. Find the bank discount and the proceeds on a simple discount note written on May 6, with a maturity date of December 2 and interest rate of 9% for $5,000. (Use Table 4.1.)

______

12. Find the bank discount and the proceeds on a simple discount note written on June 1, with a maturity date of December 28 and an interest rate of 7% for $8,000. (Use Table 4.1.)

____

13. Find the bank discount and the proceeds on a simple discount note for $7,500 at 15% interest written on July 1, with a maturity date of September 29.

____

14. Find the bank discount and the proceeds on a simple discount note for $1230 at $8\frac{1}{2}$% interest written on March 15, with a maturity date of August 6.

____

15. Alexis was trying to decide between a simple interest loan of $5000 at 12% interest for 240 days or a simple discount loan. Which loan will give her the most money to use and how much does she pay to use the bank's money?

____

16. Karen was trying to decide between a simple interest loan of $6,600 at 11% for 270 days or a simple discount loan. Which loan will give her the most money to use and how much does she pay to use the bank's money?

____

ANSWERS

1. D = $20; S = $230 **2.** D = $9.88; S = $385.12 **3.** D = $354.38; S = $6,395.62 **4.** D = $974.40; S = $7,145.60 **5.** D = $545.63; S = $9,154.37 **6.** D = $1,875; S = $73,125 **7.** D = $8,000; S = $142,000 **8.** D = $450; S = $5,550 **9.** D = $200; S = $4,800 **10.** D = $196.88; S = $2303.12 **11.** D = $262.50; S = $4737.50 **12.** D = $326.67; S = $7673.33 **13.** D = $281.25; S = $7218.75 **14.** D = $41.82; S = $1,188.18 **15.** simple interest loan ($5,000 > $4,600); $400 **16.** simple interest loan ($6,600 > $6,055.50); $544.50

FOCUS ON PROBLEM SOLVING

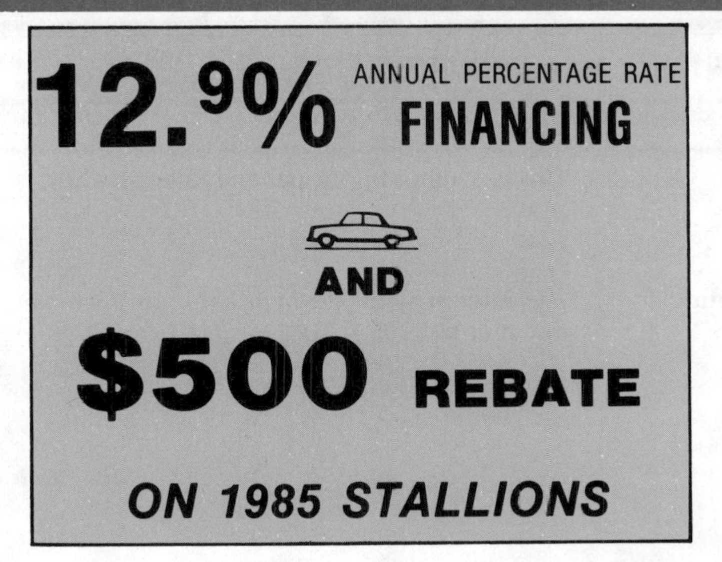

In 1969 the United States government passed a **Truth in Lending Law** that requires a creditor to tell a borrower the true yearly interest rate, called the **annual percentage rate** (abbreviated a.p.r.), on all installment loans written by a bank, a department store, a car dealer and so on. Interest (money paid for the privilege of borrowing money) is a major source of business income. Hence, it is only fair that the consumer clearly understand the borrowing rate that he or she is being charged by the creditor.

Our automobile illustration advertises an a.p.r. of 12.9%. All billings from major credit cards, such as VISA or Mastercard, will specify their a.p.r. All retail store billings will specify their a.p.r. All bank loans will specify their a.p.r. Look for the a.p.r. statement before accepting credit; it must be given to you.

The annual percentage rate may be computed using the following formula:

$$\text{a.p.r.} = \frac{2 \times \text{number of payments per year} \times \text{interest charged}}{\text{beginning balance owed} \times (\text{total number of payments} + 1)}$$

Example A woman purchases an $8,200 car and makes a down payment of $2,000. She then owes a balance of $8,200 − $2,000 = $6,200, which is to be paid back as part of 36 equal monthly payments in the amount of $214.60. What annual percentage rate is she paying?

Solution She has agreed to 12 payments per year for three years or a total of 36 payments. She owes a beginning balance of $6,200. Her interest is the difference between the amount that she must pay back and the amount that she owes. That is, her interest is (36)($214.60) − $6,200 = $7,725.60 − $6,200 = $1,525.60.

$$\begin{aligned}\text{a.p.r.} &= \frac{2 \times 12 \text{ payments per year} \times \$1{,}525.60 \text{ interest}}{\text{a beginning balance of } \$6{,}200 \times (36 + 1)} \\ &= \frac{\$36{,}614.40}{\$229{,}400} \\ &= 0.1596, \text{ or } 15.96\%\end{aligned}$$

BUILDING YOUR MATH VOCABULARY

TERM	DEFINITION	EXAMPLE
Amount, A	This is a sum of principal and interest. Thus, $A = P + I$.	If Principal = \$5,000 and Interest = \$500, A = \$5,000 + \$500 = \$5,500
Bank Discount, D	The interest withheld by a bank on a simple discount note. $D = MRT$	If Maturity Value = 3,000, Rate = 15%, and Time = 90 days, $D = \$\overset{\overset{15}{\cancel{30}}}{\cancel{3{,}000}} \times \frac{15}{\underset{1}{\cancel{100}}} \times \frac{\overset{1}{\cancel{90}}}{\underset{\underset{2}{\cancel{4}}}{\cancel{360}}}$ $= \frac{222}{4} = \$112.50$
Bankers' Rule	The common method used by banks to compute time for the purpose of determining interest. It is determined by dividing the exact number of days (money is borrowed) by 360.	On a 90 day loan, $T = \frac{90}{360} = \frac{1}{4}$
Base, B	The starting number or reference number in a percentage problem.	75% of 300 = 225, or 0.75 × 300 = 225 (↑ 300: base)
Borrower	A person who borrows money.	
Maturity Value, M	The amount of money payable to a bank when a simple discount note becomes due.	$D = \$3{,}000 \times \frac{15}{100} \times \frac{90}{360}$ (↑ \$3,000: maturity value)
Percent	One percent is $\frac{1}{100}$, or 0.01, or 1% of a quantity.	1% of 200 is 2: $\overset{2}{\cancel{200}} \times \frac{1}{\underset{1}{\cancel{100}}} = 2$
Percentage, P	A percentage is a number that represents a portion of the base, B. Percentage is computed using the formula $P = RB$.	75% of 300 = 225; 0.75 × 300 = 225 (↑ 225: percentage)
Principal	The amount of a deposit or loan.	The simple interest on \$5,000 borrowed at 10% for 2 years is I = \$5,000 × 0.10 × 2 = \$1,000 (↑ \$5,000: principal)
Proceeds, S	The amount of money that you receive from a bank when you take out a simple discount note. $S = M - D$.	Maturity value of a loan is \$5,000 and bank discount is \$500; thus S = \$5,000 − \$500 = \$4,500
Rate, R	A percent (%) of the base.	75% of 300 = 225; 0.75 × 300 = 225 (↑ 0.75: rate)

TERM	DEFINITION	EXAMPLE
Simple Interest, *I*	Money received or paid for on a loan. $I = PRT$.	If Principal = \$2,000, Rate = 12%, Time = 2 years: I = \$2,000 × 0.12 × 2 = \$480 (↑ simple interest)
Term	The term of a loan is the time period for which the loan will be outstanding.	A person borrows \$7,500 for 3 years at 12%; term of loan is 3 years.
Worth	The current rate of interest (%) being charged for a loan.	Person borrows \$7,500 for 3 years at 12%; worth of loan is 12%.

4

Chapter Review

Determine the fraction, decimal, or percent equivalents missing in problems 1–10.

	Fraction	*Decimal*	*Percent (%)*
1.	$\frac{3}{4}$		
2.			76%
3.		0.06	
4.	$\frac{9}{10}$		
5.		0.035	
6.	$1\frac{1}{2}$		
7.		0.45	
8.			20.5%
9.			300%
10.		1.75	

11. 16% of 240 is what number? ________

12. 72% of 35 is what number? ________

13. $8\frac{1}{4}$% of 400 is what number? ________

14. 40.5% of 84 is what number? ________

15. 18% of what number is 72? ________

16. 56% of what number is 30.8? ________

17. 8.5% of what number is 10.2? ________

18. $7\frac{1}{2}$% of what number is 6.3? ________

19. 18 is what percent of 60? ________

20. 72 is what percent of 180? ________

21. 0.056 is what percent of 0.28? ________

22. 0.64 is what percent of 0.00192? ________

23. 58 is what percent of 20? ________

24. 52.5 is what percent of 30? ________

25. Find the percent increase from 75 to 105. ________

26. Find the percent decrease from 92.4 to 68.4. ________

27. Find the percent decrease from 106 to 79.5. ____________

28. Find the percent increase from 120 to 300. ____________

29. A company places a staggering 125% markup on a certain electric light fixture. Assuming the company purchases the fixture for $16, what is the selling price?

____

30. A weighing machine is guaranteed accurate to within $\frac{1}{5}$% of the recorded weight. If this is true, what is the maximum possible error associated with a load of 10,000 lb?

____

31. An agent sold a shipment of grain for $800,000 for which he received $\frac{1}{8}$% commission. Compute the commission and the *net proceeds* (selling price less commission) for this grain sale.

____

32. General Gear Company has announced a wage increase of 10.5%. If Denise earns $180 per week before the wage increase, how much will she earn after the wage increase?

____

33. A lamp is listed for $62.00 less a 10% discount and then another 10% discount. Compute the sale price. (*Hint:* Do not combine the two discounts.)

____

34. Out of 32 students, 8 received a grade of B. What percent of the students received a B?

____

35. Production economics reduced the manufacturing expense of a particular item from $3.00 to $2.75. What percent savings was realized?

____

36. The population of the United States in 1920 was approximately 105,710,000. In 1940 the population had increased to approximately 131,670,000. What was the percent of increase?

____

37. A car costs $7,200. Of this, $380 is for bucket seats. What percent of the price of the car is for bucket seats?

____

38. Fifteen is 25% of what number? ____________

39. A man received $46.20 in overtime pay. This amount of money represents 12% of his gross pay. How much is his gross pay?

____

40. What amount of savings would be needed to generate $650 in interest at the end of a year if the bank is giving interest at the rate of 6.5%?

____

41. Workmen have loaded 16,000 pounds of merchandise into a truck. If the men have loaded 32% of the truck, how much does the truck hold? How much remains to be loaded?

____

42. The rate of depreciation of a certain piece of equipment is 4% per year. If the amount of depreciation is $1,440, what is the value of the equipment?

____

43. A collection agency collected a past due bill of $980 for a doctor. The agency charged a fee of $147 for its services. What rate of commission did the agency charge?

____

44. Louise Clarendon wishes to buy a house selling for $56,000. She can obtain a loan for 85% of the price of the house. How much cash does she need to raise to purchase the house?

____

45. The monthly expenses of a small business are $5,950. If the owner wishes to make a profit of 12%, how much money must he take in each month?

____

46. Thirty percent of Attleborn High School's graduates went to college the fall after graduation. If 60 people entered college, how many Attleborn High School graduates were there?

____

47. Thirty-eight of 50 club members attended a meeting. What percent of the club members attended this meeting?

____

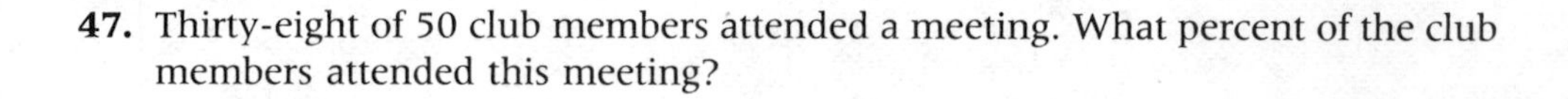

48. Barbara Pickard lives in a town in which the property tax rate is $3\frac{1}{2}$% of the house valuation. If her house is valued at $64,000, what is her property tax bill?

____

49. A store has an item that listed for $54 now selling at a reduced price of $42.12. What percent of list price was this item marked down?

____

50. This month's inventory of 248 cases is 80% of last month's inventory. What was the inventory of the previous month?

____

51. Peter earned $32 commission on the sale of a television set. If the commission rate is 4%, find the selling price of the television.

52. If the commission on a sale of $42.00 is $3.78, what percent commission is paid?

53. Cathy bought corporate stock for $7,400 and sold it for $7,992. What percent of her original investment was her profit?

54. Tobacco production declined from 2.14 billion pounds in 1976 to 1.91 billion pounds in 1977. Calculate the percent of decrease in production for the one-year period.

55. What amount is 5% of $3,000?

56. Arthur Heinz purchased a new refrigerator priced at $440. He made a down payment of 15% of the price. Find the amount of the down payment.

57. Jones and Jones, Inc. reported that profits of the year 1982 had declined by 7% to $102,500. Calculate the profits of the previous year.

58. If $P = \$16{,}000$, $R = 10\%$ and $T = 2$, find the simple interest, I, and the amount, A.

59. If $P = \$50{,}000$, $R = 8\%$ and $T = 4.25$, find the simple interest, I, and the amount, A.

60. Nellie loaned $8,700 at 12% simple interest for 90 days.

(a) How much interest did she receive?

(b) What was the amount she received at the end of the term?

61. Sidney borrowed $20,000 at 15% simple interest from April 10 to June 21.

(a) How much interest did he pay?

(b) What was the amount that he paid on June 21?

62. Find the bank discount, D, and proceeds, S, on a simple discount note of $10,000 for 60 days at 9.5%.

63. Find the bank discount, D, and the proceeds, S, on a simple discount note of $25,000 for 45 days at 18%.

____

64. Jose borrowed $3,750 at 11% for 180 days on a simple discount note.

(a) How much interest did he pay? _______________

(b) How much money did Jose receive the day he signed the note?

____

(c) How much did Jose repay in 180 days? _______________

65. Jacques signed a simple discount note for $2,800 on March 31. The note was payable on October 27 at 8.75%.

(a) How much interest did he pay? _______________

(b) How much money did he receive the day he signed the note?

____

ANSWERS

1. 0.75; 75% **2.** $\frac{19}{25}$; 0.76 **3.** $\frac{3}{50}$; 6% **4.** 0.9; 90% **5.** $\frac{7}{200}$; 3.5% **6.** 1.5; 150% **7.** $\frac{9}{20}$; 45% **8.** $\frac{41}{200}$; 0.205 **9.** 3; 3 **10.** $1\frac{3}{4}$; 175% **11.** 38.4 **12.** 25.2 **13.** 33 **14.** 34.02 **15.** 400 **16.** 55 **17.** 120 **18.** 84 **19.** 30% **20.** 40% **21.** 20% **22.** $333\frac{1}{3}$% **23.** 290% **24.** 175% **25.** 40% **26.** 25.97% **27.** 25% **28.** 150% **29.** $36 **30.** 20 pounds **31.** $1,000; $799,000 **32.** $198.90 **33.** $50.22 **34.** 25% **35.** $8\frac{1}{3}$% **36.** 24.6% **37.** 5.3% **38.** 60 **39.** $385 **40.** $10,000 **41.** 50,000 pounds **42.** $36,000 **43.** 15% **44.** $8,400 **45.** $6,761.36 **46.** 200 **47.** 76% **48.** $2,240 **49.** 22% **50.** 310 cases **51.** $800 **52.** 9% **53.** 8% **54.** 10.7% **55.** $150 **56.** $66 **57.** $110,215.05 **58.** I = $3,200; A = $19,200 **59.** I = $17,000; A = $67,000 **60.** **(a)** $261 **(b)** $8,961 **61.** **(a)** $600 **(b)** $20,600 **62.** D = $158.33; S = $9,841.67 **63.** D = $562.50; S = $24,437.50 **64.** **(a)** $206.25 **(b)** $3,543.75 **(c)** $3,750 **65.** **(a)** $142.92 **(b)** $2,657.08

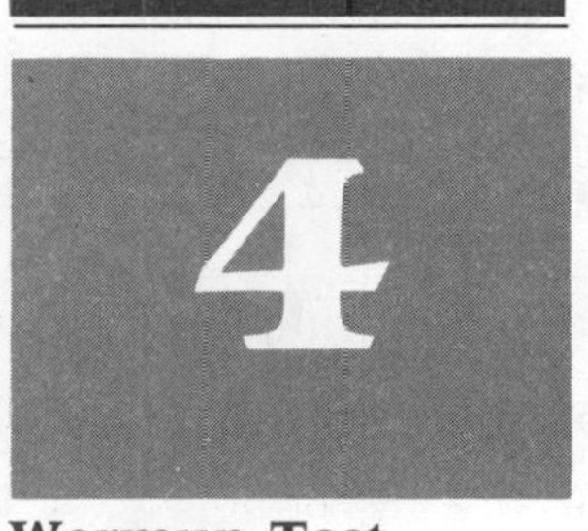

Warmup Test

1. Convert the following to percents (if necessary, carry to the ten-thousandths position).

(a) $\frac{17}{20}$ ____ **(b)** $\frac{5}{16}$ ____ **(c)** 0.32 ____ **(d)** 0.0145 ____ **(e)** 3 ____

2. Convert the following percents to fractions expressed in their simplest form.

(a) 12% ____ **(b)** $7\frac{1}{2}\%$ ____ **(c)** 225% ____ **(d)** 8.25% ____ **(e)** 0.8% ____

3. Convert the following percents to decimals.

(a) 16% ____ **(b)** 3.6% ____ **(c)** 0.045% ____ **(d)** 12.75% ____

4. Find 15% of 225. ____________

5. 160 is what percent of 250? ____________

6. 7.5% of what number is 21? ____________

7. The Santiago family has a variable-rate mortgage. Their monthly mortgage payments increased from $300 to $312. What is the percent of increase in their monthly payments?

____

8. A clothing store reduced its top-of-the-line winter coats from $235.50 to $175. What is the percent of decrease? (Round off to tenths of a percent.)

____

9. A group invested $15,000 in bonds at $12\frac{1}{2}\%$ for 4 years.

(a) What is the simple interest? ____________

(b) What is the amount accumulated? ____________

10. Find the bank discount, D, and proceeds, S, on a simple discount note of $18,000 at $10\frac{1}{4}\%$ simple interest for 150 days.

____

11. Bob shot 38 baskets in 45 attempts. What is the percentage of baskets made? (Round off to the nearest tenth of a percent.)

____

12. Last year at Drexel State 18% of the student body took accounting. If 342 students took accounting, how many students were at Drexel State last year?

____

Challenge Test

1. Convert the following to percents (if necessary, carry to the ten-thousandths position).

(a) $\frac{9}{20}$ ____ **(b)** $\frac{7}{12}$ ____ **(c)** 0.64 ____ **(d)** 0.0919 ____ **(e)** 4 ____

2. Convert the following percents to fractions expressed in their simplest form.

(a) 16% ____ **(b)** $5\frac{1}{2}$% ____ **(c)** 140% ____ **(d)** 4.75% ____ **(e)** 0.6% ____

3. Convert the following percents to decimals.

(a) 22% ____ **(b)** 7.1% ____ **(c)** 0.088% ____ **(d)** 15.65% ____

4. Find 35% of 360. ____

5. 280 is what percent of 350? ____

6. 6.4% of what number is 32? ____

7. The Barret family has been making monthly VISA payments of $80. The bank that they are dealing with wants them to start making monthly payments of $95. What is the percent of increase in their monthly payment?

____

8. Taylor and Lord reduced the price of a certain item from $98 to $84. What is the percent of decrease in price? (Round off to tenths of a percent.)

____

9. A man invested $18,000 at $15\frac{3}{4}$% for 5 years.

(a) What is the simple interest? ____

(b) What is the amount accumulated? ____

10. Find the bank discount, *D*, and the proceeds, *S*, on a simple discount note of $12,000 at $12\frac{1}{2}$% simple interest for 270 days.

____

11. A pitcher completed 150 of 160 innings that he started. What is the percent of innings completed? (Round off to the nearest tenth of a percent.)

____

12. This semester, 85% of the students who started elementary statistics passed the course. If 459 students passed this course, how many students started the course?

____

Following are three cumulative review tests for Part 1. A calculator is not to be used.

Cumulative Test

1. Convert $6\frac{1}{8}$ to an improper fraction. ________

2. Express 76.94 in words. ________

3. Place an inequality sign between: \$0.05 ____ \$0.50.

4. Round off 796.215 to the nearest hundredth ______; to the nearest tenth ______.

5. What is the reciprocal of $\frac{7}{11}$?

6. $\begin{array}{r} 96 \\ \times\ 200 \\ \hline \end{array}$

7. $63.75 + 38.46 =$ ________

8. $\begin{array}{r} \$760.50 \\ -\ 395.40 \\ \hline \end{array}$

9. $0.325 \times 1.6 =$ ________

10. $38.35 \div 9.1 =$ ________

11. $\begin{array}{r} 5\frac{1}{2} \\ +\ 2\frac{3}{5} \\ \hline \end{array}$

12. $\begin{array}{r} 7\frac{1}{4} \\ -\ 2\frac{3}{5} \\ \hline \end{array}$

13. $\frac{3}{8} \times \frac{6}{5} =$ ________

14. $\frac{8}{3} \div 3\frac{1}{5} =$ ________

15. What is the LCM of 9, 12, and 20? ________

16. Express $\frac{4}{5}$ as a percent with a percent sign (%). ________

17. Find 40% of 90. ________

18. Fifteen represents 25% of what number? ________

19. Fourteen is what percent of 70? ________

20. What is the ratio of 9 feet to 2 yards? ________

21. Given $\frac{5}{7} = \frac{50}{70}$. The means are ________; the extremes are ________.

22. If $\frac{9}{4} = \frac{x}{28}$, then x must equal ________.

23. What is the simple interest, I, on a principal of \$6,000 for two years at 10% ?

____

24. In problem 23, what is the amount at the end of the two-year period?

____

25. Clair takes out a simple discount note at a local bank for one year with a maturity value of $10,000 and interest rate of 15%. Determine her proceeds, *S*.

____

26. Matthew has sold 125% of his last year's sales total for new cars. This year he sold 60 cars. How many cars did he sell last year?

____

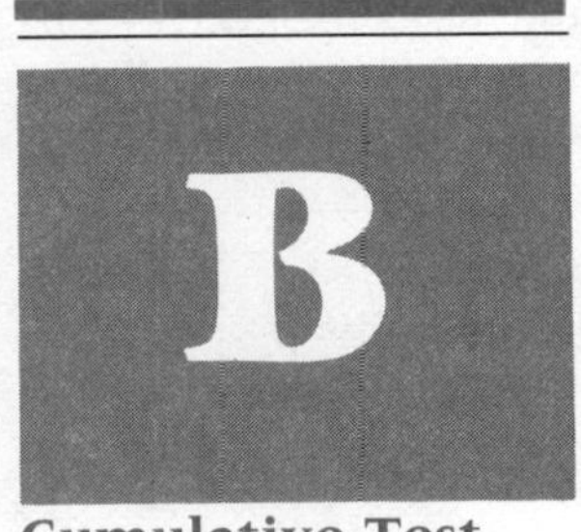

Cumulative Test

1. Convert $\frac{47}{6}$ to a mixed number. ______

2. Eight thousand six and forty-four hundredths is expressed as ______.

3. Express $\frac{7}{8}$ as a decimal. ______

4. Arrange the following decimals in order of magnitude from smallest to largest: 0.312, 0.02, 0.103, and 0.30.

______

5. Round off $4,490.765 to the nearest hundredth ______;

to the nearest tenth ______.

6. What is the reciprocal of $\frac{25}{37}$? ______

7. $129.88 + 97.51 =$ ______

8. $\begin{array}{r} 31.4 \\ \times\ 100 \\ \hline \end{array}$

9. $0.0097 \times 0.015 =$ ______

10. You gave the clerk at Rex Mart $20.00 and received change of $13.72 for a particular item. How much did the item cost you?

______

11. $\begin{array}{r} 7\frac{1}{3} \\ +\ 4\frac{1}{4} \\ \hline \end{array}$

12. $\begin{array}{r} 19\frac{2}{3} \\ -\ 6\frac{7}{8} \\ \hline \end{array}$

13. $\frac{3}{4} \times \frac{12}{7} =$ ______

14. $\frac{8}{5} \div 3\frac{1}{5} =$ ______

15. What is the LCM of 3, 4, and 9? ______

16. Express 0.10 as a percent with a percent sign (%). ______

17. Express 2.5% as a decimal. ______

18. Find 18% of 300. ______

19. Nineteen represents 20% of what number? ______

20. Twenty-three is what percent of 72? ______

21. What is the ratio of 12 nickels to 3 quarters? ______

22. If 3:7 as 18:42, the means are ______ and the extremes are ______.

23. Tricia is working on a scale drawing in which 1 inch equals 10 yards. How many inches on the scale drawing represent 25 yards?

____

24. What is the simple interest, I, on a principal of $10,500 for 6 months at 16% ?

____

25. Harold took out a simple discount note at a local bank for 1 year with a maturity value of $4,000 and interest rate of 16%. Determine the proceeds, S.

____

26. Last year attendance at State University's basketball games totaled 126,000. This year's attendance increase was 112% of last year's attendance. What was this year's increase in attendance?

____

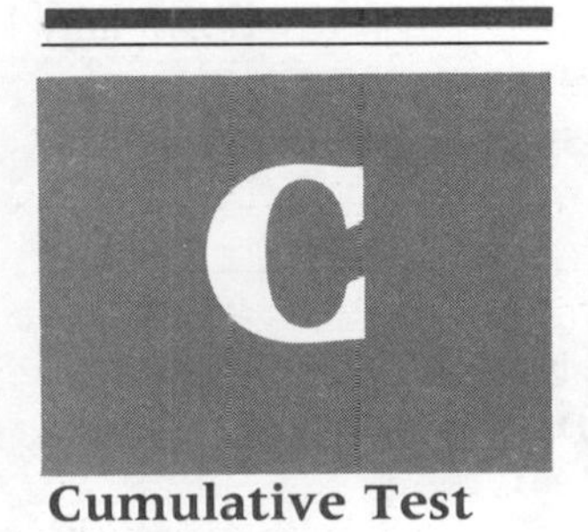

Cumulative Test

1. Convert $3\frac{4}{5}$ to an improper fraction. ________

2. Two hundred fifty and seven thousandths may be expressed as ________.

3. Express $\frac{2}{3}$ as a decimal. ________

4. Round off 8.055 to the nearest hundredth ________;

   to the nearest tenth ________.

5. What is the reciprocal of 4? ________? Of $\frac{5}{9}$? ________

6. Place an inequality sign between: \$0.10 ____ \$0.01.

7. You purchase two items that cost \$0.75 and \$0.86, respectively. Assuming that you pay for these items with a \$5 bill, what is your change?

   ____

8. $\begin{array}{r} 33{,}746.19 \\ -\ 10{,}695.21 \\ \hline \end{array}$

9. $2.15 \times 0.00018 =$ ________

10. $2.71\overline{)48.78}$

11. $18\frac{2}{3} + 17\frac{4}{5} =$ ________

12. $\begin{array}{r} 7\frac{1}{5} \\ -\ 3\frac{7}{10} \\ \hline \end{array}$

13. $\frac{8}{7} \times \frac{5}{24} =$ ________

14. $\frac{4}{5} \div 1\frac{5}{7}$ ________

15. What is the LCM of 4, 8, and 9? ________

16. Which is larger, $\frac{7}{8}$ or $\frac{9}{10}$? ________

17. Express 0.569 as a percent with a percent sign (%). ________

18. If you earned \$24,000 this year and 6.7% of this amount was withheld for social security, how much social security tax did you pay?

    ____

19. Find 56% of 340. ________

20. Eighteen is what percent of 90? ________

21. Thirty-two represents 20% of what number? ________

22. What is the ratio of 96 inches to 10 feet? ________

23. If $\frac{11}{4} = \frac{x}{20}$, then x must equal ________________.

24. What is the simple interest, I, on a principal of $8,500 for 9 months at 12%?

____

25. Meghan took out a simple discount note at a local bank for 6 months with a maturity value of $20,000 and interest rate of 18%. Determine her proceeds, S.

____

26. Last year Helen paid $1.11 for a gallon of heating oil. This year she will pay 95% of last year's price. How much is this year's price to the nearest cent?

____

Part

INDEPENDENT APPLICATIONS OF ARITHMETIC

Geometry

CHAPTER OUTLINE

5.1 PERIMETER

Historically, men and women studied geometry because of its logical beauty and precision which has fascinated generations of people. Today people study geometry because of its practical applications. Some of the people who use geometry include draftsmen, chemists, architects, engineers, carpenters, and plumbers.

We all know what points and lines are but sometimes find it difficult to define them. A **point** is represented by a "dot" and is named by placing a capital letter near it. A point has no length, width, or thickness. Thus, in reality, a point cannot be seen. On the other hand, a **straight line** or simply a **line** has length and is represented by drawing a "straight mark." A line is named either by two points placed upon it or by a small letter placed near it.

Figure 5.1

In Figure 5.1, the line may be named l or line AB, or line AC, or line BC, where A, B, and C are points on the line. A straight line extends without end in two opposite directions.

As you have already observed in preceding chapters, lines are quite useful to help us visualize numbers, fractions, and decimals. Like the words **point** and **line,** the **plane** is also undefined in mathematics. Any three points, not all lying in a straight line, determine a plane. It is important that we have a real feeling for the meaning of this word. Visualize extending a sheet of paper in all directions and you have a model for a plane. Extending a wall in your room or extending a desktop surface will also serve as a model for a plane. Thus, we see points as lying on a line and lines as lying on a plane.

A **straight-line segment** or a line segment is the part of a straight line between two points and including the two points, called the **endpoints.** Line segments are named by their endpoints. In Figure 5.1, AB is a line segment. Because the length of a line segment is finite, you can measure it. It is clear that a line segment is the shortest distance between two points? Can you name two other line segments in this figure?

Example 1 Name all the line segments of line l.

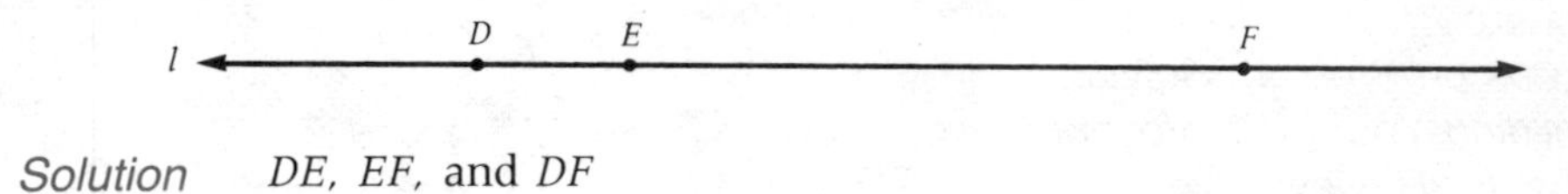

Solution DE, EF, and DF

A **ray** is part of a line extending indefinitely in one direction from a point. In Figure 5.2 you see a ray that begins at A and moves along a line through B; this is a ray from A through B, denoted $\overrightarrow{AB}$ or "ray k."

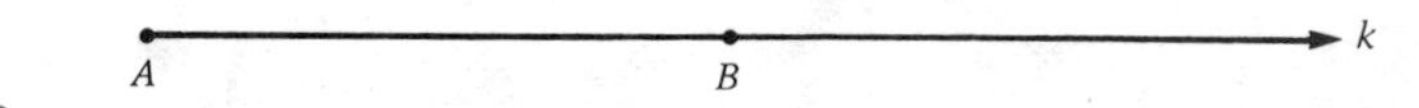

Figure 5.2

Polygons are closed figures whose sides are line segments. A **triangle** is a three-sided polygon, a **quadrilateral** is a four-sided polygon, and a **pentagon** is a five-sided polygon.

> The **perimeter,** P, of a polygon is the distance around the polygon and may be found by adding the lengths of its sides.

Example 2 Find the perimeter of the polygon $ABCDEF$.

Solution

$$\begin{aligned} P &= AB + BC + CD + DE + EF + AF \\ &= 4 \text{ in} + 5 \text{ in} + 6 \text{ in} + 4 \text{ in} + 7 \text{ in} + 3 \text{ in} \\ &= 29 \text{ in} \end{aligned}$$

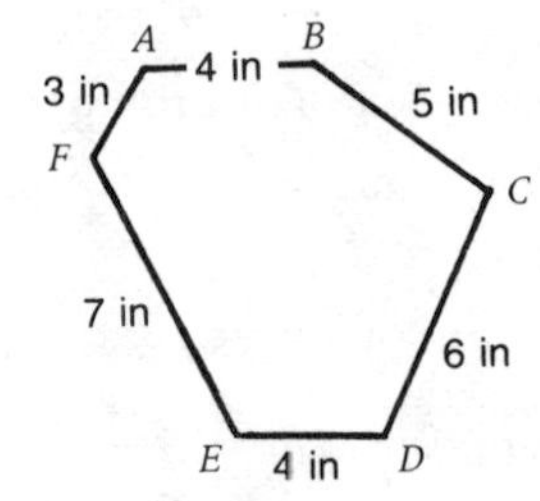

A **parallelogram** is a quadrilateral whose opposite sides are parallel and of equal length. See Figure 5.3.

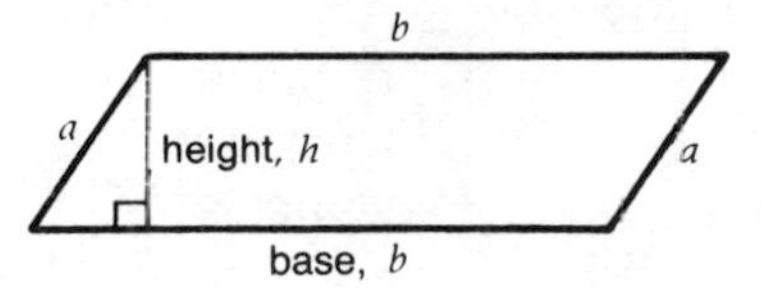

Figure 5.3

> The formula for the perimeter, P, of any parallelogram is 2 times side a plus 2 times side b, or $P = 2 \cdot a + 2 \cdot b = 2a + 2b$.

Example 3 Find the perimeter of the parallelogram $ABCD$ as shown.

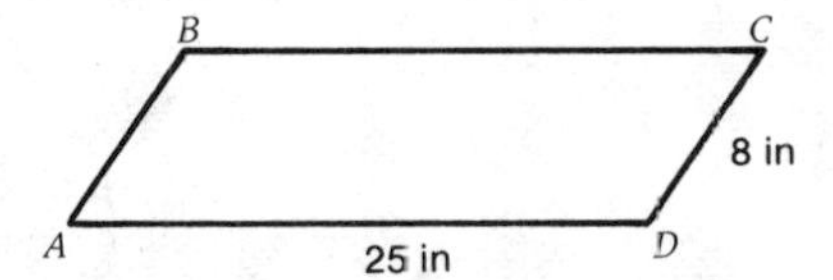

Solution

$$\begin{aligned} P &= 2 \cdot a + 2 \cdot b \\ &= 2 \cdot 8 \text{ in} + 2 \cdot 25 \text{ in} \\ &= 16 \text{ in} + 50 \text{ in} \\ &= 66 \text{ in} \end{aligned}$$

A **rectangle** is a quadrilateral with opposite sides of equal length and four right angles (90° angles). See Figure 5.4.

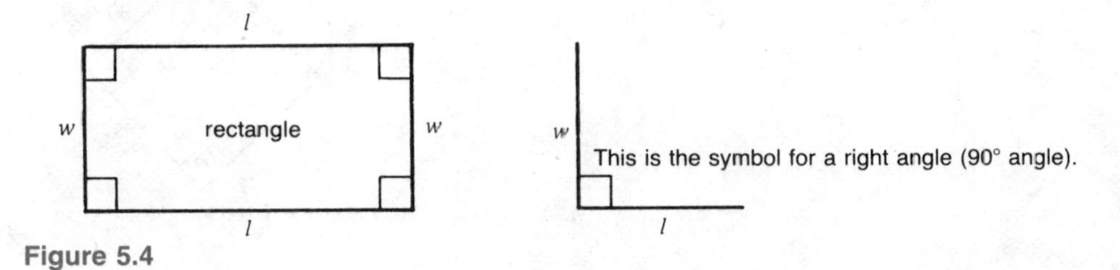

Figure 5.4

> The formula for the perimeter, P, of any rectangle is 2 times its length + 2 times its width, or $P = 2 \cdot l + 2 \cdot w = 2l + 2w$.

Example 4 Find the perimeter of rectangle *ABCD*, as shown.

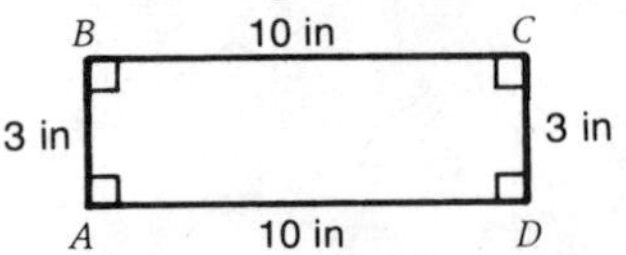

Solution

$$\begin{aligned} P &= 2 \cdot l + 2 \cdot w \\ &= 2 \cdot 10 \text{ in} + 2 \cdot 3 \text{ in} \\ &= 20 \text{ in} + 6 \text{ in} \\ &= 26 \text{ in} \end{aligned}$$

Example 5 A home gardener wishes to build a fence around his 30-foot by 50-foot rectangular garden. How much fencing does he need? If the cost of fencing is $3.00 per foot, how much will it cost him to fence in this garden?

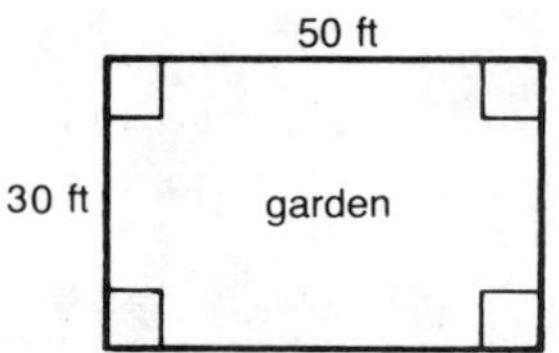

Solution

$$\begin{aligned} P &= 2 \cdot l + 2 \cdot w \\ &= 2 \cdot 50 \text{ ft.} + 2 \cdot 30 \text{ ft} \\ &= 100 \text{ ft} + 60 \text{ ft} \\ &= 160 \text{ ft} \end{aligned}$$

$$\begin{aligned} \text{Cost} &= (\$3.00)(160) \\ &= \$480 \end{aligned}$$

A **square** is a rectangle with all four sides of equal length.

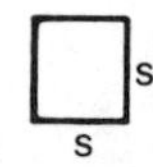

> The formula for the perimeter, P, of any square is 4 times the length of any side, s, or $P = 4 \cdot s = 4s$.

Example 6 A baseball diamond is a square with a distance of 90 feet between successive bases. What is the distance around the diamond?

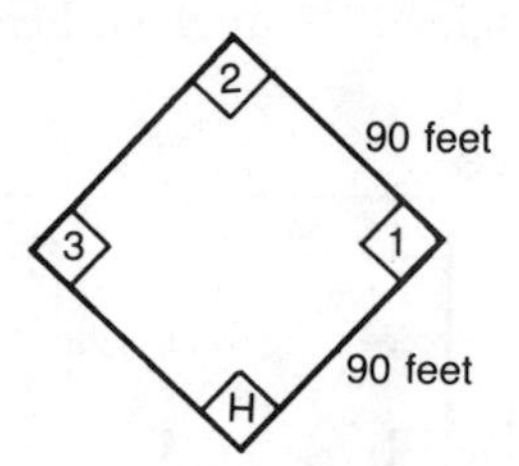

Solution

$$\begin{aligned} P &= 4 \cdot s \\ &= 4 \cdot 90 \text{ ft} \\ &= 360 \text{ ft} \end{aligned}$$

A **triangle** is a polygon composed of three line segments as shown in Figure 5.5.

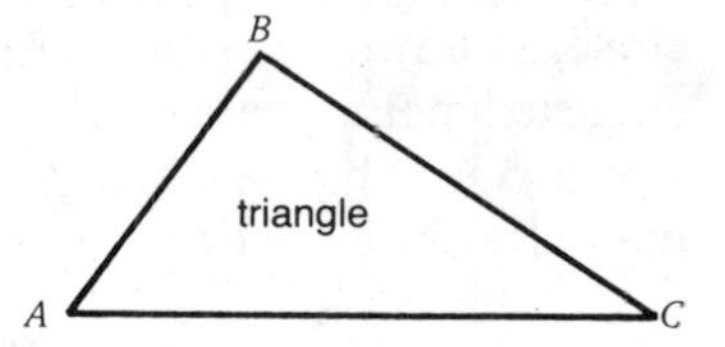

Figure 5.5

> The formula for the perimeter, P, of any triangle is the sum of its three sides, or $P = AB + BC + CA$.

Example 7 Find the perimeter of a triangle whose sides are 8 in, 9 in, and 10 in.

Solution
$$\begin{aligned} P &= 8 \text{ in} + 9 \text{ in} + 10 \text{ in} \\ &= 27 \text{ in} \end{aligned}$$

Example 8 A sailmaker is making a triangular sail for a boat with the dimensions as pictured and wants to bind the outer edge of the sail. How much binding does he need?

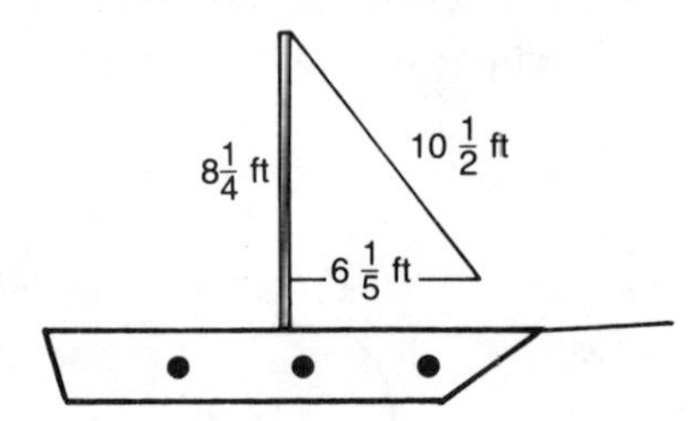

Solution
$$\begin{aligned} P &= 10\tfrac{1}{2} \text{ ft} + 8\tfrac{1}{4} \text{ ft} + 6\tfrac{1}{5} \text{ ft} \\ &= 10\tfrac{10}{20} \text{ ft} + 8\tfrac{5}{20} \text{ ft} + 6\tfrac{4}{20} \text{ ft} \\ &= 24\tfrac{19}{20} \text{ ft} \end{aligned}$$

Example 9 Find the perimeter of the four-sided figure ABCD shown.

B
$3\frac{1}{4}$ ft
$2\frac{5}{8}$ ft
D
$1\frac{7}{8}$ ft
C
A
$1\frac{3}{4}$ ft

Solution
$$\begin{aligned} \text{Perimeter} &= AB + BC + CD + DA \\ &= 3\tfrac{1}{4} \text{ ft} + 2\tfrac{5}{8} \text{ ft} + 1\tfrac{7}{8} \text{ ft} + 1\tfrac{3}{4} \text{ ft} \\ &= 3\tfrac{2}{8} \text{ ft} + 2\tfrac{5}{8} \text{ ft} + 1\tfrac{7}{8} + 1\tfrac{6}{8} \text{ ft} \\ &= 7\tfrac{20}{8} \text{ ft} \\ &= 9\tfrac{1}{2} \text{ ft} \end{aligned}$$

The Circle

The perimeter of a circle is called the **circumference,** C. A **diameter** of a circle, d, is a line segment that joins two points on a circle and passes through the center of the circle. The **radius,** r, of a circle is a line segment extending from the center of the circle to a point on the circle. A diameter, d, is always twice as long as a radius, r (see Figure 5.6). That is, $d = 2 \times r$. Is it clear that you can draw an unlimited number of diameters in any circle?

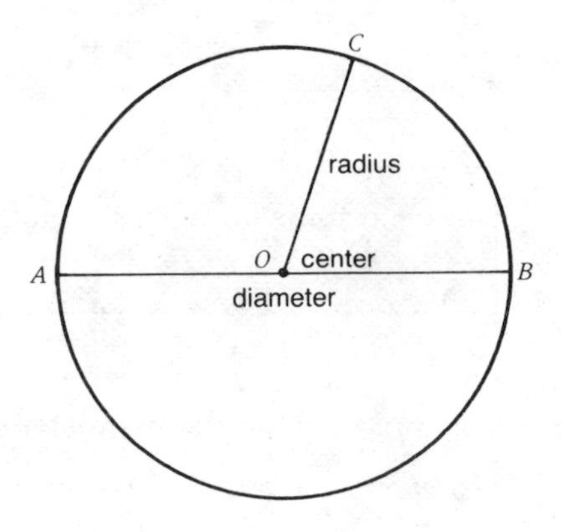

Figure 5.6

If you measure the circumference (use a string) and the diameter of a saucer, a dinner plate, and a serving dish from your kitchen cabinet and then form the ratio $\frac{C}{d}$, you will obtain a number approximately equal to (denoted $\doteq$), 3.14. See Figure 5.7, where we show examples in which we have actually done this experiment.

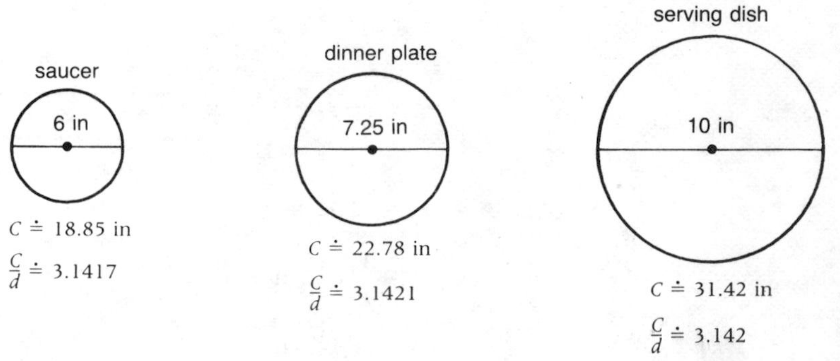

$C \doteq 18.85$ in
$\frac{C}{d} \doteq 3.1417$

$C \doteq 22.78$ in
$\frac{C}{d} \doteq 3.1421$

$C \doteq 31.42$ in
$\frac{C}{d} \doteq 3.142$

Figure 5.7

An approximate value for this ratio is 3.14159265 (accurate to 8 decimal places). In our experiment the most accurate approximation was 3.1417. If we were to form the same ratio by dividing the circumference of any circle by its diameter, we would get approximately the same number. This number is called "pi," symbolized by the Greek letter π. There are two standard approximations for π. They are: (1) 3.14 as a decimal and (2) $\frac{22}{7}$ as a fraction. To see that the two approximations are very close you have merely to divide 22 by 7. The result correct to two decimal places is 3.14. *It is the usual practice in mathematics to use 3.14 as an approximation of π unless you find $\frac{22}{7}$ more convenient.*

> The formula for the circumference, C, of a circle is π times its diameter or $C = \pi \cdot d = 2 \cdot \pi \cdot r$ or $\pi d = 2\pi r$.

Example 10 (a) If the diameter, d, of a circle is 10 inches, how long is the radius, r?
(b) If the radius, r, of a circle is 5.31 feet, how long is the diameter, d?

Solution (a) $r = \frac{d}{2} = \frac{10 \text{ in}}{2} = 5 \text{ in}$

(b) $d = 2 \cdot r = 2(5.31 \text{ ft}) = 10.62 \text{ ft}$

Example 11 Find the circumference, C, of a circle whose diameter is 6.02 in.

Solution $C = \pi \cdot d = (3.14)(6.02 \text{ in}) = 18.90 \text{ in}$

Example 12 An automobile wheel has a radius of 8 inches. How far does the car travel in 100 revolutions of the wheel?

Solution 1 revolution equals the circumference of the wheel

$$\begin{aligned}
\text{step 1: } C = 1 \text{ revolution} &= 2 \cdot \pi \cdot r \\
&= 2(3.14)(8 \text{ in}) \\
&= 50.24 \text{ in} \\
\text{step 2: } 100 \text{ revolutions} &= (100)(50.24 \text{ in}) \\
&= 5{,}024 \text{ in}
\end{aligned}$$

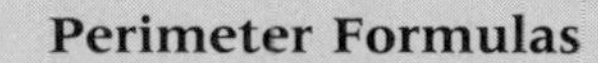

Perimeter Formulas

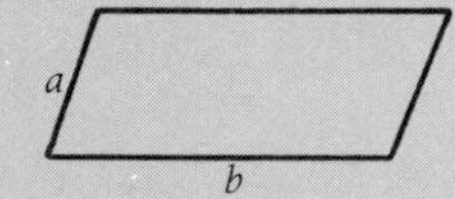

Parallelogram: $P = 2 \cdot a + 2 \cdot b$

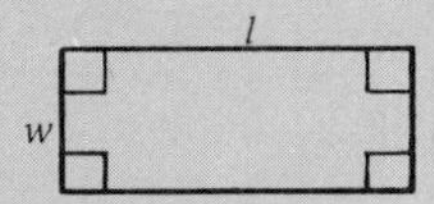

Rectangle: $P = 2 \cdot l + 2 \cdot w$

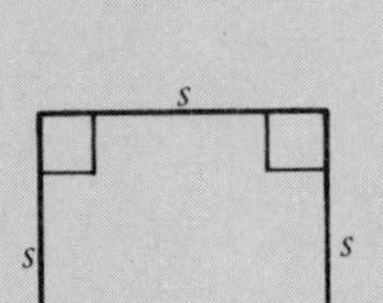

Square $P = 4 \cdot s$

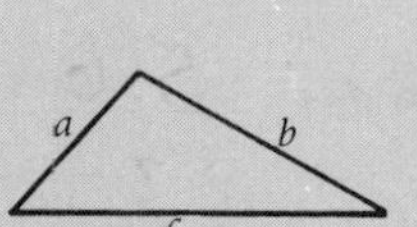

Triangle: $P = a + b + c$

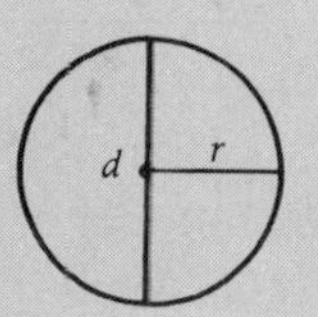

Circle: $C = \pi \cdot d$ or $2 \cdot \pi \cdot r$

5.1 EXERCISES

Round off all decimal answers in this section to the nearest hundredth. In problems 1–12, find the perimeter of the given figure.

1. 12 in; 4 in — 32

2. 5 in; 5 in — 20

3.

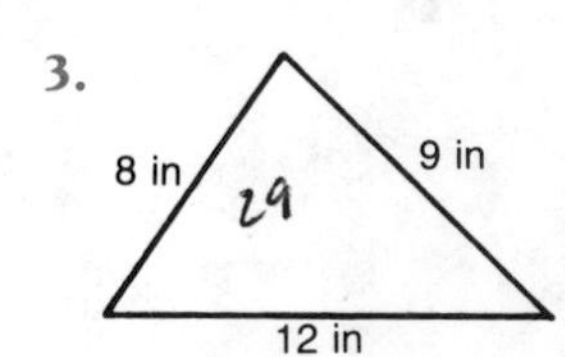

4. 15 in — C=47.10

5.

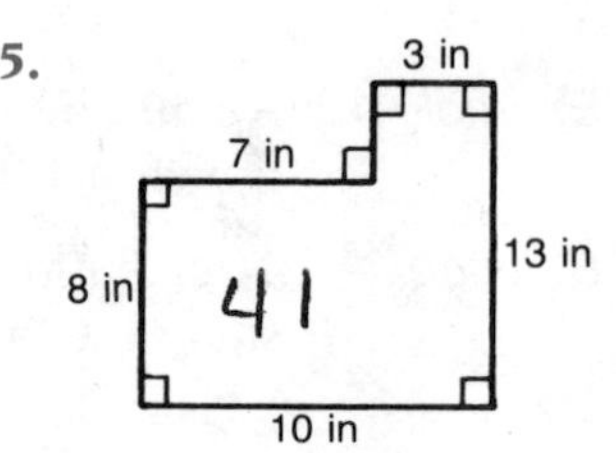

6.

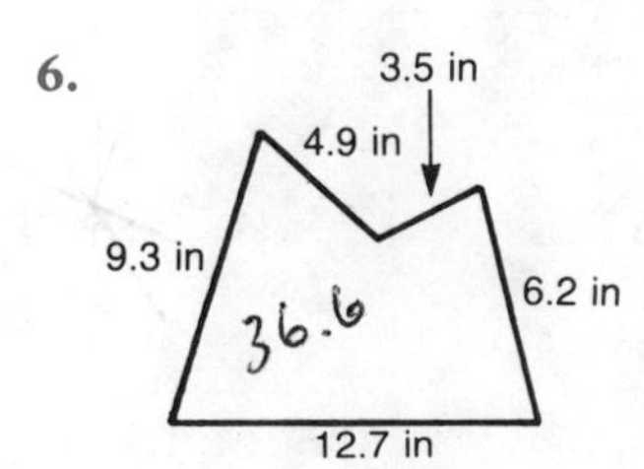

7.

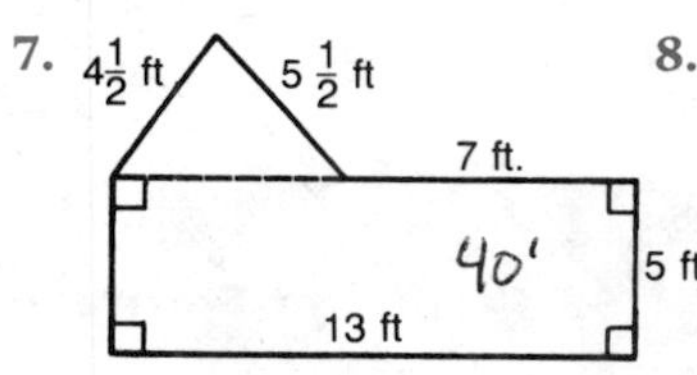

8.

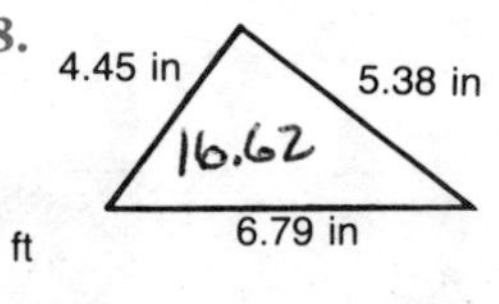

9.

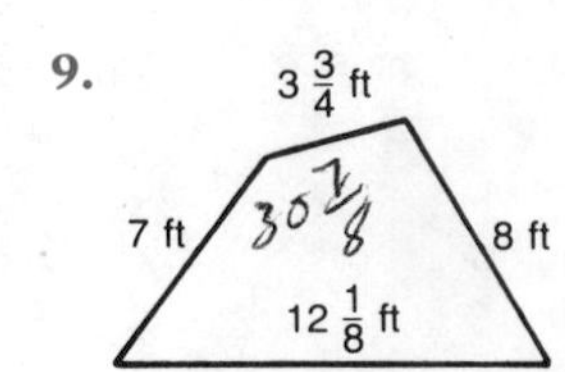

10.

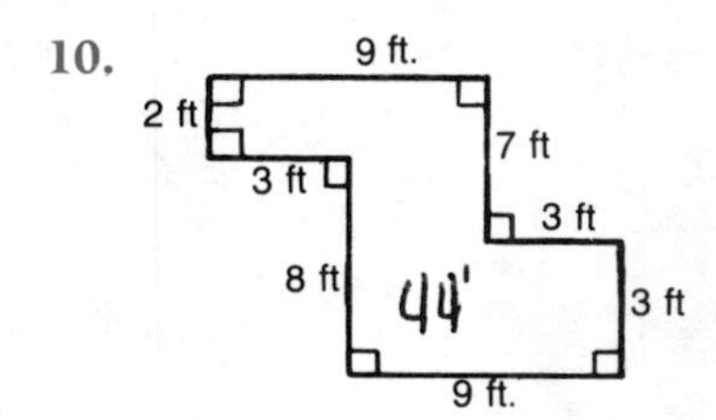

11.

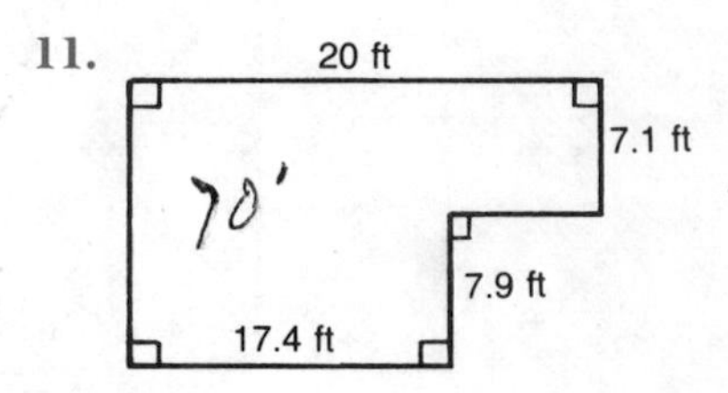

12.

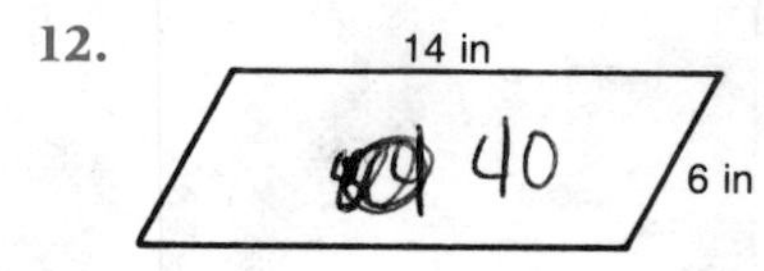

13. If $r = 2$ inches, then $d =$ 4 and $C =$ 12.56.

14. If $r = 2.8$ inches, then $d =$ 5.6 and $C =$ 17.58.

15. If $d = 6$ inches, then $r =$ 3 and $C =$ 18.84.

16. If d = 6.44 inches, then $r =$ 3.22 and C = 20.22.

17. (true or false) $C = 2 \times \pi \times r$. True

18. (true or false) An unlimited number of radii (this is the plural of radius) can be drawn in any circle.

true

19. (true or false) In any circle, $d \div 2 = r$. ______

20. What is the perimeter, P, of a square

(a) 6 inches on a side? ______

(b) 8 inches on a side? ______

(c) 10 inches on a side? ______

21. If the width of a rectangle is 2.61 inches and the length is 14.07 inches, what is the perimeter?

______

22. If the width of a rectangle is $1\frac{7}{8}$ meters and the length is $3\frac{5}{6}$ meters, what is the perimeter?

______

23. You have a square 3 inches on a side. What is the perimeter? Double the length of each side of the square. Now what is the perimeter?

______

24. A farmer wishes to enclose a rectangular pasture 300 meters by 150 meters. The cost of fencing is \$2.50 per meter. What will be his cost to enclose this pasture?

______

25. Find the perimeter of a triangle whose sides are 8 centimeters, 7.5 centimeters, and 7.75 centimeters, respectively.

______

26. Find the perimeter of the parallelogram whose unequal sides are 32 inches and 18 inches.

______

27. Find the perimeter of the parallelogram whose unequal sides are 17.6 ft and 6.8 ft.

______

28. Find the perimeter of a triangle whose sides are 6.23 inches, 7.15 inches, and 8.09 inches.

______

29. A polygon has sides with lengths of $12\frac{1}{3}$ ft, $13\frac{2}{5}$ ft, and $12\frac{7}{8}$ ft. What kind of polygon is this? What is its perimeter?

______

30. What is the circumference of a circle whose diameter is $5\frac{1}{4}$ feet? Use $\pi = \frac{22}{7}$.

______

31. What is the circumference of a circle whose diameter is $6\frac{1}{8}$ feet? Use $\pi = \frac{22}{7}$.

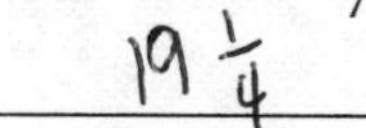

32. What is the circumference of a circle whose radius is 3.74 inches?

33. Find the perimeter of the polygon *ABCDE,* as shown. All measurements are in yards.

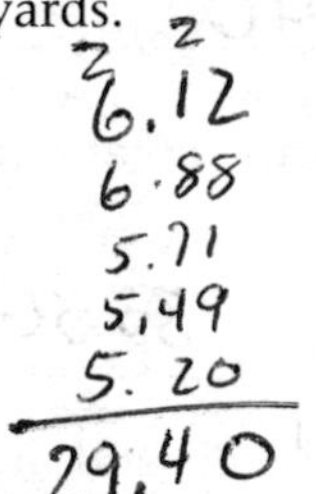

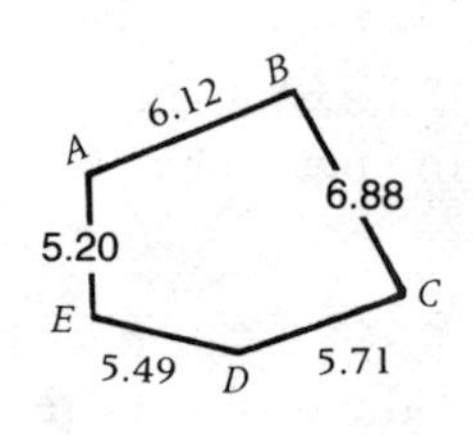

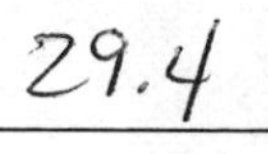

34. Given polygon *ABCD* with $AB = 4.71$ inches, $BC = 3.99$ inches, $CD = 3.60$ inches and $DA = 4.25$ inches. What kind of a polygon is this?

What is the perimeter of this polygon?

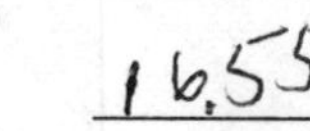

35. All line segments in the cross shown are 1 inch. What is the perimeter of the cross?

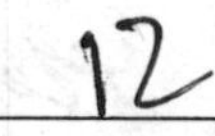

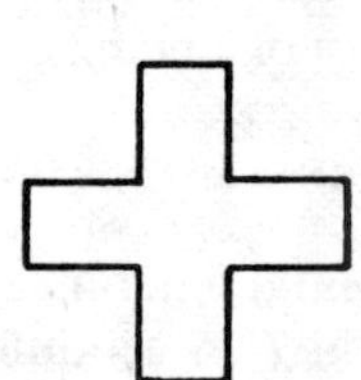

36. Find the approximate length of the equator if the radius of the earth is taken as 4,000 miles. Express your answer to the nearest whole number.

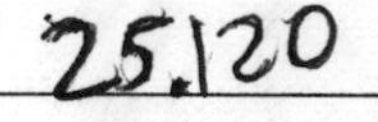

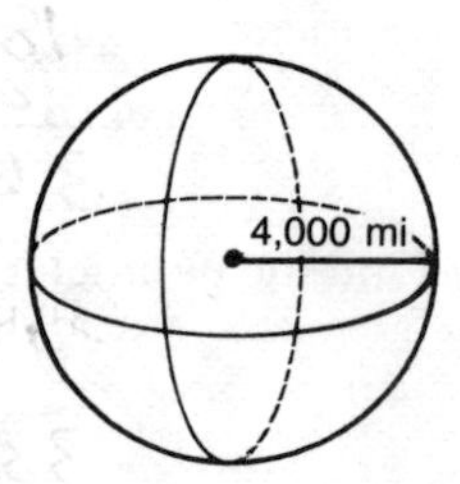

37. A bicycle has a wheel diameter of 26 inches. How far does it travel through 100 revolutions of the wheel?

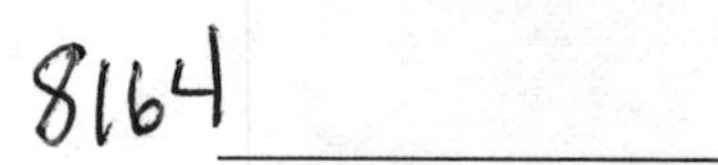

38. Find the circumference of a circle "inscribed" in a square with a 12-inch side.

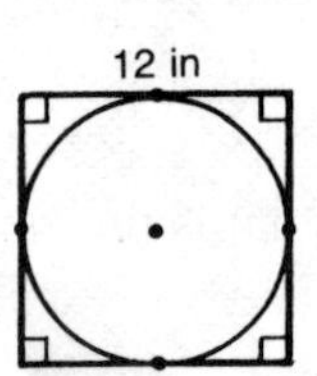

39. What is the diameter of a tree if a string passed around it is 62 inches long? Give your answer to the nearest hundredth. Assume that the cross section of the tree is circular and $\pi = 3.14$.

____

40. A semicircle is half of a circle. If the length of a semicircle is 2π inches, find the radius of the circle.

____

41. A semicircle is half of a circle. If the length of a semicircle is 16π ft, find the radius of the circle.

____

ANSWERS

1. 32 in **2.** 20 in **3.** 29 in **4.** 47.1 **5.** 46 in **6.** 36.6 in **7.** 40 ft **8.** 16.62 in **9.** $30\frac{7}{8}$ ft **10.** 44 ft **11.** 70 ft **12.** 40 in **13.** $d = 4$ in; $C = 12.56$ in **14.** $d = 5.6$ in; $C = 17.58$ in **15.** $r = 3$ in; $C = 18.84$ in **16.** $r = 3.22$ in; $C = 20.22$ in **17.** true **18.** true **19.** true **20. (a)** 24 in; **(b)** 32 in; **(c)** 40 in **21.** 33.36 in **22.** $11\frac{5}{12}$ meters **23.** 12 in; 24 in **24.** $2,250 **25.** 23.25 cm **26.** 100 in **27.** 48.8 ft **28.** 21.47 in **29.** triangle; $38\frac{73}{120}$ ft **30.** $16\frac{1}{2}$ ft **31.** $19\frac{1}{4}$ ft **32.** 23.49 in **33.** 29.4 yd **34.** quadrilateral; 16.55 in **35.** 12 in **36.** 25,120 miles **37.** 8,164 in **38.** 37.68 in **39.** 19.75 in **40.** $r = 2$ in **41.** $r = 16$ ft

5.2 AREA

The universally accepted unit of measure for area is a square. A surface enclosed by a square each of whose sides is one inch is called a *square inch* (see Figure 5.8). Other common units of area are the square foot, the square centimeter, and the square mile. Can you name another common unit of area?

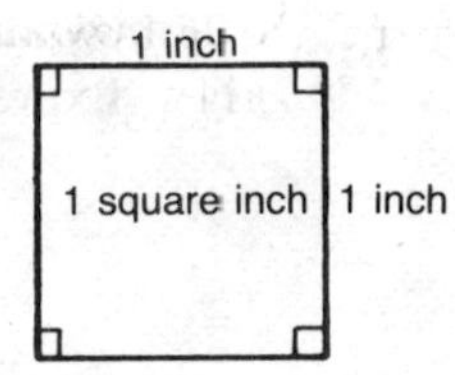

Figure 5.8

The area of a plane figure is the number of times the unit of area is contained in the figure.

Rectangle

Example 1 Find the area of a rectangle whose width is 3 inches and whose length is 5 inches. Draw a picture.

Solution

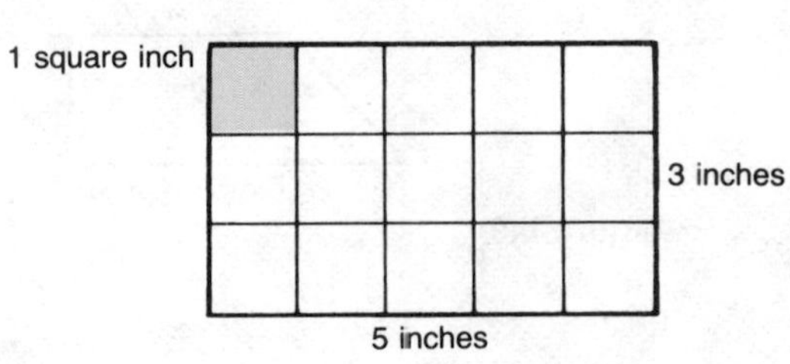

Our sketch shows that there are 3 rows of 5 square inches, or 15 square inches.

> To find the area, A, of any rectangle multiply the length, l, by the width, w, or $A = l \times w = l \cdot w = lw$.

Example 2 How much surface area is on the top of a desk 2.5 feet by 4 feet?

Solution

$$\begin{aligned} A &= l \cdot w \\ &= (2.5\text{ ft})(4\text{ ft}) \\ &= 10\text{ square feet} \end{aligned}$$

A square is a special rectangle in which all four sides are equal.

> To find the area, A, of any square multiply the length of a side, s, by itself, or $A = s \times s = s^2$.

Example 3 Find the area of a square 8 inches on a side.

Solution

$$\begin{aligned} A &= s \times s \\ &= (8\text{ in})(8\text{ in}) \\ &= 64\text{ square inches} \end{aligned}$$

Example 4 Can you place 25 squares, each 2 inches on a side, upon a rectangle whose dimensions are 8 inches by 10 inches? The squares are to be placed side by side.

Solution Since each square will have an area of 4 square inches, 25 such squares will have cover an area of 100 square inches. However, the rectangle will have an area of only 80 square inches. Thus you cannot complete the placement.

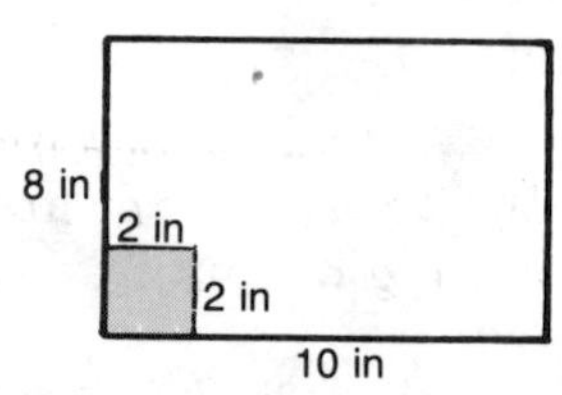

Parallelogram By cutting off one corner of a parallelogram and attaching it to its opposite side, as pictured in Figure 5.9, the parallelogram can be transformed into a rectangle. Thus the area of a parallelogram is equal to the area of a rectangle.

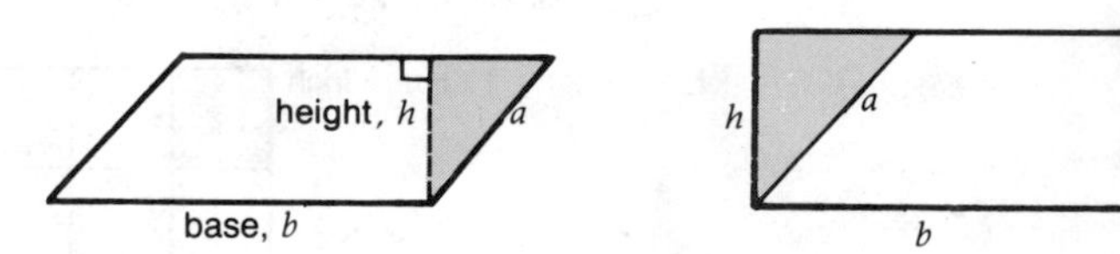

Figure 5.9

> The area, A, of a parallelogram can be found by multiplying the base, b, by the height, h, or $A = b \times h = b \cdot h = bh$.

Example 5 Find the area of the parallelogram whose base is 12 inches and whose height is 9 inches.

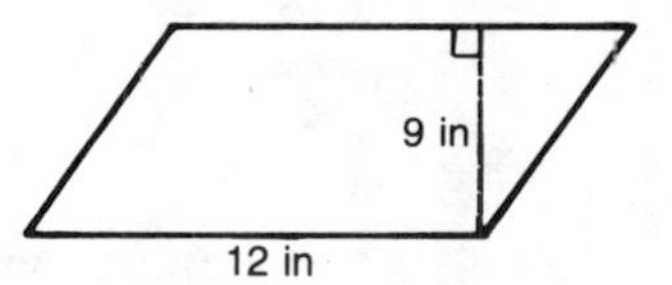

Solution

$$A = b \cdot h$$
$$= 12 \text{ in} \times 9 \text{ in}$$
$$= 108 \text{ square inches}$$

Triangle The area of a triangle may be considered half of the area of a rectangle. The area of each rectangle in Figure 5.10 is $b \times h$. Can you see that the area of the shaded triangle is $\frac{1}{2}$ the area of the rectangle or the area of the triangle is $\frac{1}{2} \times b \times h$?.

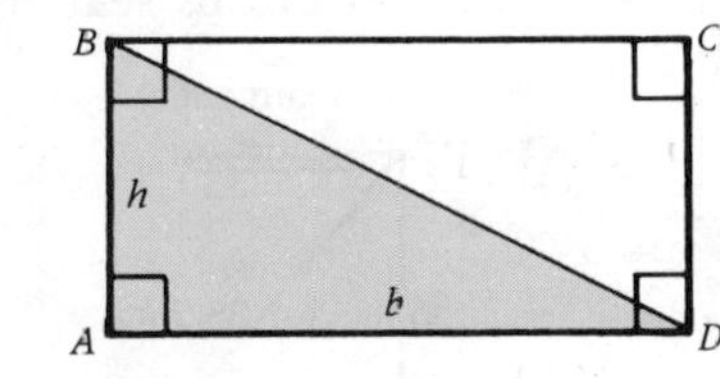

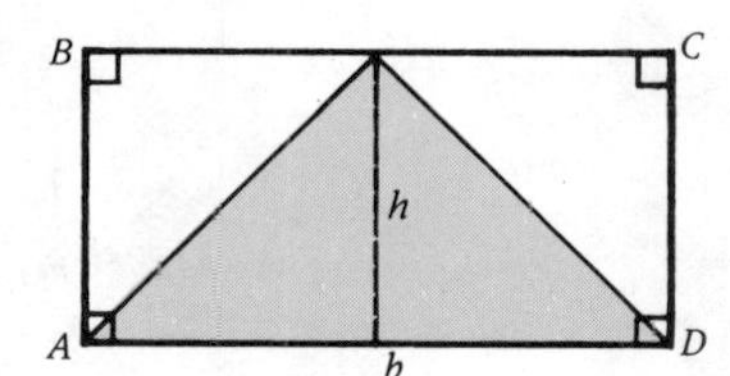

Figure 5.10

The area, A, of a triangle can be found by multiplying $\frac{1}{2}$ times its base, b, times its height, h, or $A = \frac{1}{2} \times b \times h = \frac{1}{2} \cdot b \cdot h = \frac{1}{2}bh$.

The **base,** b, of a triangle may be any side of the triangle that is convenient. Usually we set up the triangle so that the base side is "horizontal" as in Figure 5.10. The **height,** h, is the vertical distance from the base to the top of the triangle.

Example 6 Find the area of a triangle with a base of 2.72 decimeters and a height of 1.06 decimeters.

Solution

$$A = \frac{1}{2} \times b \times h$$
$$= (.5)(2.72 \text{ decimeters})(1.06 \text{ decimeters})$$
$$= 1.44 \text{ square decimeters}$$

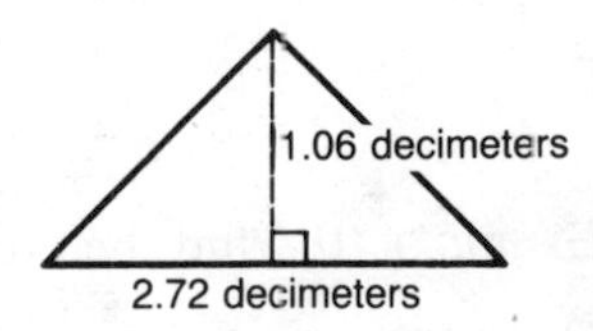

In order to find the height of some triangles it is necessary to extend the base. In such a case the height of the triangle meets the base extended. This is illustrated in our next example.

Example 7 Find the area of triangle ABC pictured as shown.

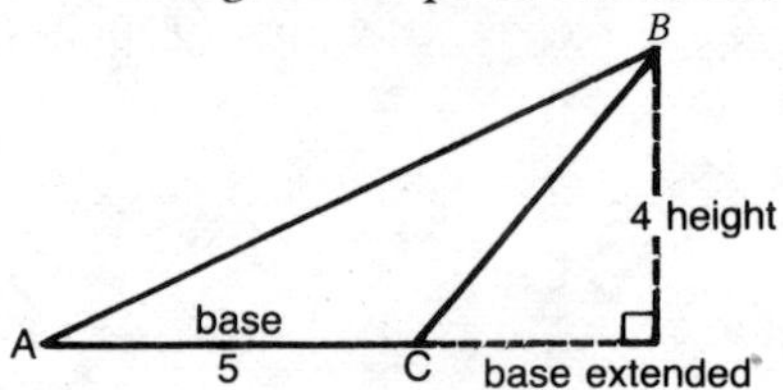

Solution The base is 5 and the height is 4.

Thus $A = \frac{1}{2} \times b \times h$

$= \frac{1}{2} \times 5 \times 4$

$= 10$ square units

Example 8 A fishing boat is lost at sea. A Coast Guard cutter is given a triangular area, as shown, in which to conduct its search.
(a) What is the area of this search?
(b) What is the perimeter of this search area?

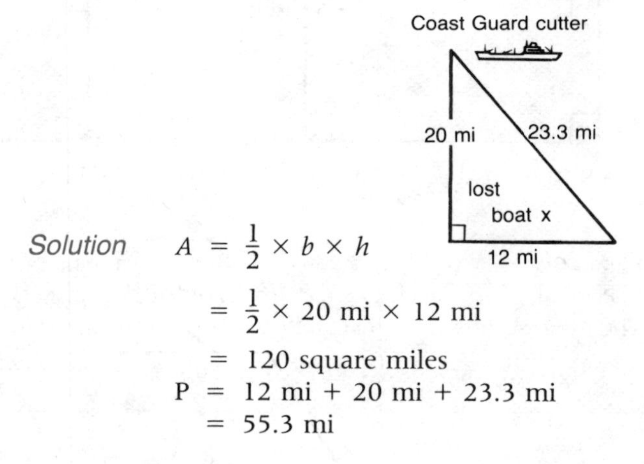

Solution $A = \frac{1}{2} \times b \times h$

$= \frac{1}{2} \times 20 \text{ mi} \times 12 \text{ mi}$

$= 120$ square miles

$P = 12 \text{ mi} + 20 \text{ mi} + 23.3 \text{ mi}$

$= 55.3 \text{ mi}$

Circle Because the formula for the area of a circle is based upon a branch of mathematics called calculus, the derivation of this formula is beyond the scope of this text.

> The formula for the area, A, of any circle is $\pi \times$ radius $\times$ radius or $A = \pi \cdot r^2 = \pi r^2$.

Example 9 Find the area of a circle with a radius of 3 meters.

Solution $A = \pi \cdot r^2 = (3.14)(3 \text{ meters})^2$

$= (3.14)(9 \text{ square meters})$

$= 28.26$ square meters

Example 10 Find the area of a circle with a diameter of 10 inches.

Solution $r = \frac{d}{2} = \frac{10 \text{ inches}}{2} = 5 \text{ inches}$

$A = \pi r^2 = (3.14)(5 \text{ inches})^2$

$= (3.14)(25 \text{ square inches})$

$= 78.5$ square inches

Example 11 A circular house is the most efficient house to build, in the sense that it encloses the most area with the least perimeter. The shape of the foundations of two houses are given below. One is square with a base of 33.67 ft and the other is circular with a radius of 19 ft.

(a) Find the area of each house rounded to the nearest unit.
(b) Find the perimeter of each house.
(c) Which house has the greatest area?
(d) Which house has the smaller perimeter? How much smaller is it?

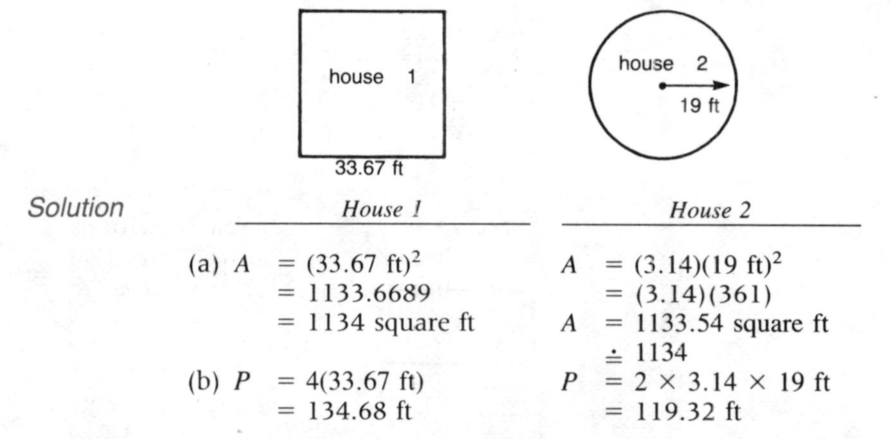

Solution

	House 1	*House 2*
(a)	$A = (33.67 \text{ ft})^2$	$A = (3.14)(19 \text{ ft})^2$
	$= 1133.6689$	$= (3.14)(361)$
	$= 1134$ square ft	$A = 1133.54$ square ft
		$\doteq 1134$
(b)	$P = 4(33.67 \text{ ft})$	$P = 2 \times 3.14 \times 19$ ft
	$= 134.68$ ft	$= 119.32$ ft

(c) Both houses have the same area (to the nearest square foot).
(d) The circular house has the smaller perimeter by 134.68 ft − 119.32 ft = 15.36 ft.

Trapezoid A **trapezoid** is a quadrilateral with at least one pair of parallel sides. The parallel sides are called the **bases** of the trapezoid. See Figure 5.11.

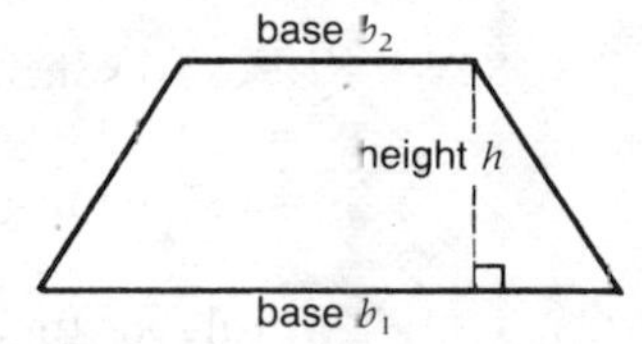

Figure 5.11

By taking two trapezoids of the same size and shape, turning one of them over and attaching it to the other, as shown in Figure 5.12, we form a parallelogram whose base is $b_1 + b_2$ and whose height is h. The area of our trapezoid is then one-half the area of the parallelogram.

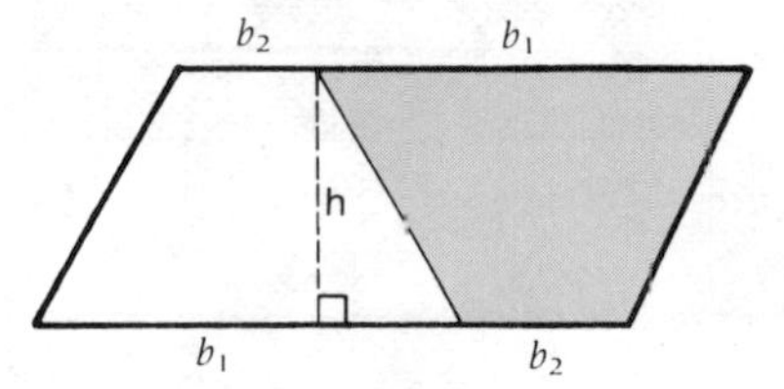

Figure 5.12

> The area, A, of a trapezoid can be found by multiplying $\frac{1}{2}$ times the sum of its two bases, $b_1 + b_2$, times its height, h, or $A = \frac{1}{2} \times (b_1 + b_2) \times h = \frac{1}{2} \cdot (b_1 + b_2) \cdot h = \frac{1}{2}(b_1 + b_2)h$.

Example 12 Find the area of the trapezoid shown.

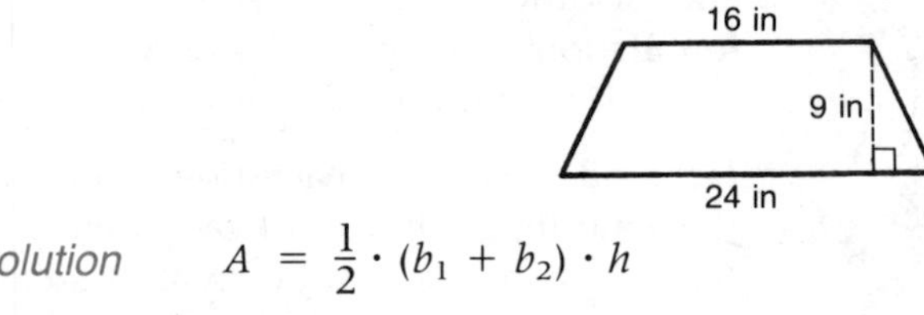

Solution

$$A = \frac{1}{2} \cdot (b_1 + b_2) \cdot h$$
$$= \frac{1}{2}(24 \text{ in} + 16 \text{ in})(9 \text{ in})$$
$$= \frac{1}{2}(40 \text{ in})9 \text{ in}$$
$$= 180 \text{ square inches}$$

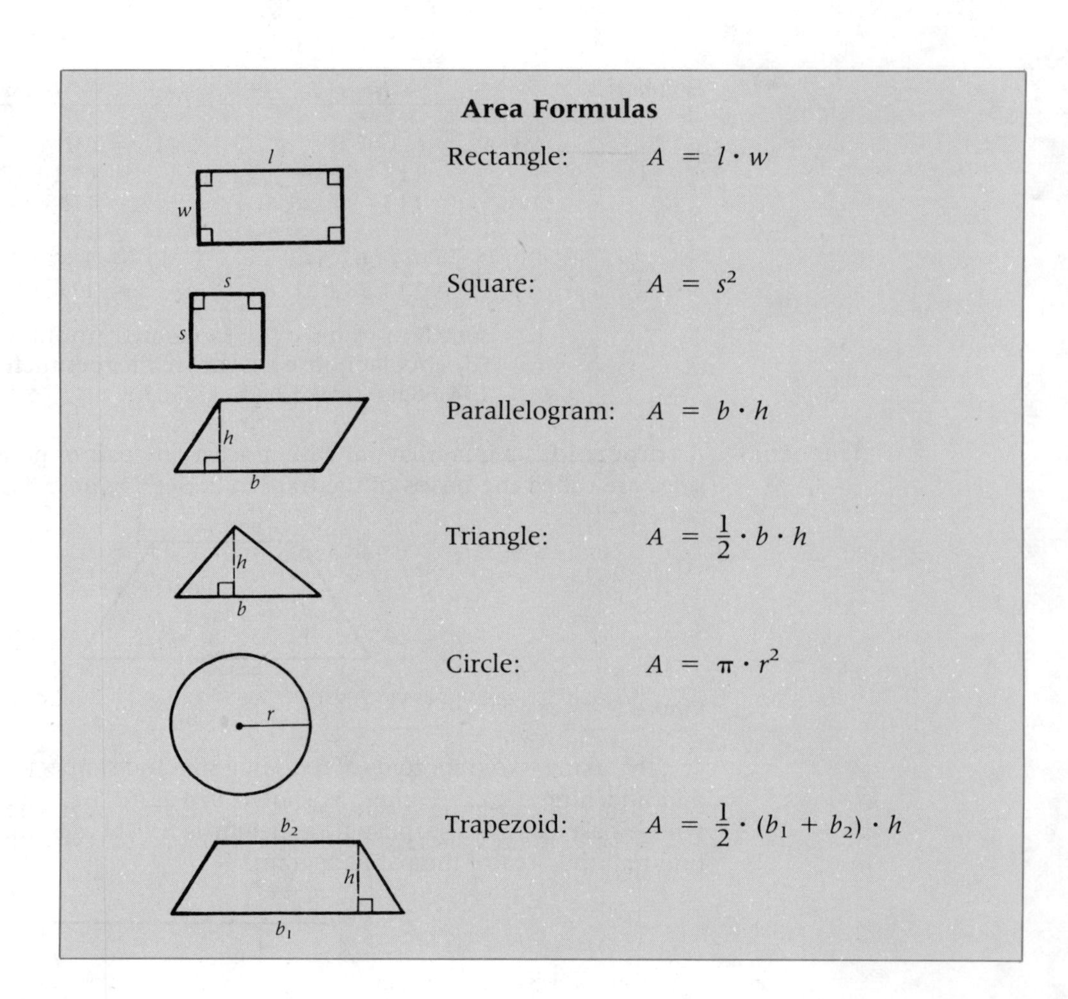

Area Formulas

Rectangle: $A = l \cdot w$

Square: $A = s^2$

Parallelogram: $A = b \cdot h$

Triangle: $A = \frac{1}{2} \cdot b \cdot h$

Circle: $A = \pi \cdot r^2$

Trapezoid: $A = \frac{1}{2} \cdot (b_1 + b_2) \cdot h$

5.2 EXERCISES

Round off all decimal answers in this section to the nearest hundredth. In problems 1–12 find the area of the given figure.

1. 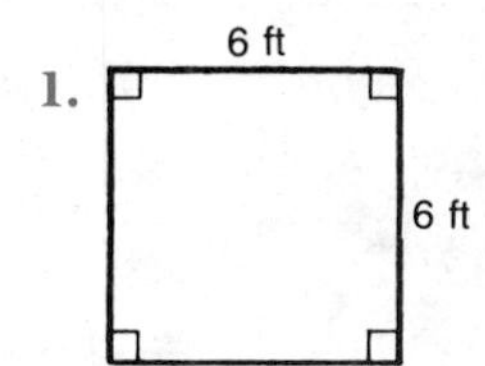

2. 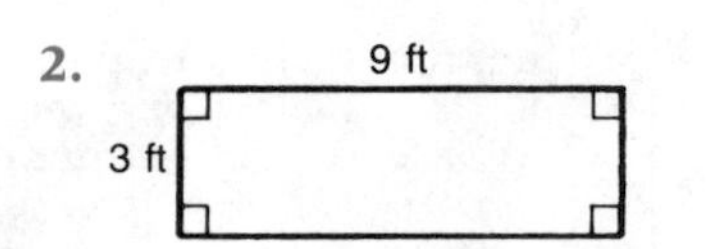

3. 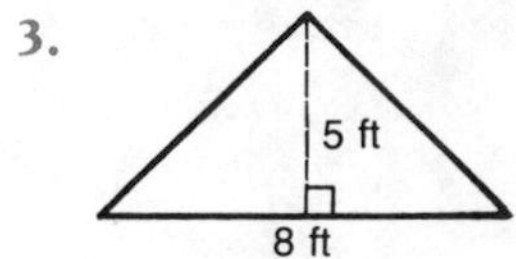

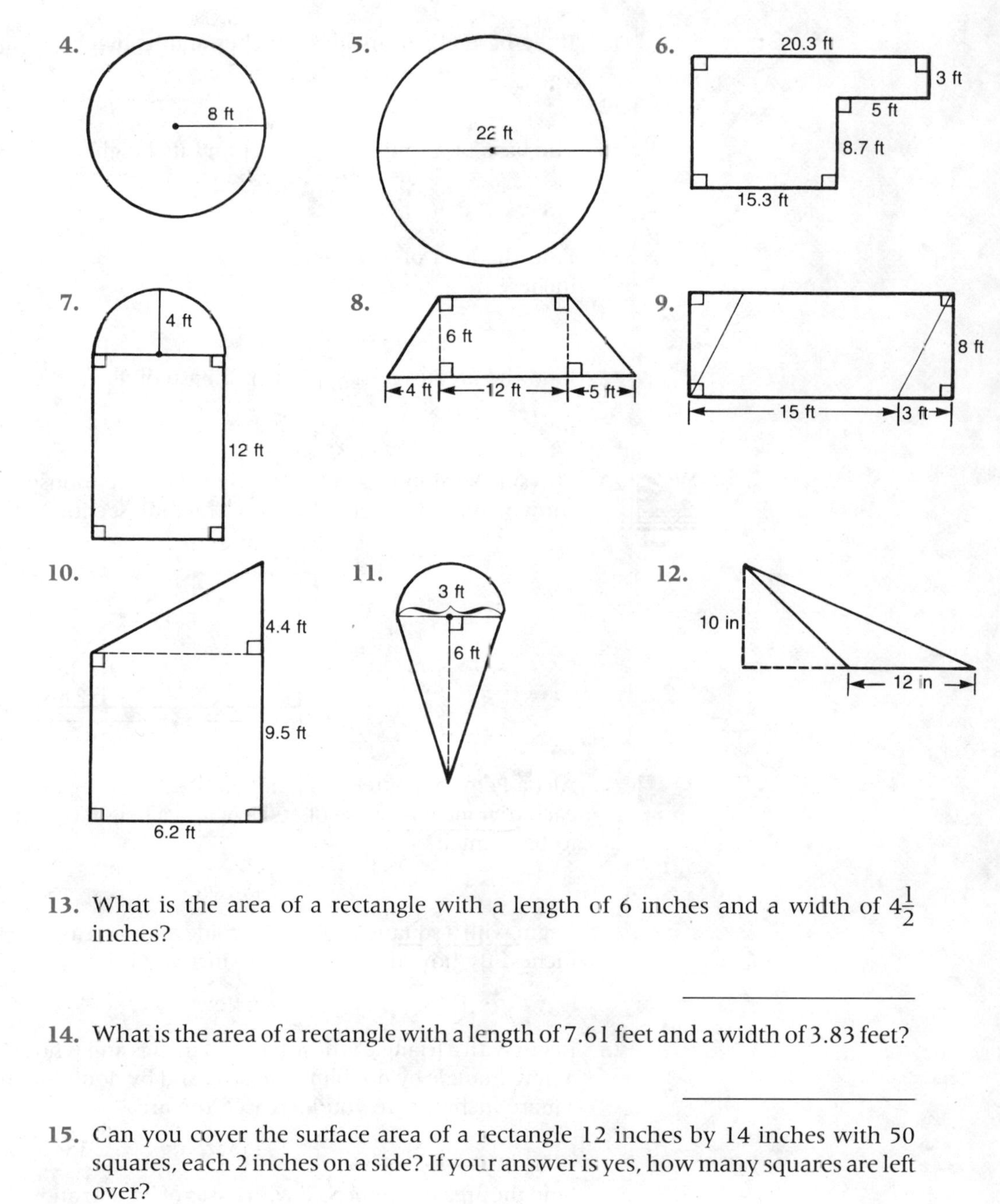

13. What is the area of a rectangle with a length of 6 inches and a width of $4\frac{1}{2}$ inches?

____

14. What is the area of a rectangle with a length of 7.61 feet and a width of 3.83 feet?

____

15. Can you cover the surface area of a rectangle 12 inches by 14 inches with 50 squares, each 2 inches on a side? If your answer is yes, how many squares are left over?

____

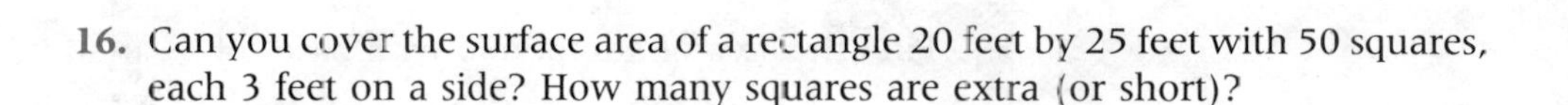

16. Can you cover the surface area of a rectangle 20 feet by 25 feet with 50 squares, each 3 feet on a side? How many squares are extra (or short)?

____

17. A football field is 120 yards long and 55 yards wide. How many square yards of polyturf are required to cover the field?

____

18. A square has a side of length 10 feet and a circle has a diameter of 10 feet. Which figure has the largest area? How much larger is it?

____

19. If the base of a triangle is 8 inches and its height is 9 inches, what is its area?

20. The base of a triangle is 50 feet and its height is 40 feet. What is its area?

21. Find the area of a triangle with a base of 13.02 inches and a height of 6.88 inches.

22. Find the area of a triangle with a base of $4\frac{1}{4}$ feet and a height of $3\frac{2}{5}$ feet.

23. The side section of a south-facing solar greenhouse attached to a house is as shown. What is the total area of this side section?

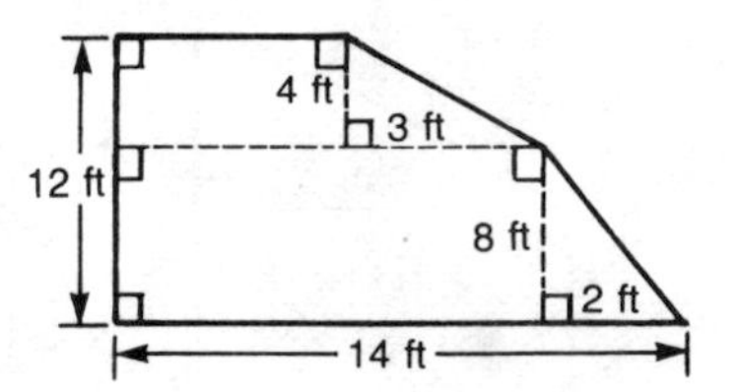

24. Alcoa Paint has agreed to paint a roof consisting of 4 triangles of equal area. If each triangle has a base of 100 feet and a height of 60 feet, what is the total area to be painted?

25. Begin with a square 5 inches on a side and increase the length of each side to 10 inches. By how much have you increased the area?

26. Begin with a triangle whose base is 4 inches and whose height is 2 inches. Create a new triangle by doubling the base and by doubling the height. By how many square inches have you increased the area?

27. Find the area of a trapezoid with bases of 24.2 in and 17.8 in and height of 10 in.

28. Find the area of a trapezoid with bases of $8\frac{1}{2}$ ft and $6\frac{1}{4}$ ft and height of 4 ft.

29. Find the area of a circle whose radius is 4 inches.

30. Find the area of a circle whose diameter is 8 inches.

31. Find the area of a circle with a radius of 2.84 inches.

32. Find the area of a circle with a radius of $1\frac{1}{2}$ feet. Use $\pi = \frac{22}{7}$.

33. A machinist has a washer with a large diameter of 2.4 inches and the hole of the washer has a diameter of 0.8 inch. What is the surface area of the washer?

____

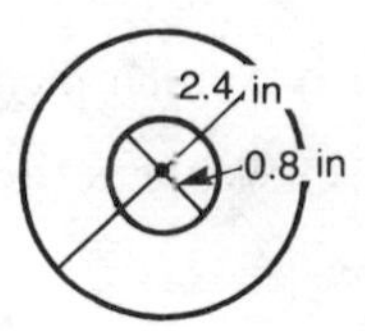

34. A circular rod is to be milled from a beam whose cross section is a square 2 inches on a side. See the figure of the cross section to the right below. What is the area of the material that is wasted in doing this procedure?

____

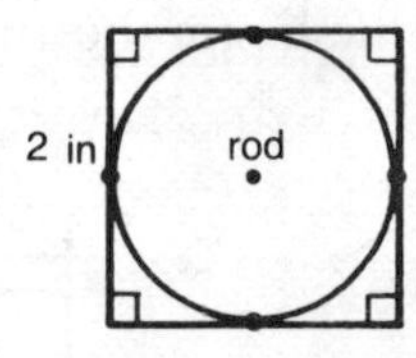

35. If you double the radius of a circle whose radius to start is 4 inches, how much greater is the area of the new circle?

____

36. A floor is 20 feet by 80 feet. How many square floor tiles 2 feet on a side are needed to cover this floor?

____

37. A semicircle is half of a circle. What is the area of a semicircle with a radius of 8 yards?

____

38. A semicircle is half of a circle. What is the area of a semicircle with a diameter of 12 feet?

____

39. A dentist has just signed a long-term lease on one floor of a building with space for office, waiting room, and so on, as pictured. Find the total area that the dentist is renting.

____

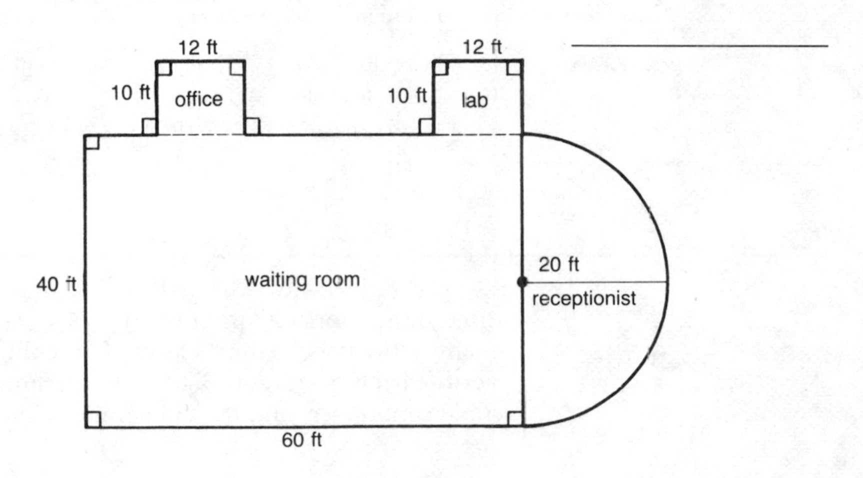

40. Two plots of land, one a square and the other a circle are being surveyed. The side of the square is 63.24 ft and the radius of the circle is 35.69 ft.

(a) Find the area of each piece of land. __________

(b) Which piece of land has the greatest area? __________

(c) What is the difference in area rounded to the nearest square foot?

____

(d) Find the perimeter of each piece of land. __________

(e) Which piece of land has the least perimeter? __________

41. In 1978 it cost $35 a square foot to build a house. How much did it cost to build the house shown?

____

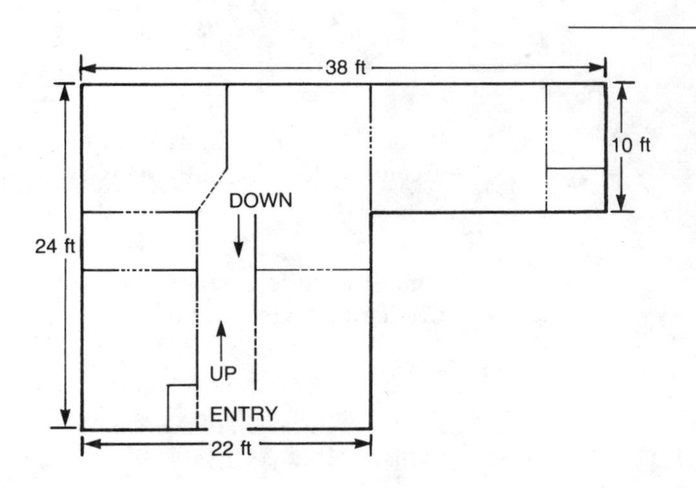

ANSWERS

1. 36 sq ft **2.** 27 sq ft **3.** 20 sq ft **4.** 200.96 sq ft **5.** 379.94 sq ft **6.** 194.01 sq ft **7.** 121.12 sq ft **8.** 99 sq ft **9.** 144 sq ft **10.** 72.54 sq ft **11.** 12.53 sq ft **12.** 60 sq in **13.** 27 sq in **14.** 29.15 sq ft **15.** yes; 8 squares **16.** no; $5\frac{5}{9}$ squares short **17.** 6,600 sq yd **18.** square; 21.5 sq ft larger **19.** 36 sq in **20.** 1,000 sq ft **21.** 44.79 sq in **22.** $7\frac{9}{40}$ sq ft **23.** 146 sq ft **24.** 12,000 sq ft **25.** 75 sq in **26.** 12 sq in **27.** 210 sq in **28.** $29\frac{1}{2}$ sq ft **29.** 50.24 sq in **30.** 50.24 sq in **31.** 25.33 sq in **32.** $7\frac{1}{14}$ sq ft **33.** 4.02 sq in **34.** 0.86 sq in. **35.** 150.72 sq in **36.** 400 tiles **37.** 100.48 sq yds **38.** 56.52 sq ft **39.** 3,268 sq ft **40. (a)** square = 3,999.30 sq ft; circle = 3,999.66 sq ft **(b)** circle **(c)** 0 **(d)** square = 252.96; circle = 224.13 **(e)** circle **41.** $24,080

5.3 VOLUME

The universally accepted unit of measure for volume is a cube. A cube is a three-dimensional figure consisting of six sides that are identical squares. A die is an example of a cube. If the six faces of the cube are square inches, then the cube is called a **cubic inch.** See Figure 5.13. Other common units of volume are the cubic foot, the cubic centimeter, and the cubic mile. Can you think of another unit of volume?

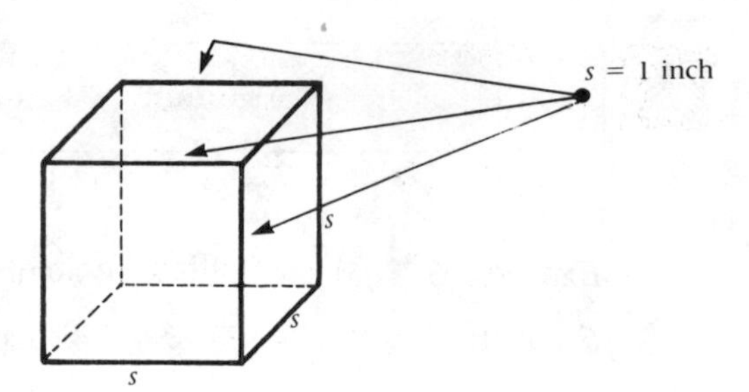

Figure 5.13

Rectangular Solid

A **rectangular solid** is a three-dimensional figure with all right angles and shape, similar to an ordinary carton. See Figure 5.14.

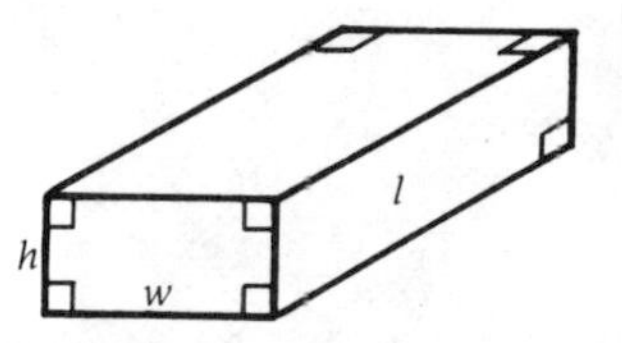

Figure 5.14

> The volume, V, of a rectangular solid is its length, l, times its width, w, times its height, h, or $V = l \times w \times h = l \cdot w \cdot h = lwh$.

Example 1 Find the volume, V, of a rectangular solid whose length is 4 inches, width is 3 inches, and height is 2 inches.

Solution 1 The area of the rectangular base is (4 in)(3 in) = 12 square inches. Upon this rectangular base you can construct 2 layers of 12 cubic inches.

Volume = V = (2)(12 cubic inches)
= 24 cubic inches

Solution 2 $V = l \cdot w \cdot h$
= (4 inches)(3 inches)(2 inches)
= 24 cubic inches

Example 2 Halifax Electronics is shipping rectangular crates 2 feet by 3 feet by 5 feet in a truck 20 feet by 10 feet by 12 feet. How many of these crates can this truck hold for shipment?

Solution
step 1: First we will find the volume of each crate.
V = (2 ft)(3 ft)(5 ft) = 30 cubic feet
step 2: Find the volume of the truck.
V = (20 ft)(10 ft)(12 ft) = 2,400 cubic feet
step 3: 2,400 ÷ 30 = 80 crates can be packed in the truck.

A *cube* is a special rectangular solid.

> The volume of a cube, V, is $V = s \times s \times s = s^3$

Example 3 Find the volume of a cube 4 inches on a side.

Solution $V = s \times s \times s = (4 \text{ in})(4 \text{ in})(4 \text{ in})$
$= 64$ cubic inches

Cylinder A **cylinder** is a three-dimensional figure similar in shape to a soup can. See Figure 5.15.

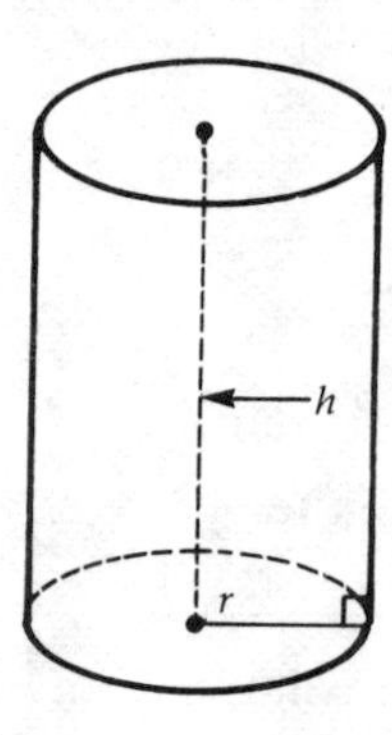

Figure 5.15

> The volume, V, of a cylinder is $\pi \times r \times r \times h$, or $V = \pi \cdot r^2 \cdot h = \pi r^2 h$.

Example 4 Find the volume of a cylinder with a radius, r, of 5 inches and a height, h, of 8 inches.

Solution

$$\begin{aligned} V &= \pi \times r^2 \times h \\ &= (3.14)(5 \text{ in})^2(8 \text{ in}) \\ &= (3.14)(25 \text{ sq in})(8\text{in}) \\ &= 628 \text{ cubic inches} \end{aligned}$$

(Remember we must first square 5 in)

Example 5 Denver Gas is selling a cylindrical water heater with a radius of 10 inches and a height of 72 inches.

(a) What is the volume of this heater?
(b) If 231 cubic inches equals 1 gallon, how many gallons of water, to the nearest gallon, does this heater hold?

Solution $V = \pi \times r^2 \times h$

(a) $$\begin{aligned} V &= (3.14)(10 \text{ in})^2(72 \text{ in}) \\ &= (3.14)(100 \text{ sq in})(72 \text{ in}) \\ &= 22{,}608 \text{ cubic inches} \end{aligned}$$

(b) $$\begin{aligned} \text{Number of Gallons} &= 22{,}608 \div 231 \\ &= 97.8 \text{ gallons} \doteq 98 \text{ gallons} \end{aligned}$$

Cone A **cone** is a three-dimensional figure similar in shape to an ice-cream cone. See Figure 5.16.

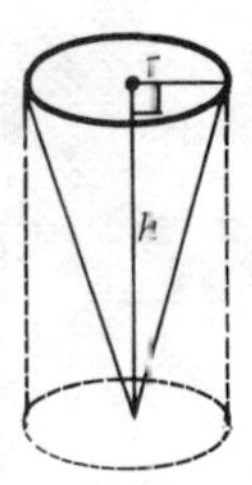

Figure 5.16

> The volume, V, of a cone is $\frac{1}{3}$ times the volume of a cylinder, or $V = \frac{1}{3} \times \pi \times r^2 \times h$ $= \frac{1}{3} \cdot r^2 \cdot h = \frac{1}{3}r^2h$.

Example 6 Determine the volume of a cone with a radius, r, of 5 feet and a height, h, of 8 feet.

Solution

$$V = \frac{1}{3} \times \pi \times r^2 \times \text{h}$$
$$= \left(\frac{1}{3}\right)(3.14)(5 \text{ ft})^2(8 \text{ ft})$$
$$= \frac{1}{3}(3.14)(25 \text{ sq ft})(8 \text{ ft})$$
$$= \frac{628 \text{ cubic feet}}{3}$$
$$= 209.33 \text{ cubic feet}$$

Example 7 A landscaper has a load of loam delivered to his next job. The truck driver dumped the loam in a conical pile 6 feet high and 15 feet in diameter. What is the volume of the loam?

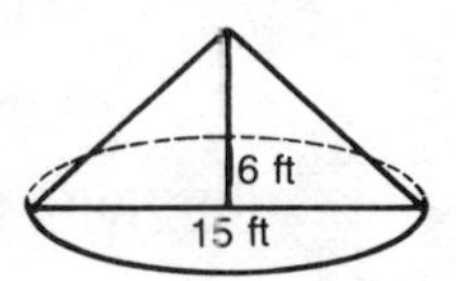

Solution

$$V = \frac{1}{3} \times \pi \times r^2 \times h$$
$$= \left(\frac{1}{3}\right) (3.14)(7.5 \text{ ft})^2 (6 \text{ ft})$$
$$= \left(\frac{1}{3}\right)(3.4)(56.25 \text{ sq ft})(6 \text{ ft})$$
$$= 353.25 \text{ cubic feet}$$

Sphere A **sphere** is a three-dimensional figure similar in shape to a baseball. See Figure 5.17.

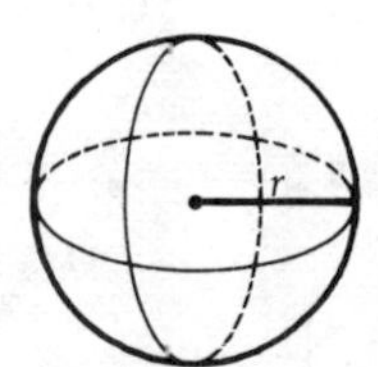

Figure 5.17

> The volume, V, of a sphere is $\frac{4}{3} \times \pi \times$ radius $\times$ radius $\times$ radius, or $V = \frac{4}{3} \times \pi \times r^3 = \frac{4}{3} \cdot \pi \cdot r^3 = \frac{4}{3}\pi r^3$.

Example 8 Determine the volume, V, of a sphere whose radius, r, is 2 inches. Let $\pi = \frac{22}{7}$ for convenience. Round to nearest hundredth.

Solution

$$V = \frac{4}{3} \times \pi \times r^3$$
$$= \frac{4}{3} \times \frac{22}{7} \times (2 \text{ in})^3$$
$$= \frac{704}{21} \text{ cubic inches}$$
$$= 33.52 \text{ cubic inches}$$

Example 9 Litton Chemical is building spherical storage tanks because they hold maximum volume for a given amount of material.

(a) If the radius of each tank is 30 ft, what is the volume of each tank?

____

(b) If one cubic foot equals 7.5 gallons, how many gallons will each tank hold?

____

Solution

(a) $V = \frac{4}{3} \times \pi \times r^3$
$$= \left(\frac{4}{3}\right)(3.14)(30 \text{ ft})^3$$
$$= \left(\frac{4}{3}\right)(3.14)(27{,}000 \text{ cu ft})$$
$$= 113{,}040 \text{ cubic feet}$$

(b) Number of Gallons $= 113{,}040 \div 7.5$
$= 15{,}072$ gallons

Example 10 A tank for storing grain is shaped as follows. It is cylindrical with a cone on its bottom. The cylinder has a radius of 6 ft and a height of 15 ft. The cone has a height of 6 ft. What is the volume of the tank?

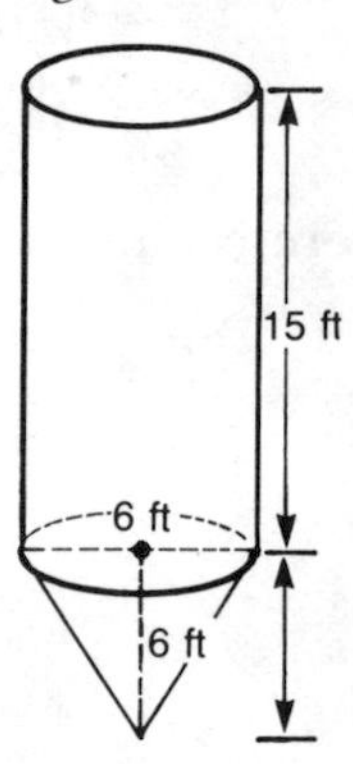

Solution

Volume of Tank = Volume of Cylinder + Volume of Cone

Volume of Cylinder: $V = \pi \times r^2 \times h$
$= (3.14)(6 \text{ ft})^2(15 \text{ ft})$
$= (3.14)(36 \text{ sq ft})(15 \text{ ft})$
$= 1{,}695.6$ cubic feet

Volume of Cone: $V = \frac{1}{3} \times \pi \times r^2 \times h$
$= \left(\frac{1}{3}\right)(3.14)(6 \text{ ft})^2\ (6 \text{ ft})$
$= \left(\frac{1}{3}\right)(3.14)(36 \text{ sq ft})(6 \text{ ft})$
$= 226.08$ cubic feet

Volume of Tank: $V = 1{,}695.6$ cubic feet $+ 226.08$ cubic feet
$= 1{,}921.68$ cubic feet

Volume Formulas

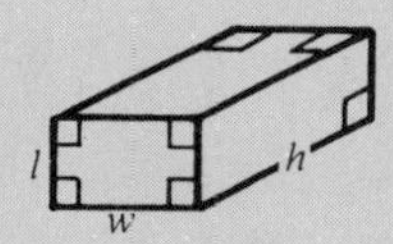

Rectangular solid: $V = l \cdot w \cdot h$

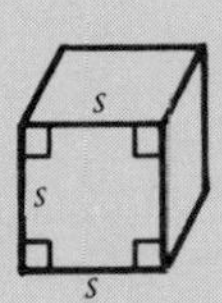

Cube: $V = s^3$

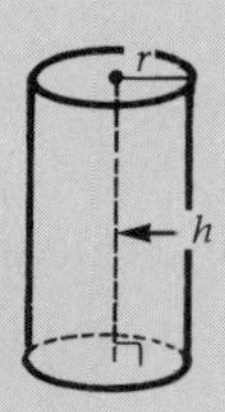

Cylinder: $V = \pi \cdot r^2 \cdot h$

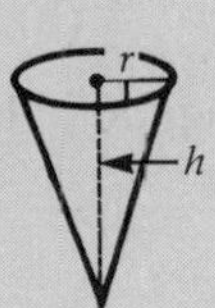

Cone: $V = \frac{1}{3} \cdot \pi \cdot r^2 \cdot h$

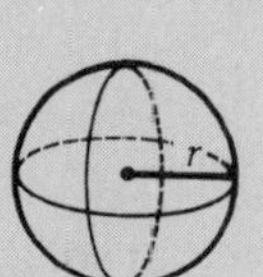

Sphere: $V = \frac{4}{3} \cdot \pi \cdot r^3$

5.3 EXERCISES

Round off all decimal answers in this section to the nearest hundredth. In problems 1–8 find the volume of the given figure.

1.

8 ft
2 ft
4 ft

2.

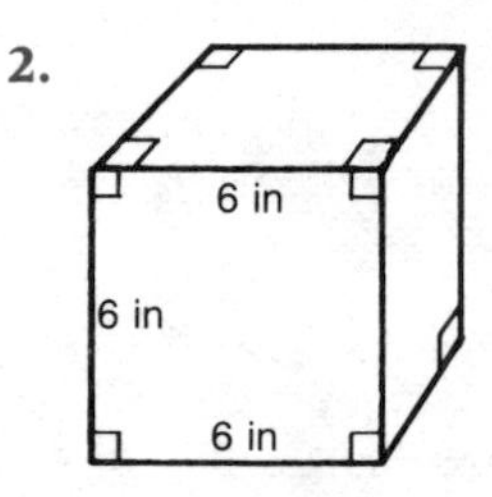

3.

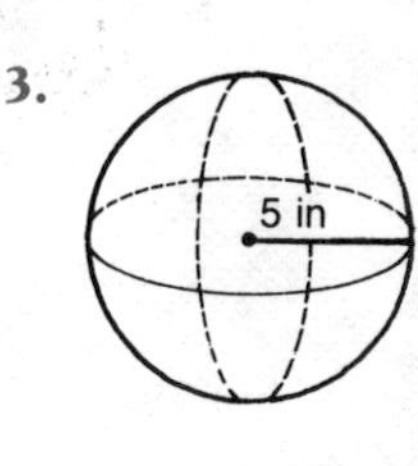

4.

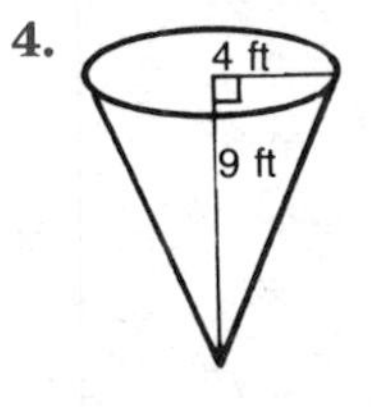

5.

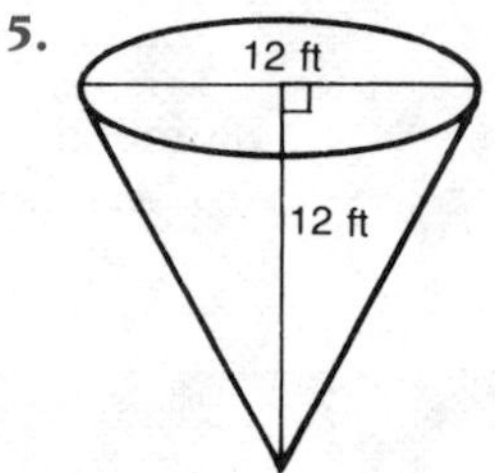

6.

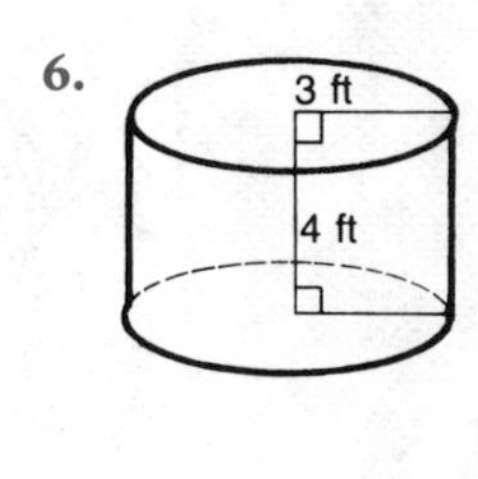

7.

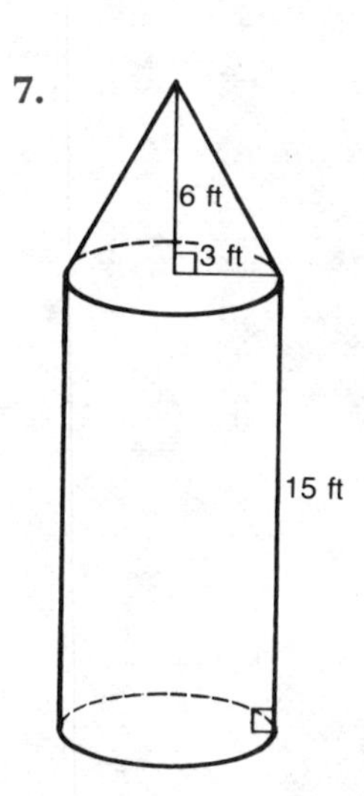

8.

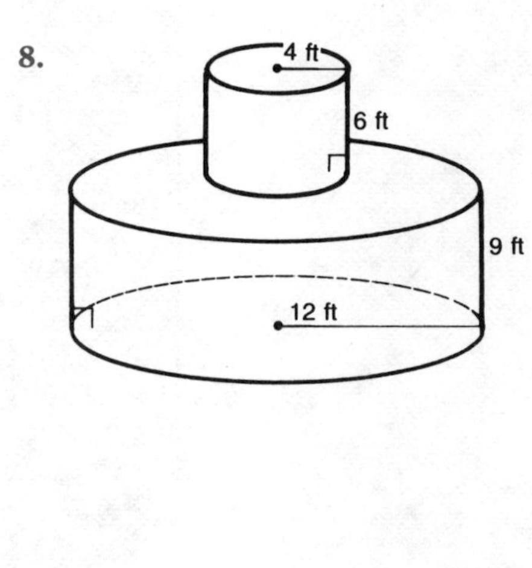

9. Find the volume of a cube whose side measure is:

(a) 2 ________________; **(b)** 3 ________________

10. Find the volume of a cube whose side measure is:

(a) 3.1 ________________; **(b)** 1.75 ________________

11. What is the volume of a rectangular solid whose length is 3 inches, width is 5 inches, and height is 10 inches?

____

12. What is the volume of a rectangular solid whose length is 8 ft, width is 11 ft, and height is 5 ft?

____

13. What is the volume of a rectangular solid with length of 6.2 inches, width of 4.0 inches, and height of 5.16 inches?

____

14. What is the volume of a rectangular solid with a length of $3\frac{3}{4}$ meters, width of $3\frac{3}{4}$ meters, and height of 5 meters?

15. A "rule of thumb" for heating a room all winter with a wood-burning stove is: "It takes as much wood as the room will hold." If a cord of wood equals a rectangular solid 8 ft by 4 ft by 4 ft, how many cords of wood would it take to heat a living room that is 8 ft by 20 ft by 14 ft?

16. A box is 2 ft by 1 ft by 1.5 ft. One cubic foot of sand weighs 145 pounds. How many cubic feet of sand will it take to fill this box?

What is the weight of the sand?

17. A rectangular water storage tank is 10 ft long, 4 ft wide, and 3 ft high. What is its volume and how many gallons will it hold (7.5 gallons per cubic foot)?

18. Find the volume of a cylinder whose radius, r, equals 6 inches and whose height, h, equals 12 inches.

19. Determine the volume of a cylinder with a radius of 5.26 feet and a height of 8.82 feet.

20. Find the volume of a cone with the same circular base and the same height as the cylinder of problem 18.

21. Find the volume of a cone with the same circular base and the same height as the cylinder of problem 19.

22. Find the volume of a cylinder with a radius of $5\frac{3}{5}$ centimeters and a height of 8 centimeters. Use $\pi = \frac{22}{7}$.

23. Find the volume of a cone with a circular base and a radius of $4\frac{3}{4}$ inches and a height of 6 inches. Use $\pi = \frac{22}{7}$.

24. A circular swimming pool is 24 ft in diameter and 6 ft deep. What is its volume and how many gallons of water does it take to fill it (7.5 gallons per cubic foot)?

25. A can of soup is cylindrical with a radius of 1 inch and a height of 3.5 inches. What is the volume of the can?

26. How many cubic feet are in a cubic yard?

27. A sphere has a radius of 2 inches, what is its volume?

28. If a sphere has a radius of 10 ft, what is its volume?

29. If a sphere has a radius of $2\frac{1}{4}$ inches, what is its volume? $\left(\text{Use } \pi = \frac{22}{7}\right)$.

30. When a circle with a diameter of 4 inches is rotated about its diameter, what is the volume generated by this rotation?

31. If the volume of a cone is 49.7 cubic inches, what is the volume of a cylinder with the same circular base?

32. If the volume of a cone is 38.6 cubic feet, what is the volume of a cylinder with the same circular base?

33. A conical filter of a coffee maker has a diameter of 4 inches and a height of 5 inches. What is its volume?

34. You have a circle whose area is 6.5 square inches and you raise this circle vertically to a height of 10 inches. What is the volume generated through this vertical movement?

35. A basketball is 9.5 inches in diameter. How many cubic inches of air does it hold?

36. A sphere is inside and touching the sides of a cube whose edges are all 4 inches. What is the difference between the volume of the cube and the volume of the sphere?

37. A cylinder and a cone have the same circular base. What will the height of the cone have to be compared to the height of the cylinder if the volume of the cone and the volume of the cylinder are equal?

ANSWERS

1. 64 cu ft **2.** 216 cu in **3.** 523.33 cu in **4.** 150.72 cu ft **5.** 452.16 cu ft **6.** 113.04 cu ft **7.** 480.42 cu ft **8.** 4,370.88 cu ft **9. (a)** 8 **(b)** 27 **10. (a)** 29.79 **(b)** 5.36 **11.** 150 cu in **12.** 440 cu ft **13.** 127.97 cu in **14.** $70\frac{5}{16}$ cu m **15.** 17.5 cords **16.** 3 cu ft; 435 pounds **17.** 120 cu ft; 900 gallons **18.** 1,356.48 cu in **19.** 766.25 cu ft **20.** 452.16 cu in **21.** 255.42 cu ft **22.** $788\frac{12}{25}$ cu cm **23.** $141\frac{23}{28}$ cu in **24.** 2,712.96 cu ft; 20,347.2 gallons **25.** 10.99 cu in **26.** 27 **27.** 33.49 cu in **28.** 4,186.67 cu ft **29.** $47\frac{41}{56}$ cu in **30.** 33.49 cu in **31.** 149.10 cu in **32.** 115.80 cu ft **33.** 20.93 cu ft **34.** 65 cu in **35.** 448.69 cu in **36.** 30.51 cu in **37.** 3:1

FOCUS ON PROBLEM SOLVING

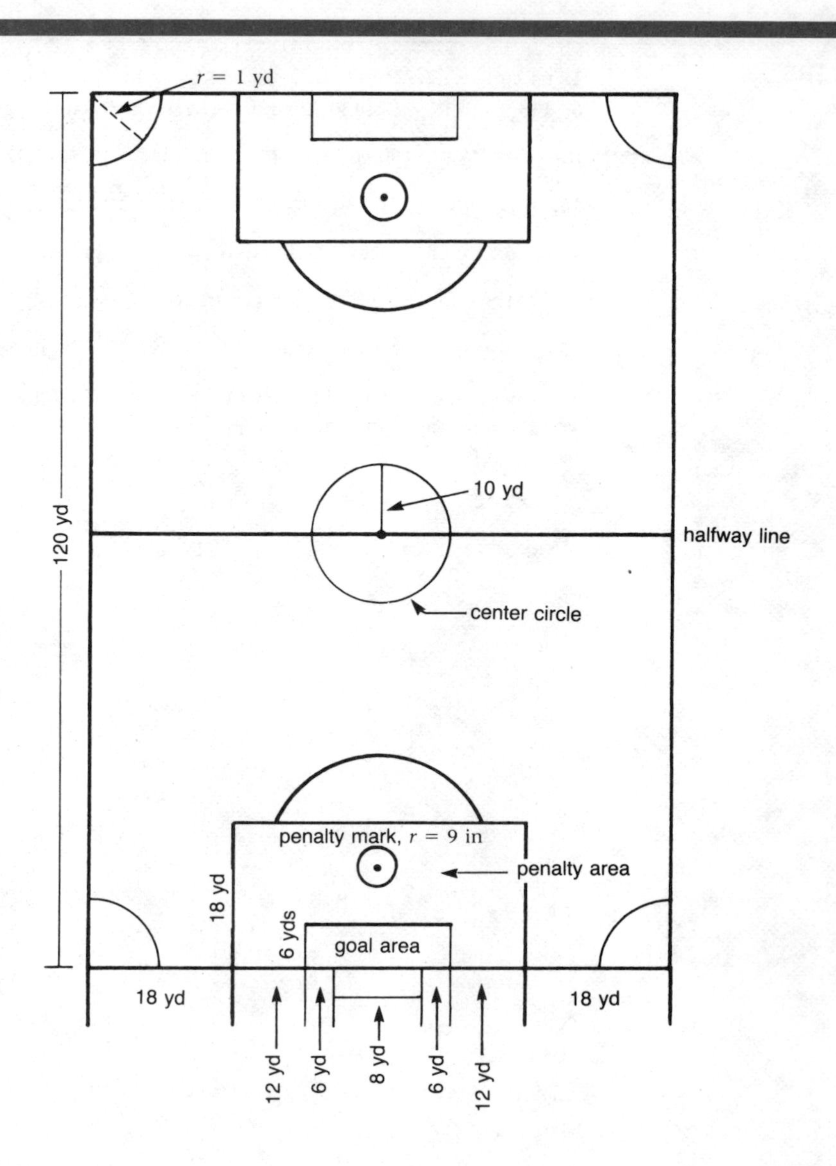

Soccer has long been an international sport played and admired by millions of people. Among the many skills demanded of a soccer athlete are endurance, agility, and speed. Fortunately, people need only a modest amount of equipment to play soccer, and it may be played in an area approximately the size of a football field.

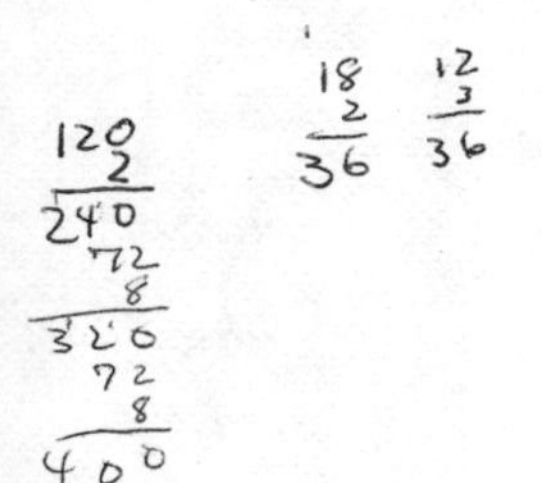

Example The preceding diagram represents the field of play for the game of soccer.

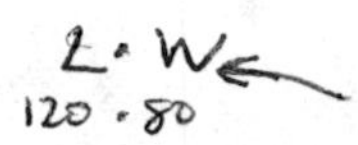

(a) What is the perimeter of the entire field?
(b) What is the area of the field?
(c) What is the perimeter of the penalty area (not including the arc of the circle at its top)?
(d) What is the perimeter of the goal area?
(e) What is the area of the penalty mark?
(f) What is the length of the arc of the circle at each corner?
(g) What is the area of the field of play outside of the penalty area?

Solution

(a) The width is 18 yd + 12 yd + 6 yd + 8 yd + 6 yd + 12 yd + 18 yd = 80 yards.
So $P = 2 \times w + 2 \times l = 2 \times 80 + 2 \times 120 =$ 400 yards.

(b) $A = l \times w = 120$ yd $\times$ 80 yd = 9,600 square yards

(c) Its dimensions are 44 yards by 18 yards. Thus
$P = 2 \times 44$ yd $+ 2 \times 18$ yd = 124 yards

(d) Its dimensions are 20 yards by 6 yards. Thus
$P = 2 \times 20$ yd $+ 2 \times 6$ yd = 52 yards

(e) $A = \pi \times r^2 = (3.14)(9 \text{ inches})^2 =$ 254.34 square inches

(f) Length of arc $= \frac{1}{4} \times \pi \times d = \frac{1}{4}$ yd $\times \pi \times 2$ yd = 1.57 yards

(g) The area of each penalty area is 18 yd × 44 yd = 792 square yards.
The answer is:
9,600 yd − 2(792) yd = 8,016 square yards

Soccer is fast becoming a major sport in the United States.
(Courtesy of the Minnesota Strikers)

BUILDING YOUR MATH VOCABULARY

TERM	DEFINITION	EXAMPLE
Area, A	A unit of area is a surface enclosed by a square each of whose sides is of unit length.	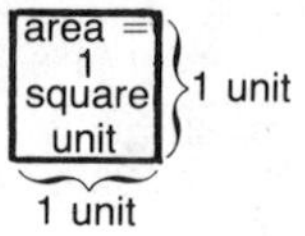 area = 1 square unit; 1 unit; 1 unit
Base, b	Any side of a triangle can be its base, but we usually select the "horizontal" side for convenience.	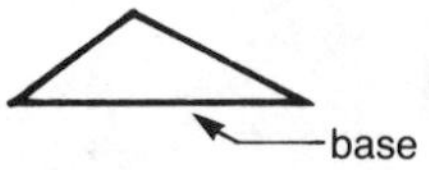 base
Circle	A circle is a closed curve all of whose points are equidistant from a fixed point called its center.	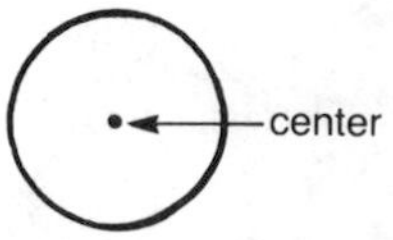 center
Circumference, C	The perimeter of the circle.	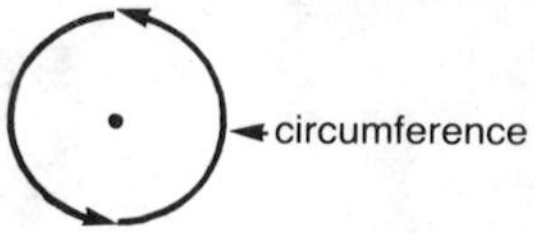 circumference
Cone	A three-dimensional figure similar in shape to an ice cream cone.	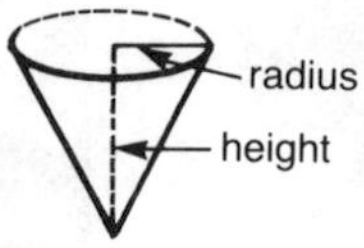 radius; height
Cylinder	A three-dimensional figure similar in shape to a soup can.	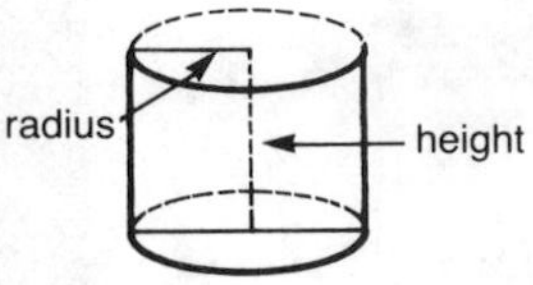 radius; height
Diameter, d	A line segment drawn from one point on a circle to another point on the circle and passing through the center of the circle.	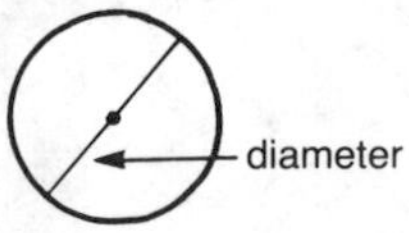 diameter
Height, h	The height of a triangle is the distance from the base side to the top of the triangle.	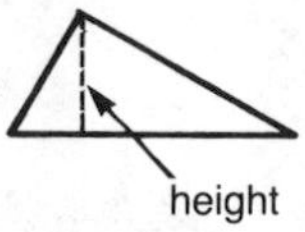 height
Line	An undefined term in geometry. A model of a line is a "straight mark" obtained by folding a piece of paper.	
Line segment	A line segment is part of a line between two points and including the two points.	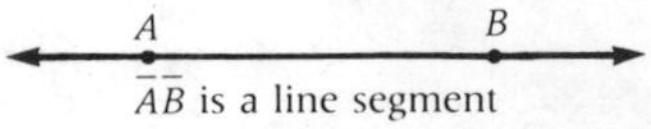 A B; $\overline{AB}$ is a line segment
Parallelogram	A quadrilateral whose opposite sides are parallel and of equal length.	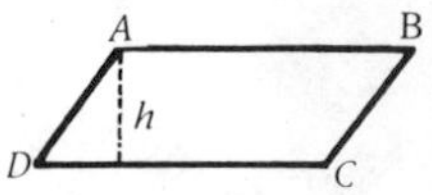 A B h D C
Perimeter, P	The perimeter of a polygon is the distance around the polygon and may be found by adding the lengths of its sides.	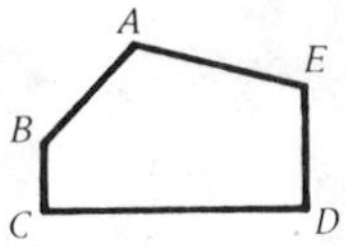 A E B C D; $P = AB + BC + CD + DE + EA$
Pi (π)	This is a ratio of $\frac{C}{d}$.	$\pi \doteq 3.14$ or $\frac{22}{7}$

TERM	DEFINITION	EXAMPLE
Point	An undefined term in geometry. A "dot" is a model for a point.	•
Plane	An undefined term in geometry. Extending a piece of paper in all directions serves as a model for a plane.	
Polygon	A closed figure all of whose sides are line segments.	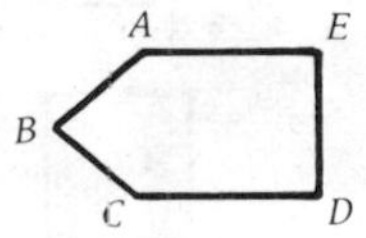 A E B C D
Quadrilateral	A four-sided polygon.	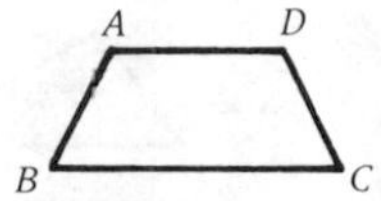 A D B C
Radius, *r*	A line segment extending from the center of a circle to any point on the circle. The radius is $\frac{1}{2}$ the length of the diameter, *d*, of a circle. The plural of radius is radii.	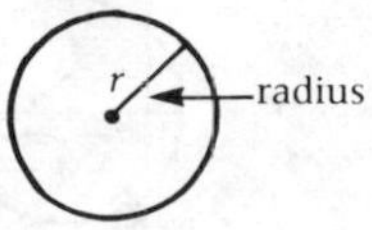 *r* radius
Ray	A ray is part of a line extending indefinitely in one direction from a point.	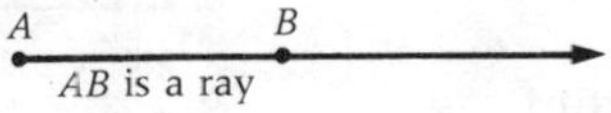 A B; *AB* is a ray
Rectangle	A quadrilateral in which the opposite sides are equal in length and all angles are right angles.	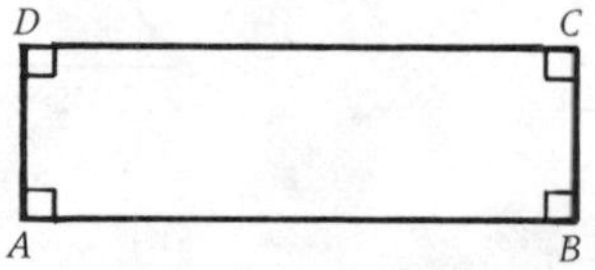 D C A B
Rectangular solid	A three-dimensional figure similar in shape to a box.	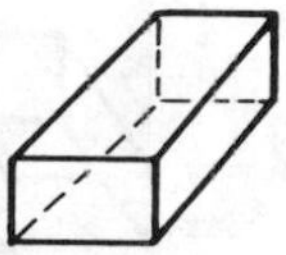
Square	A rectangle with four equal sides.	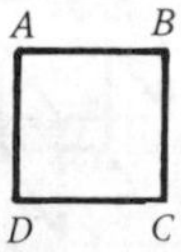 A B D C
Sphere	A three-dimensional figure similar in shape to a basketball.	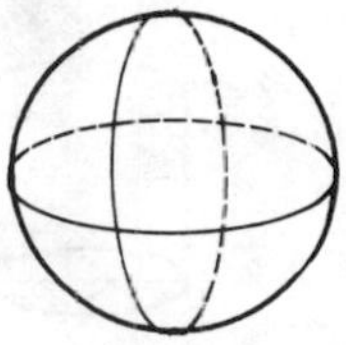
Trapezoid	A quadrilateral with at least one pair of parallel sides.	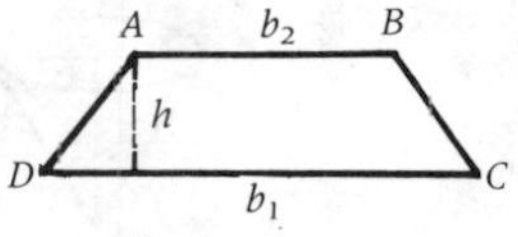 A b_2 B *h* D b_1 C
Triangle	A three-sided polygon.	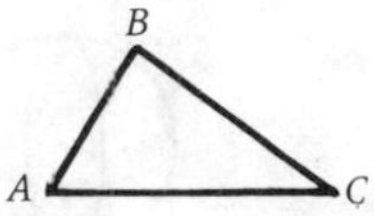 B A C
Volume	A unit of volume is a cube, each of whose sides is of unit length.	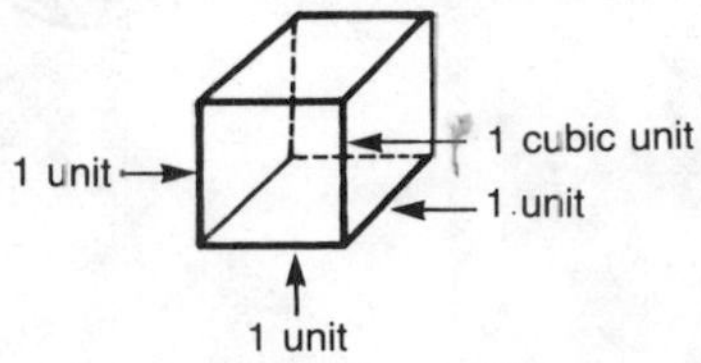 1 unit; 1 cubic unit; 1 unit; 1 unit

Formulas

Figure	Perimeter	Area
Rectangle	$P = 2 \cdot l + 2 \cdot w$	$A = l \cdot w$
Square	$P = 4 \cdot s$	$A = s \cdot s = s^2$
Triangle	$P = a + b + c$	$A = \frac{1}{2} \cdot b \cdot h$
Circle	$P = \pi \cdot d = 2 \cdot \pi \cdot r$	$A = \pi \cdot r^2$
Parallelogram	$P = 2 \cdot a + 2 \cdot b$	$A = b \cdot h$
Trapezoid		$A = \frac{1}{2} \cdot (b_1 + b_2) \cdot h$

Figure	Volume
Rectangular Solid	$V = l \cdot w \cdot h$
Cube	$V = s \cdot s \cdot s = s^3$
Cylinder	$V = \pi \cdot r^2 \cdot h$
Cone	$V - \frac{1}{3} \cdot \pi \cdot r^2 \cdot h$
Sphere	$V = \frac{4}{3} \cdot \pi \cdot r^3$

Chapter Review

Round off all decimal answers to the hundredths place.

1. Find the perimeter of a triangle with sides of 3.2 inches, 4.8 inches, and 5.2 inches.

2. How many cubic inches are in a cubic foot?

3. How many square inches are in a square foot?

4. The two sides that enclose the right angle (90° angle) in a right triangle are of lengths 9 inches and 12 inches. The third side is 15 inches.

(a) What is the perimeter? ________ **(b)** What is the area? ________

5. If the radius of a circle is 20 meters find

(a) its circumference ____________ and **(b)** its area ____________.

6. If AB = 4.1 ft, BC = 3.8 ft, CD = 2.85 ft, DE = 3.71 ft, and EA = 2.60 feet, what is the perimeter of polygon ABCDE?

7. A cylinder has a radius of 8 inches and a height of $4\frac{1}{4}$ inches. Find its volume.

8. A cylinder has a radius of 10 inches and a height of $3\frac{2}{3}$ inches. Find its volume.

9. A cone has the same circular base and height as the cylinder in problem 7. What is its volume?

10. A cone has the same circular base and height as the cylinder in problem 8. What is its volume?

11. A rectangle has length of 9 yards and width of 6 yards.

(a) What is its perimeter? ________ **(b)** What is its area? ________

12. A square is 6.5 inches on a side.

(a) What is its perimeter? ________ **(b)** What is its area? ________

13. If a cubic foot of water weighs 62.5 pounds, what is the weight of water in a rectangular tank 2 ft by 3 ft by 4 ft filled with water?

14. If a cubic foot of water weighs 62.5 pounds, what is the weight of water in a rectangular tank 5 ft by 1 ft by 6 ft filled with water?

15. What is the volume of a sphere whose radius is 2.5 inches?

____

16. What is the volume of a sphere whose diameter is 7 inches?

____

17. (true or false) $\pi = \frac{C}{d}$. ________________

18. (true or false) $\pi = 22.7$. ________________

19. A parallelogram is pictured below.

(a) What is its perimeter? __________ **(b)** What is its area? __________

20. Find the area of the trapezoid with bases of 8.4 in and 6.3 in and a height of 4.2 in.

____

21. Find the area of trapezoid *ABCD*. ________________

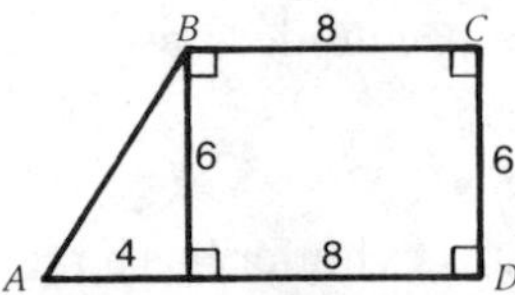

22. Find the cost of digging a cylindrical well 5 feet in diameter and 23 feet deep if the cost per cubic foot is $17.50.

____

23. A circle with a diameter of 9 meters is rotated about its diameter to generate a sphere. What is the volume of the sphere?

____

24. You have a rectangle with a width of 2 inches and a length of 4 inches. If this rectangle is rotated about its length to create a cylinder, what is the volume of the cylinder?

____

25. Determine the total surface area of a cube whose side dimension is 10 inches.

____

26. If a plane passes through the center of a sphere, what is the shape of its intersection with the surface of the sphere?

____

27. What is the difference between the volumes of a sphere with a 4-ft radius and a sphere with a 2-ft radius?

____

28. A circle with a $\frac{1}{2}$-inch radius and a center located at the center of a 2-inch square is cut out of the square by a clothing manufacturer. How much material is wasted in doing this?

____

29. Determine the total surface area of a cube whose side dimension is 2 inches.

____

30. Determine the total surface area of a cube whose side dimension is 3 inches.

____

31. The surface area of a sphere is given by the formula: $S = 4 \cdot \pi \cdot r^2$. If $r = 4$ ft, find the surface area.

____

32. The surface area of a sphere is given by the formula: $S = 4 \cdot \pi \cdot r^2$. If $r = 3$ ft, find the surface area.

____

ANSWERS

1. 13.2 in **2.** 1,728 **3.** 144 **4. (a)** $P = 36$ in **(b)** $A = 54$ sq in **5. (a)** $C = 125.6$ meters **(b)** $A = 1{,}256$ sq m **6.** 17.06 ft **7.** 854.08 cu in **8.** 1,151.33 cu in **9.** 284.69 cu in **10.** 383.78 cu in **11. (a)** 30 yd **(b)** 54 sq yd **12. (a)** 26 in **(b)** 42.25 sq in **13.** 1,500 pounds **14.** 1,875 pounds **15.** 65.42 cu in **16.** 179.50 cu in **17.** true **18.** false **19.(a)** 84 in **(b)** 192 sq in **20.** 30.87 sq in **21.** 60 **22.** $7,899.06 **23.** 381.51 cu m **24.** 50.24 cu in **25.** 600 sq in **26.** a circle **27.** 234.45 cu ft **28.** 3.215 sq in **29.** 24 sq in **30.** 54 sq in **31.** 200.96 sq ft **32.** 113.04 sq ft

Warmup Test

Round off all decimal answers to the nearest hundredth.

1. Completions:

(a) π is approximately equal to ________ (leave as a decimal).

(b) If the diameter of a circle is 8.5 yards, then the radius is ____________.

(c) If the area of a semicircle is 3π, then the area of the whole circle is ________ (leave your answer in terms of π).

(d) If the volume of a cone is 5π, then the volume of a cylinder with the same circular base is ________ (leave your answer in terms of π).

2. (a) What is the perimeter of a square $3\frac{1}{2}$ inches on a side? ____________

(b) What is the area of this same square? ____________

3. (a) What is the perimeter of triangle *ABC*? ____________

(b) What is the area of this triangle? ____________

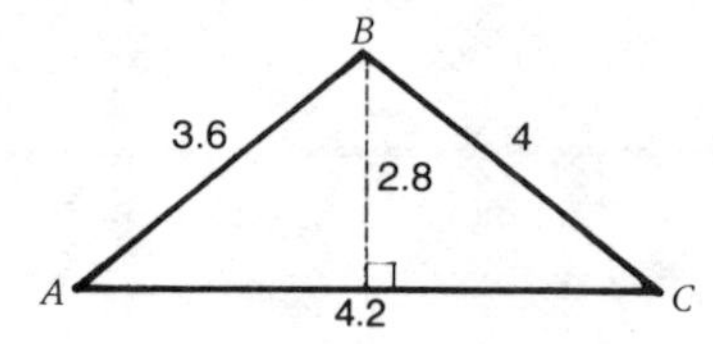

4. A rectangular solid has dimensions of 4, 5, and 6. What is its volume?

____

5. How many crates, each holding 10 cubic units, can be emptied into the rectangular solid of problem 4?

____

6. If a rectangle has a length of 4 meters and a width of $1\frac{1}{4}$ meters, find

(a) its perimeter ________________ **(b)** its area ________________.

7. If a parallelogram has unequal sides of 3.5 ft and 4.2 ft, find its perimeter.

____

8. If a parallelogram has a base of 15 yd and height of 6.3 yd, find its area.

____

9. Find the area of the trapezoid with bases 10.65 in and 9.7 in and a height of 4.3 in (round off to thousandths).

____

10. A cylinder with a 15-inch height rests on a circle whose radius is 3 inches.

(a) What is the area of the base circle? ____________

(b) What is the volume of the cylinder? ____________

11. What is the volume of a sphere whose diameter is 20 inches?

12. A circle of radius 8 feet has a square constructed inside it such that the four corners of the square lie on the circle. If the side of the square is 11.4 feet, what is the difference in areas?

13. Carpeting costs $10.50 per square yard. Disregarding waste, how much will it cost to cover the area shown with carpeting?

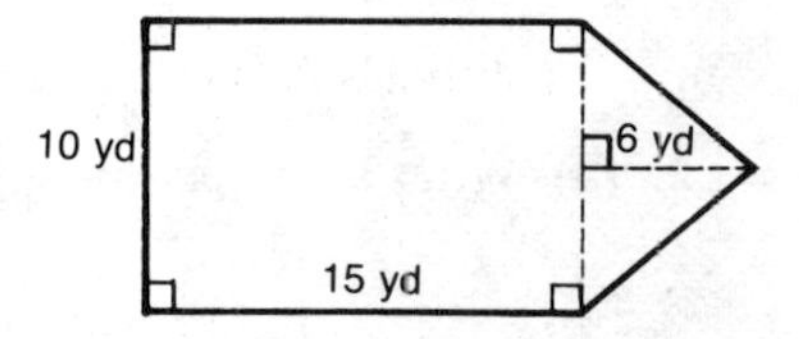

14. A cylindrical water tank has a base radius of 12 feet and a height of 18 feet. What is the volume?

How many gallons will it hold (7.5 gallons per cubic foot)?

15. A home gardener wishes to fence in her 20-ft by 50-ft garden. If fencing costs $5.50 per foot, what is the cost to fence in her garden?

Challenge Test

Round off all decimal answers to the nearest hundredth.

1. Completions:

(a) π, expressed in fraction form, is approximately equal to ______________

(b) If the radius of a circle is 3.6 ft, then the diameter is ______________

(c) If the area of a semicircle is 3π, then the area of the whole circle is ______________ (leave your answer in terms of π).

(d) If the volume of cone is 8π, then the volume of a cylinder with the same circular base is __________ (leave your answer in terms of π).

2. (a) What is the perimeter of a square $5\frac{1}{4}$ inches on a side?

____

(b) What is the area of this same square? (Round off to hundredths.)

____

3. (a) What is the perimeter of triangle *ABC*? ______________

(b) What is the area of this triangle? ______________

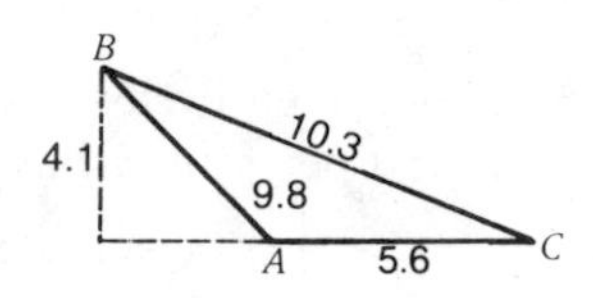

4. A rectangular solid has dimensions of 6 in, 10 in, and 14 in. What is its volume?

____

5. How many crates, each holding 12 cubic inches, can be emptied into the rectangular solid of problem 4?

____

6. If a rectangle has a length of 6.2 yd and a width of 3.4 yd, find

(a) its perimeter ______________ **(b)** its area ______________.

7. If a parallelogram has unequal sides of 8.6 feet and 9.3 feet, find its perimeter.

____

8. If a parallelogram has a base of 24 in and a height of $9\frac{1}{2}$ in, find its area.

____

9. Find the area of the trapezoid with bases of 12.84 in and 8.62 in and a height of 5 in.

____

10. A cylinder with a 20-inch height rests on a circle whose radius is 7 inches.

(a) What is the area of the base circle? ________________

(b) What is the volume of the cylinder? ________________

11. What is the volume of a sphere whose diameter is 12 inches?

____

12. A circle of radius 10 feet has a square constructed inside it such that the four corners of the square lie on the circle. If the side of the square is 14.2 feet, what is the difference in areas?

____

13. Carpeting costs $12.75 per square yard. Disregarding waste, how much will it cost to cover the area shown with carpeting?

____

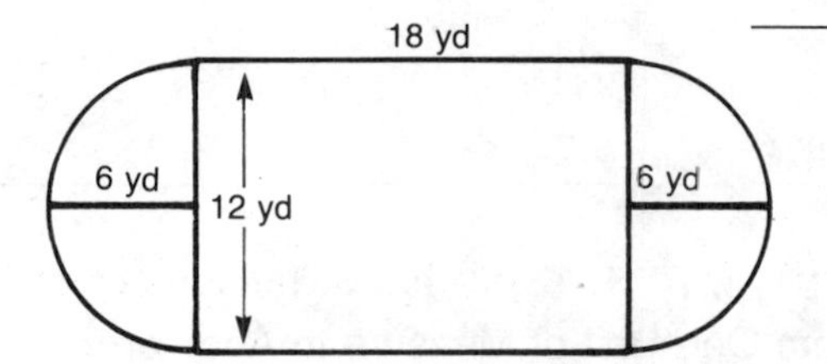

14. A conical water tank has a base radius of 9 ft and a height of 12 ft. What is the volume? How many gallons will it hold (7.5 gallons per cubic foot)?

____

15. A farmer wishes to fence in a 40-ft by 60-ft corral. If fencing costs $16.25 per foot, what is the cost to fence in the corral?

____

Measurement

CHAPTER OUTLINE

6.1 ENGLISH SYSTEM

Each country has its own standard system of measurement. The United States uses the English system. This is unlike all other major countries which use the metric system. At the present time, the United States is undergoing a transition from the English system to the metric system, but this change will not be completed for many years. During this transition period it is necessary that we understand the English system of measurement.

Converting from One Unit of Measure to Another

It is important for us to learn how to convert from one unit of measure to another within the English system. The following tables will help us in making the conversions. The tables state the relationships between units of length, units of weight, units of volume, units of time and household measures.

Table 6.1

Units of Length

1 foot (ft)	=	12 inches (in)
1 yard (yd)	=	3 feet
1 mile (mi)	=	5,280 feet
1 mile	=	1,760 yards

Table 6.2

Units of Weight

1 pound (lb)	=	16 ounces (oz)
1 ton	=	2,000 lb

Table 6.3

Units of Volume

1 pint (pt)	=	16 fluid ounces (fl oz)
1 quart (qt)	=	2 pints
1 gallon (gal)	=	4 quarts

Table 6.4

Units of Time

1 minute (min)	=	60 seconds
1 hour (h)	=	60 minutes
1 day (da)	=	24 hours
1 week (wk)	=	7 days
1 year (yr)	=	365 days
1 year (yr)	=	52 weeks
1 year	=	12 months (mo)

Table 6.5

Household Measures

1 tablespoon (tbsp)	=	3 teaspoons (tsp)
1 cup	=	16 tablespoons
1 cup	=	8 fluid ounces
1 pint	=	2 cups

The following examples show how we use proportions to convert from one unit of measure to another in the English system.

Example 1a 12 feet = ________ yards.

Solution Let x be the number of yards in 12 feet.
Since 1 yard = 3 feet, we can set up a proportion.

$$\frac{12 \text{ feet}}{x \text{ yards}} = \frac{3 \text{ feet}}{1 \text{ yard}}$$

$$3x = 12 \quad \text{(Rule for Proportions)}$$

$$\frac{\overset{1}{\cancel{3}}x}{\underset{1}{\cancel{3}}} = \frac{12}{3} \quad \text{(Divide by 3)}$$

$$x = 4 \text{ yards}$$

Example 1b 16 yards = ________ feet.

Solution Let x be the number of feet in 16 yards.
Since 1 yard = 3 feet, we can set up a proportion.

$$\frac{16 \text{ yards}}{x \text{ feet}} = \frac{1 \text{ yard}}{3 \text{ feet}}$$

$$x = 16 \times 3 \quad \text{(Rule for Proportions)}$$

$$x = 48 \text{ feet}$$

Example 2a 5,000 pounds = ________ tons.

Solution Let x be the number of tons in 5,000 pounds.
Since 1 ton = 2,000 pounds we can set up a proportion.

$$\frac{5{,}000 \text{ lb}}{x \text{ tons}} = \frac{2{,}000 \text{ lb}}{1 \text{ ton}}$$

$$2{,}000x = 5{,}000 \quad \text{(Rule for Proportions)}$$

$$\frac{\overset{1}{\cancel{2000}}x}{\underset{1}{\cancel{2000}}} = \frac{5000}{2000} \quad \text{(Divide by 2,000)}$$

$$x = 2\tfrac{1}{2} \text{ tons}$$

Example 2b $3\frac{1}{4}$ tons = ________ pounds.

Solution Let x be the number of pounds in $3\frac{1}{4}$ tons.

Since 1 ton = 2,000 pounds, we can set up a proportion.

$$\frac{3\frac{1}{4}}{x \text{ lb}} = \frac{1 \text{ ton}}{2{,}000 \text{ lb}}$$

$$x = 3\tfrac{1}{4} \times 2{,}000 \quad \text{(Rule for Proportions)}$$

$$x = \frac{13}{\underset{1}{\cancel{4}}} \times \overset{500}{\cancel{2{,}000}}$$

$$x = 6{,}500 \text{ lb}$$

Example 3 $6\frac{1}{2}$ gallons = ________ quarts.

Solution Let x be the number of quarts in $6\frac{1}{2}$ gallons.

Since 1 gallon = 4 quarts we can set up a proportion.

$$\frac{6\frac{1}{2}\text{ gallons}}{x\text{ quarts}} = \frac{1\text{ gallon}}{4\text{ quarts}}$$

$$x = 6\frac{1}{2} \times 4 \quad \text{(Rule for Proportions)}$$

$$x = \frac{13}{\cancel{2}_1} \times \cancel{4}^{2} = 26\text{ quarts}$$

Example 4 $3\frac{1}{4}$ days = ________ hours.

Solution Let x be the number of hours in $3\frac{1}{4}$ days.

Since 1 day = 24 hours, we can set up a proportion.

$$\frac{3\frac{1}{4}\text{ days}}{x\text{ hours}} = \frac{1\text{ day}}{24\text{ hours}}$$

$$x = 3\frac{1}{4} \times 24 \quad \text{(Rule for Proportions)}$$

$$x = \frac{13}{\cancel{4}_1} \times \cancel{24}^{6} = 78\text{ hours}$$

Example 5 6 cups = ________ pints.

Let x be the number of pints in 6 cups.
Since 1 pint = 2 cups we can set up a proportion.

$$\frac{6\text{ cups}}{x\text{ pints}} = \frac{2\text{ cups}}{1\text{ pint}}$$

$$2x = 6$$

$$\frac{\cancel{2}x}{\cancel{2}_1} = \frac{6}{2} \quad \text{(Divide by 2)}$$

$$x = 3\text{ pints}$$

Example 6 $4\frac{1}{3}$ pounds = ________ ounces.

Solution Let x be the number of ounces in $4\frac{1}{3}$ pounds.

Since 1 pound = 16 ounces we set up a proportion.

$$\frac{4\frac{1}{3}\text{ lb}}{x\text{ oz}} = \frac{1\text{ lb}}{16\text{ oz}}$$

$$x = 4\frac{1}{3} \times 16$$

$$x = \frac{13}{3} \times 16 = \frac{208}{3} = 69\frac{1}{3}\text{ lbs.}$$

Example 7 8 gallons = ________ pints.

Solution step 1: Change 8 gallons to quarts.
Let x be the number of quarts in 8 gallons.
Since 1 gallon = 4 quarts, we have

$$\frac{8 \text{ gal}}{x \text{ qt}} = \frac{1 \text{ gal}}{4 \text{ qt}}$$

$$x = 4 \times 8 = 32 \text{ qt}$$

step 2: Change 32 quarts to pints.
Let y be the number of pints in 32 quarts.
Since 1 quart = 2 pints, we have

$$\frac{32 \text{ qt}}{y \text{ pt}} = \frac{1 \text{ qt}}{2 \text{ pt}}$$

$$y = 32 \times 2 = 64 \text{ pt}$$

Therefore 8 gallons = 64 pints.

Example 8 A tailor needs 24 in. of calico material that cost $3.15 per yard. How much does he pay for the calico material?

Solution step 1: Change 24 inches to yards.
Let x = the number of yards in 24 inches.
Since 1 yard = 36 inches, we have

$$\frac{24 \text{ in}}{x \text{ yd}} = \frac{36 \text{ in}}{1 \text{ yd}}$$

$$36x = 24$$

$$\frac{\overset{1}{\cancel{36}}x}{\underset{1}{\cancel{36}}} = \frac{\cancel{24}}{\cancel{36}}$$

$$x = \frac{2}{3} \text{ yard}$$

step 2: Find the tailor's cost.

$$\text{Cost} = \frac{2}{3} \text{ yd.} \times \$3.15 = \$2.10$$

6.1 EXERCISES

Convert each of the following to the indicated units.

1. 48 in = ________ ft
2. 20 qt = ________ gal
3. 15 tsp = ________ tbsp
4. 84 days = ________ wk
5. 6,500 lb = ________ tons
6. 64 fl oz = ________ pt
7. $5\frac{1}{4}$ h = ________ min
8. 3 cups = ________ tbsp
9. $10\frac{1}{2}$ lb = ________ oz
10. 72 min = ________ h

11. 4 mi = ______ yd **12.** 150 s = ______ min

13. 12,750 lb = ______ tons **14.** $5\frac{1}{2}$ tons = ______ lb

15. 6 tons = ______ lb **16.** $\frac{1}{4}$ ton = ______ oz

17. 10,560 ft = ______ mi **18.** 198 in = ______ yd

19. $25\frac{1}{2}$ ft = ______ yd **20.** 612 in = ______ ft

21. 3.5 yd = ______ ft **22.** $5\frac{1}{4}$ mi = ______ yd

23. 8.2 yd = ______ in **24.** 2.4 mi = ______ ft

25. 60 h = ______ da **26.** 78 wk = ______ yr

27. 21 mo = ______ yr **28.** 260 s = ______ min

29. $3\frac{1}{3}$ h = ______ s **30.** 2 wk = ______ h

31. $1\frac{1}{5}$ yr = ______ days **32.** 4.5 yr = ______ mo

33. 10 pt = ______ gal **34.** 40 tbsp = ______ cups

35. 12 fl oz = ______ tbsp **36.** 22 qt = ______ gal

37. 21 tsp = ______ tbsp **38.** 15 cups = ______ pt

39. 4 gal = ______ fl oz **40.** 6 cups = ______ fl oz

41. A seamstress needs 16 in of lace that costs $7.20 per yard. How much does she pay for the material?

______

42. A tentmaker needs 30 in of material that costs $10.56 per yard. How much does she pay for the material?

______

ANSWERS

1. 4 ft **2.** 5 gal **3.** 5 tbsp **4.** 12 wk **5.** $3\frac{1}{4}$ tons **6.** 4 pt **7.** 315 min
8. 48 tbsp **9.** 168 oz **10.** $1\frac{1}{5}$ h **11.** 7,040 yd **12.** $2\frac{1}{2}$ min **13.** $6\frac{3}{8}$ tons
14. 11,000 lb **15.** 12,000 lb **16.** 8,000 oz **17.** 2 mi **18.** $5\frac{1}{2}$ yd **19.** $8\frac{1}{2}$ yd
20. 51 ft **21.** 10.5 ft **22.** 9,240 yd **23.** 295.2 in **24.** 12,672 ft **25.** $2\frac{1}{2}$ da
26. $1\frac{1}{2}$ yr **27.** $1\frac{3}{4}$ yr **28.** $4\frac{1}{3}$ min **29.** 12,000 s **30.** 336 h **31.** 438 da
32. 54 mo **33.** $1\frac{1}{4}$ gal **34.** $2\frac{1}{2}$ cups **35.** 24 tbsp **36.** $5\frac{1}{2}$ gal **37.** 7 tbsp
38. $7\frac{1}{2}$ pt **39.** 512 fl oz **40.** 48 fl oz **41.** $3.20 **42.** $8.80

6.2 ARITHMETIC OF MEASUREMENT NUMBERS

Simplifying Measurements

If we were given the height of an NBA basketball player to be 79 inches or told that a cable television installation took 100 minutes, we would convert the 79 inches to 6 feet 7 inches and the 100 minutes to 1 hour 40 minutes. *We would have simplified the measurements by converting the smaller units of measure to larger units of measure.*

Example 1 Simplify 22 inches.

Solution Since 12 inches = 1 foot,
22 inches = 12 inches + 10 inches
= 1 foot + 10 inches
Measurements are written without the plus sign:
Thus 22 inches = 1 foot 10 inches

Example 2 Simplify 100 minutes.

Solution Since 60 minutes = 1 hour,
100 minutes = 60 minutes + 40 minutes
= 1 hour + 40 minutes
= 1 hour 40 minutes

Example 3 Simplify 225 minutes.

Solution Since 60 minutes = 1 hour,
225 minutes = 180 minutes + 45 minutes
(60 min × 3 = 180 min)
= 3 hours + 45 minutes
= 3 hours 45 minutes

Example 4 Simplify 33 quarts.

Solution Since 4 quarts = 1 gallon,
33 quarts = 32 quarts + 1 quart (4 qts. × 8 = 32 qts.)
= 8 gallons + 1 quart
= 8 gallons 1 quart

Example 5 Simplify 3 hours 72 minutes.

Solution Since 60 minutes = 1 hour,
3 hours 72 minutes = 3 hours + 60 minutes + 12 minutes
= 3 hours + 1 hour + 12 minutes
We can only add like units of measurement
= 4 hours 12 minutes

Example 6 Simplify 12 feet 65 inches.

Solution Since 12 inches = 1 foot,
12 feet 65 inches = 12 feet + 60 inches + 5 inches
(12 in × 5 = 60 in)
= 12 feet + 5 feet + 5 inches
= 17 feet 5 inches

ADDITION

In adding measurement numbers we can only add units of like measurement: feet to feet, quarts to quarts, pounds to pounds, hours to hours, and so forth.

To Add Measurement Numbers

1. Place the numbers to be added in columns with like units of measurement below like units of measurement.
2. Add the numbers in each column.
3. Simplify when possible.

Example 7 Add: 7 feet 3 inches + 4 feet 11 inches

Solution

```
   7 feet   3 inches
+  4 feet  11 inches
--------------------
  11 feet  14 inches
            |______|
            1 foot  2 inches
  |_____________|
  12 feet   2 inches
```

Example 8 Add: 8 weeks 5 days + 6 weeks 3 days

Solution

```
   8 weeks  5 days
+  6 weeks  3 days
------------------
  14 weeks  8 days
            |____|
             1 week  1 day
    |__________|
  15 weeks  1 day
```

Example 9 Add: 7 gal 3 qt 1 pt + 3 gal 2 qt 1 pt + 1 gal 3 qt 1 pt

Solution

```
   7 gal  3 qt  1 pt
   3 gal  2 qt  1 pt
+  1 gal  3 qt  1 pt
--------------------
  11 gal  8 qt  3 pt
          |__|  |__|
       2 gals  1 qt.  1 pt.
   |_______|   |________|
+13 gal  1 qt  1 pt
```

Example 10 Add: 3 mo 7 wk 6 da + 5 mo 3 wk 4 da + 6 mo 3 wk 3 da.
(Assume there are 4 weeks in a month)

Solution

```
     3 mo   7 wk   6 da
     5 mo   3 wk   4 da
  +  6 mo   3 wk   3 da
-----------------------
    14 mo  13 wk  13 da
  |______| |____| |____|
1 yr 2 mo  3 mo 1 wk  1 wk 6 da
  |___________________________|
1 yr  5 mo  2 wk  6 da
```

Subtraction In subtracting measurement numbers we can subtract only units of like measurement.

> **To Subtract Measurement Numbers**
>
> **1.** Place the numbers to be subtracted in columns with like units of measurement below like units of measurement.
> **2.** Subtract the numbers in each column, borrowing when necessary from the next nonzero unit of measure to the left.
> **3.** Simplify when possible.

Example 11 Subtract: 12 hours 35 minutes − 8 hours 22 minutes

Solution

```
  12 hours  35 minutes
−  8 hours  22 minutes
   4 hours  13 minutes
```

Example 12 Subtract: 10 yd 1 ft − 6 yd 2 ft

Solution

```
   9      4
  10 yd   1 ft
−  6 yd   2 ft
   3 yd   2 ft
```

1 yd = 3 ft; we borrow 3 ft from 10 yd and add the 3 ft to 1 ft to give us 4 ft

Example 13 Subtract: 3 years 8 months 2 weeks from 12 years 5 months 3 weeks. (Assume there are 4 weeks in a month)

Solution

```
  11          17
  12 years     5 months  3 weeks
−  3 years     8 months  2 weeks
   8 years     9 months  1 week
```

1 yr = 12 mo; we borrow 12 mo from 12 years and add 12 mo to 5 mo to give us 17 mo.

Example 14 Subtract: 6 h 12 min 42 sec from 9 hr 25 s.

Solution

```
        59
   8    60       85
   9 h   0 min   25 s
−  6 h  12 min   42 s
   2 h  47 min   43 s
```

1. 1 h = 60 min; we borrow 60 mins from 9 h and add 60 min to 0 min to give us 60 min.
2. 1 min = 60 s; we borrow 60 s from 60 min and add 60 s to 25 to give us 85 s.

Multiplication

> **To Multiply a Measurement Number by a Nonmeasurement Number**
>
> **1.** Multiply each part of the measurement number by the nonmeasurement number.
> **2.** Simplify when possible.

Example 15 A carpenter is working on a tract of three houses, each of which has the same basic floor plan. Each house requires a piece of shelving 4 ft 7 in in length. How much shelving does he need for the three houses?

Solution

$$\begin{array}{rr} 4 \text{ ft} & 7 \text{ in} \\ & \times 3 \\ \hline 12 \text{ ft} & 21 \text{ in} \end{array}$$

1 ft 9 in

13 ft 9 in

Example 16 Multiply: 6 mo 3 wk 2 da × 5

Solution

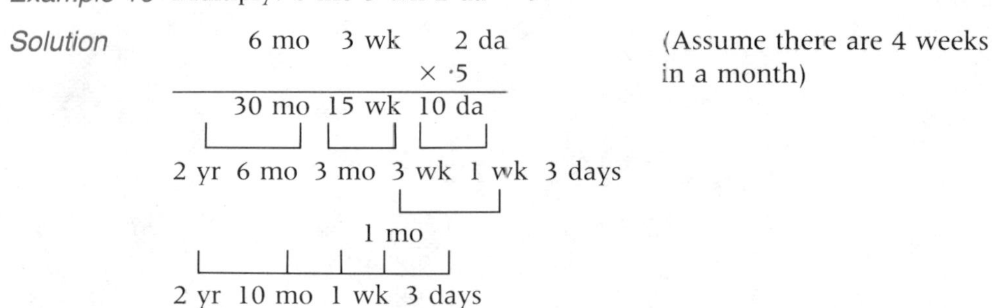

(Assume there are 4 weeks in a month)

Sometimes it is possible to multiply measurement numbers by measurement numbers.

Example 17 Find the area of the pictured triangle.

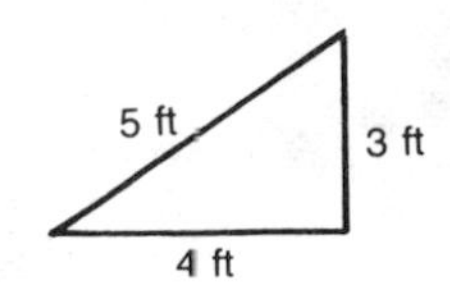

Solution

$$\begin{aligned} A &= \frac{1}{2} \cdot b \cdot h \\ &= \frac{1}{2} \cdot 4 \text{ ft} \cdot 3 \text{ ft} \\ &= 6 \text{ sq ft} \end{aligned}$$

Example 18 Multiply 6 hours by 4 inches.

Solution Multiplying 6 hours by 4 inches would give us 24 hour-inches, which has no meaning. This multiplication would not be performed.

Division

To Divide a Measurement Number by a Nonmeasurement Number

1. Divide the largest-unit in the measurement number by the nonmeasurement number.
2. If it does not divide evenly, convert the remainder into the next smaller unit and add it to the units already there.
3. Divide the result of step 2 by the nonmeasurement number.
4. Repeat steps 2 and 3 until the division is complete.

Example 19 A carpenter has a piece of board 9 ft 6 in long. He wishes to divide it into three equal parts. What is the length of each part?

Solution Divide 9 ft 6 in by 3.

$$\begin{array}{l} \quad\ \boxed{3\text{ ft}} \\ \boxed{3}\overline{)\boxed{9\text{ ft}}\,6\text{ in}} \\ \quad\ \underline{9\text{ ft}} \\ \quad\ 0 \end{array}$$

$$\begin{array}{l} \quad\ 3\text{ ft}\,\boxed{2\text{ in}} \\ \boxed{3}\overline{)\ 9\text{ ft}\,\boxed{6\text{ in}}} \\ \qquad\qquad \underline{6\text{ in}} \\ \qquad\qquad 0 \end{array}$$

$$\text{Thus}\quad 3\overline{)9\text{ ft }6\text{ in}}^{\ 3\text{ ft }2\text{ in}}$$

Example 20 Divide 7 min 10 s by 5.

Solution

$$\begin{array}{l} \quad\ \boxed{1\text{ min}} \\ \boxed{5}\overline{)\boxed{7\text{ min}}\,10\text{ s}} \\ \quad\ \underline{5\text{ min}} \\ \quad\ \boxed{2\text{ min}} \end{array}$$

Since 2 min = 120 s add: 120 s + 10 s = 130 s

$$\begin{array}{l} \quad\ 1\text{ min }\boxed{26\text{ s}} \\ \boxed{5}\overline{)7\text{ min}\,\boxed{130\text{ s}}} \\ \qquad\qquad \underline{130\text{ s}} \\ \qquad\qquad\quad 0 \end{array}$$

$$\text{Thus}\quad 5\overline{)7\text{ min }10\text{ s}}^{\ 1\text{ min }26\text{ s}}$$

Example 21 Divide 7 yd 2 ft 4 in by 4.

Solution

$$\begin{array}{l} \quad\ \boxed{1\text{ yd}} \\ \boxed{4}\overline{)\boxed{7\text{ yd}}\,2\text{ ft }4\text{ in}} \\ \quad\ \underline{4\text{ yd}} \\ \quad\ \boxed{3\text{ yd}} \end{array}$$

3 yd = 9 ft; add 9 ft + 2 ft = 11 ft

$$\begin{array}{l} \quad\ 1\text{ yd }\boxed{2\text{ ft}} \\ \boxed{4}\overline{)7\text{ yd}\,\boxed{11\text{ ft}}\,4\text{ in}} \\ \qquad\qquad \underline{8\text{ ft}} \\ \qquad\qquad \boxed{3\text{ ft}} \end{array}$$

3 ft = 36 in; add 36 in + 4 in = 40 in

$$\begin{array}{l} \quad\ 1\text{ yd }2\text{ ft}\,\boxed{10\text{ in}} \\ \boxed{4}\overline{)7\text{ yd }2\text{ ft}\,\boxed{40\text{ in}}} \\ \qquad\qquad\quad \underline{40\text{ in}} \\ \qquad\qquad\qquad 0 \end{array}$$

$$\text{Thus}\quad 4\overline{)7\text{ yd }2\text{ ft }\ 4\text{ in}}^{\ 1\text{ yd }2\text{ ft }10\text{ in}}$$

Example 22 Divide 4 gal 3 qt 1 pt by 6.

Solution

$$\begin{array}{r|l} & \boxed{0\text{ gal}} \\ \boxed{6} & \boxed{4\text{ gal}}\ 3\text{ qt}\ 1\text{ pt} \\ & \underline{0\text{ gal}} \\ & \boxed{4\text{ gal}} \end{array}$$

4 gal = 16 qt; add 16 qt + 3 qt = 19 qt

$$\begin{array}{r|l} & 0\text{ gal}\ \boxed{3\text{ qt}} \\ \boxed{6} & 4\text{ gal}\ \boxed{19\text{ qt}}\ 1\text{ pt} \\ & \quad\quad\ \underline{18\text{ qt}} \\ & \quad\quad\ \boxed{1\text{ qt}} \end{array}$$

1 qt = 2 pt; add 2 pt + 1 pt = 3 pt

$$\begin{array}{r|l} & 0\text{ gal}\ 3\text{ qt}\ \frac{1}{2}\text{ pt} \\ 6 & 4\text{ gal}\ 3\text{ qt}\ 3\text{ pt} \\ & \quad\quad\quad\quad\ \underline{3\text{ pt}} \\ & \quad\quad\quad\quad\ 0 \end{array}$$

When the nonmeasurement number does not divide evenly into the smallest unit of measure, express your answer in fractional form.

Sometimes it is possible to divide measurement numbers by measurement numbers.

Example 23 How many 5 feet long tomato stakes can a home gardener cut from a 20-foot board?

Solution

$$\begin{array}{r|l} & 4 \\ 5\text{ ft} & 20\text{ ft} \end{array}$$

The answer is a nonmeasurement number.

6.2 EXERCISES

In problems 1–14 simplify the given measurement numbers.

1. 18 in ______________________

2. 75 min ______________________

3. 12 da ______________________

4. 23 qt ______________________

5. 6 ft 19 in ______________________

6. 8 lb 45 oz ______________________

7. 2 wk 9 da ______________________

8. 5 tons 3,125 lb ______________________

9. 35 min 80 s ______________________

10. 10 yd 4 ft 15 in ______________________

11. 5 gal 6 qt 3 pt ______________________

12. 4 da 47 h 70 min ______________________

13. 3 yr 51 wk 10 da ______________________

14. 7 gals 7 qts 2 pts ______________________

In problems 15–48 perform the indicated operations and simplify.

15. 7 ft 3 in
+ 4 ft 10 in

16. 3 gal 2 qt
+ 1 gal 3 qt

17. 8 ft 9 in
9 ft 11 in
+ 10 ft 6 in

18. 12 lb 9 oz
10 lb 3 oz
+ 16 lb 14 oz

19. 6 yd 2 ft 11 in
+ 5 yd 1 ft 6 in

20. 3 h 45 min 38 s
+ 2 h 50 min 26 s

21. 4 gal 3 qt 1 pt
6 gal 2 qt 1 pt
+ 2 gal 3 qt

22. 5 yr 46 wk 6 da
3 yr 22 wk 3 da
+ 1 yr 14 wk 5 da

23. 10 ft 8 in
− 3 ft 6 in

24. 18 gal 3 qt
− 6 gal 2 qt

25. 8 wk 4 da
− 6 wk 5 da

26. 13 mi 2,000 ft
− 8 mi 3,500 ft

27. 3 gal 3 qt 1 pt
− 1 gal 2 qt 1 pt

28. 6 yd 2 ft 3 in
− 5 yd 1 ft 9 in

29. 7 yr 10 wk 2 da
− 5 yr 22 wk 6 da

30. 19 h 30 s
− 12 h 35 min 45 s

31. 26 yd 2 ft
− 17 yd 2 ft 10 in

32. 15 h
− 12 h 32 min 40 s

33. 6 wk 3 da 2 h
− 2 wk 6 da 15 h

34. 4 gal 1 qt
− 3 gal 3 qt 1 pt

35. 6 ft 7 in
× 4

36. 3 wk 4 da
× 6

37. 12 lb 10 oz
× 8

38. 2 gal 3 qt 1 pt
× 4

39. $\begin{array}{r} 3 \text{ h } 25 \text{ min } 40 \text{ s} \\ \times\ 6 \\ \hline \end{array}$

40. $\begin{array}{r} 9 \text{ yd } 2 \text{ ft } 6 \text{ in} \\ \times\ 11 \\ \hline \end{array}$

41. $5\overline{)10 \text{ tons } 500 \text{ lb}}$

42. $7\overline{)14 \text{ h } 35 \text{ min}}$

43. $6\overline{)8 \text{ ft } 6 \text{ in}}$

44. $8\overline{)2 \text{ tons } 400 \text{ lb}}$

45. $4\overline{)5 \text{ yd } 2 \text{ ft } 8 \text{ in}}$

46. $3\overline{)7 \text{ da } 23 \text{ h } 6 \text{ min}}$

47. $10\overline{)12 \text{ min } 5 \text{ s}}$

48. $4\overline{)1 \text{ gal } 5 \text{ qts } 1 \text{ pt}}$

ANSWERS

1. 1 ft 6 in **2.** 1 h 15 min **3.** 1 wk 5 da **4.** 5 gal 3 qt **5.** 7 ft 7 in **6.** 10 lb 13 oz **7.** 3 wk 2 da **8.** 6 tons 1,125 lb **9.** 36 min 20 s **10.** 11 yd 2 ft 3 in **11.** 6 gal 3 qt 1 pt **12.** 6 da 10 min **13.** 4 yr 3 da **14.** 9 gal **15.** 12 ft 1 in **16.** 5 gal 1 qt **17.** 29 ft 2 in **18.** 39 lb 10 oz **19.** 12 yd 1 ft 5 in **20.** 6 h 36 min 4 s **21.** 14 gal 1 qt **22.** 10 yr 32 wk **23.** 7 ft 2 in **24.** 12 gal 1 qt **25.** 1 wk 6 da **26.** 4 mi 3,780 ft **27.** 2 gal 1 qt **28.** 1 yd 6 in **29.** 1 yr 39 wk 3 da **30.** 6 h 24 min 45 s **31.** 8 yd 2 ft 2 in **32.** 2 h 27 min 20 s **33.** 3 wk 3 da 11 h **34.** 1 qt 1 pt **35.** 26 ft 4 in **36.** 21 wk 3 da **37.** 101 lb **38.** 11 gal 2 qt **39.** 20 h 34 min **40.** 108 yd 6 in **41.** 2 tons 100 lb **42.** 2 h 5 min **43.** 1 ft 5 in **44.** 550 lb **45.** 1 yd 1 ft 5 in **46.** 2 da 15 h 42 min **47.** 1 min $12\frac{1}{2}$ s **48.** 2 qt $\frac{3}{4}$ pt

6.3 METRIC SYSTEM

If you were to take a motor trip from the United States to Canada, upon crossing the border you would probably see highway signs that give the following kind of information: Montreal 166 km (166 kilometers), speed limit 80 km (80 kilometers).

If you stopped to fill your gas tank, the price of gasoline might be listed at 30 cents per liter. If you stopped to photograph a deer browsing in a meadow you might attach a 400mm (400-millimeter) telephoto lens to a 35mm (35-millimeter) camera. If you listened to the radio, the high temperature for that particular day might be given as 20°C (20 degrees Celsius). The immediate questions that enter your mind are as follows: How many miles are equivalent to 166 km? To 80 km? How many liters are there in a gallon? How many Fahrenheit degrees are equivalent to 20° Celsius? What does it mean when a camera is described as a 35mm camera?

Why the different systems of measurement? The United States is the only major nation in the world that is not using what is commonly called the **metric system** (International System of Units) of measurement. The United States still uses the **English system** of measurement. Currently, there are programs planned in the United States to convert from the English system of measurement to the metric system.

Why change our system of measurement? There are two main reasons:

(1) A uniform system of measurement will then exist among all the major nations of the world. So when exporting and importing goods it will not be necessary to convert from one system to the other. For example, goods can be shipped in standard-sized containers.

(2) The metric system of measurement is based on the decimal system and is thus consistent with the number system that we use. Computations within the metric system involve multiples and divisions of (powers of) 10, whereas computations within the English system necessitate the need to remember a body of numerical facts relating one unit of measure to another. For example, in order to convert pints to gallons it is necessary to know that there are 2 pints in a quart and 4 quarts in a gallon. To convert feet to inches it is necessary to know there are 12 inches in a foot.

We will first examine the basic units of measure in the metric system and show how to convert from one unit of measure to another. Then we will show how to convert from the English system of measurement to the metric system, and vice versa.

BASIC METRIC UNITS

1. **Meter:** The meter (denoted m) is the basic unit of length in the metric system. It is slightly longer than a yard.

1 meter = 1.094 yard = 39.37 inches

1 yard = 36 inches

2. **Gram:** The gram (denoted g) is the basic unit of weight in the metric system. A gram is roughly equivalent to the weight of an ordinary paper clip.

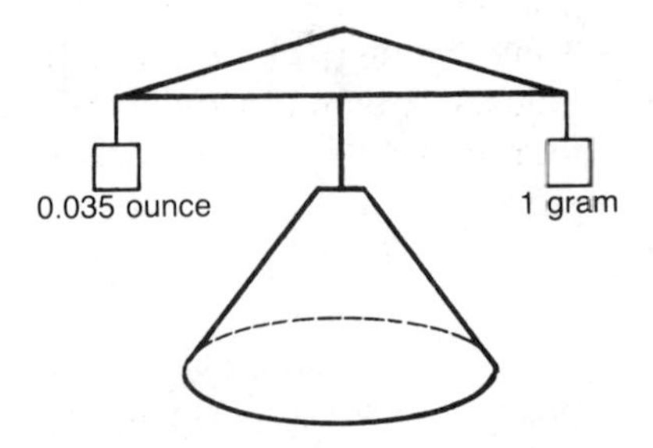

3. **Liter:** The liter (denoted l) is the basic unit of liquid volume in the metric system. A liter is a little more than a quart.

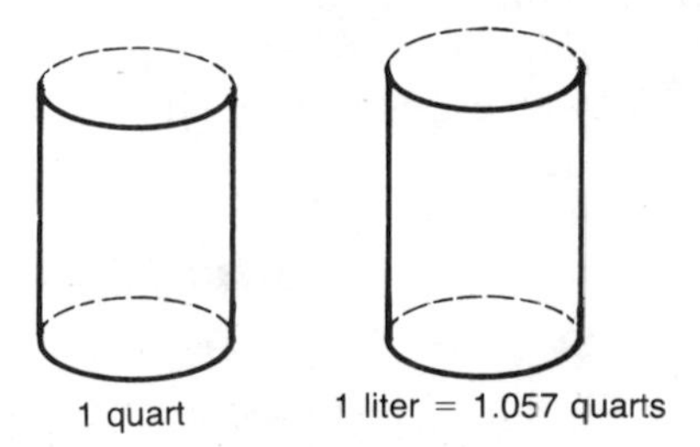

4. **Celsius Degree:** The Celsius degree (°C) is the basic unit of temperature in the metric system.

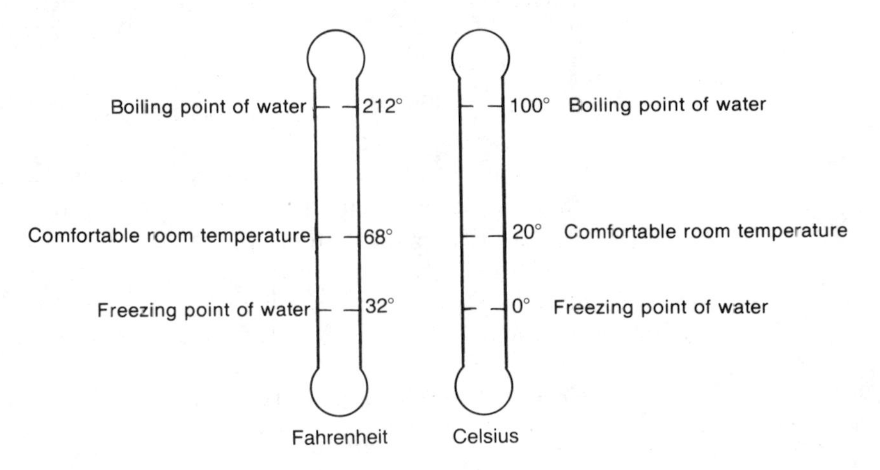

To denote larger or smaller portions of the basic units of measure we can attach prefixes to the words representing the basic units. Some commonly used metric prefixes are given in Table 6.6.

Table 6.6
Common Metric Prefixes

Metric Prefix:	mega	kilo	hecto	deca	Basic Unit	deci	centi	milli	micro
Symbol:	M	k	h	da		d	c	m	μ
Multiple of Basic Unit:	1,000,000 or 10^6	1,000 or 10^3	100 or 10^2	10 or 10^1	1 or 10^0	$\frac{1}{10} = 0.1$ or 10^{-1}	$\frac{1}{100} = 0.01$ or 10^{-2}	$\frac{1}{1,000} = 0.001$ or 10^{-3}	$\frac{1}{1,000,000} = 0.000001$ or 10^{-6}

From the table we can establish that:

1 *kilo*meter (km) = 1,000 meters (m), or $\frac{1}{1{,}000}$ km = 1 m

1 *micro*gram (μg) = $\frac{1}{1{,}000{,}000}$ gram (g), or 1,000,000 μg = 1 g

1 *hecto*liter (hl) = 100 liters (l), or $\frac{1}{100}$ hl = 1 l

1 *centi*meter (cm) = $\frac{1}{100}$ meter (m), or 100 cm = 1 m

Note: You should memorize the metric prefixes, their symbols, and their relative positions with respect to one another. This will help you to convert from one unit of measure to another.

Converting from One Unit of Measure to Another

Conversion from one unit of measure to another in the metric system involves multiplying or dividing by powers of 10. This process can also be performed by movement of the decimal point. The following examples will illustrate methods of converting from a larger unit of measure to a smaller unit of measure.

Example 1 Convert 3 meters to centimeters.

Solution 1 From Table 6.6 we can see that we are changing from a larger unit of measure to a smaller one. Using the table,

1 m = 100 cm
Thus 3 m = 3 × 100 cm = 300 cm

Solution 2 From the table we can see that to change from meters to centimeters we move 2 places to the right.

meters	decimeters	centimeters
m	dm	cm

To change from meters to centimeters we move the decimal point 2 places to the right:

3 m = 3.00. cm = 300 cm

Example 2 Convert 5.6 kilograms to grams.

Solution 1 Using the table,

1 kg = 1000 g
Thus 5.6 kg = 5.6 × 1,000 g = 5,600 g

Solution 2 From the table we can determine that to change from kilograms to grams we move 3 places to the right.

kilograms	hectograms	decagrams	basic unit
kg	hg	dag	g

To change from kilograms to grams we move the decimal point 3 places to the right.

5.6 kg = 5.600. g = 5,600 g

Example 3 Convert 10.6 deciliters to milliliters.

Solution 1 Using the table,

$$1 \text{ dl} = \frac{1}{10}\text{l}$$

$$1 \text{ l} = 1000 \text{ m}$$

Thus $1 \text{ dl} = \frac{1}{10}(1{,}000 \text{ mm}) = 100 \text{ mm}$

Then $10.6 \text{ dl} = 10.6 \times 100 \text{ ml} = 1{,}060 \text{ ml}$

Solution 2 To change from deciliters to milliliters it can be shown that we move 2 places to the right.

deciliters	centiliters	milliliters
dl	cl	ml

To change from deciliters to milliliters we move the decimal point 2 places to the right.

10.6 dl = 10.60. ml = 1,060 ml

The easiest of the two solutions used in each of the preceding examples was the one that involved movement of the decimal point.

> To change from a larger unit of measure to a smaller unit of measure by moving the decimal point count the number of places the smaller unit is *to the right* of the larger unit and move the decimal point an *equivalent number of places to the right.*

Example 4 Convert 7.5 kilometers to decimeters.

Solution Decimeters are 4 places to the right of kilometers.

km hm dam m dm

Thus 7.5 km = 7.5000. dm = 75,000 dm

Example 5 Convert 6 hectograms to milligrams

Solution Milligrams are 5 places to the right of hectograms.

hg dag g dg cg mg

Thus 6 hg = 6.00000. mg = 600,000 mg

Example 6 Convert 8.7 kiloliters to decaliters.

Solution Decaliters are 2 places to the right of kiloliters.

kl hl dal

Thus 8.7 kl = 8.70. dal = 870 dal

The next group of examples will illustrate the methods of converting from smaller units of measure to larger units of measure.

Example 7 Convert 300 centimeters to meters.

Solution 1 From the table we can see that we are converting from a smaller unit of measure to a larger unit of measure. Using the table,

$$1 \text{ cm} = \frac{1}{100} \text{ m}$$

Thus $300 \text{ cm} = 300 \times \frac{1}{100} \text{ m} = 3 \text{ m}$

Solution 2 From the table we can see that to change from centimeters to meters we move 2 places to the left.

meters	decimeters	centimeters
m	dm	cm

To change from centimeters to meters we move the decimal point 2 places to the left.

300 cm = 3.00. m = 3m

Example 8 Convert 92.7 decigrams to kilograms.

Solution 1 Using the table,

$$1 \text{ dg} = \frac{1}{10} \text{ g}$$

$$1 \text{ g} = \frac{1}{1{,}000} \text{ kg}$$

Thus $1 \text{ dg} = \frac{1}{10} \times \frac{1}{1{,}000} \text{ kg} = \frac{1}{10{,}000} \text{ kg}$

Then $92.7 \text{ dg} = 92.7 \times \frac{1}{10{,}000} \text{ kg} = 92.7 \times 0.0001$

$= 0.00927 \text{ kg}$

Solution 2 From the table we can see that to change from decigrams to kilograms we move 4 places to the left.

kilograms	hectograms	decagrams	grams	decigrams
kg	hg	dag	g	dg

To change from decigrams to kilograms we move the decimal point 4 places to the left.

92.7 dg = .0092.7 kg = 0.00927 kg

Example 9 Convert 2,000 decaliters to hectoliters.

Solution 1 Using the table,

$$1 \text{ dal} = 10 \text{l}$$

$$1 \text{ l} = \frac{1}{100} \text{ hl}$$

Thus $1 \text{ dal} = 10 \times \frac{1}{100} \text{ hl} = \frac{1}{10} \text{ hl}$

Then $2000 \text{ dal} = 2000 \times \frac{1}{10} \text{ hl} = 200 \text{ hl}$

Solution 2 From the table we can see that to change from decaliters to hectoliters we move one place to the left.

hectoliters decaliters

hl dal

To change from decaliters to hectoliters we move the decimal point one place to the left.

2,000 dal = 200.0. hl = 200 hl

> To change from a smaller unit of measure to a larger unit of measure by moving the decimal point count the number of places the larger unit is *to the left* of the smaller unit and move the decimal point an *equivalent number of places to the left.*

Example 10 Convert 8.4 meters to kilometers.

Solution Kilometers are 3 places to the left of meters.

km hm dam m

Thus 8.4 m = .008.4 km = 0.0084 km

Example 11 Convert 25,700 milligrams to hectograms.

Solution Hectograms are 5 places to the left of milligrams.

hg dag g dg cg mg

Thus 25,700 mg = .25,700. hg = 0.257 hg

Example 12 Convert 300 decaliters to kiloliters.a6

Solution Kiloliters are 2 places to the left of decaliters.

kl hl dal

Thus 300 dal = 3.00. kl = 3kl

Example 13 Find the area of the following rectangle in centimeters and in meters.

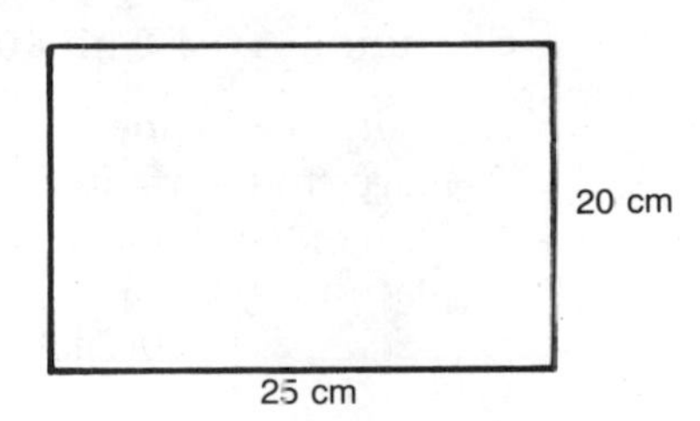

Solution *Area in centimeters:*

$A = l \times w$

$= (25\text{ cm})(20\text{ cm})$

$= 500$ square centimeters

Area in meters:

Convert the length and width of rectangle to meters.

25 cm = 0.25 m and 20 cm = 0.20 m

Find the area:

$A = l \times w$

$= (0.25\text{ m})(0.20\text{ m})$

$= 0.05$ square meters

Is it possible to use the method of moving the decimal point to convert larger units of area to smaller ones and vice versa? Observe:

1 m = 10 dm
1 m × 1 m = 10 dm × 10 dm
Thus 1 sq m = 100 sq dm

To move from one unit of area to the *next smaller* unit of area means multiplying by 100 or moving the decimal point 2 places to the right.

1 sq m = 1.00. sq dm = 100 sq dm

To move from one unit to the *next larger* unit of area means dividing by 100 or moving the decimal point 2 places to the left.

100 sq dm = 1.00. sq m = 1 sq m

Observe once more:

1 m = 100 cm
1 m × 1 m = 100 cm × 100 cm
Thus 1 sq m = 10,000 sq cm

To move 2 units of area to the right means multiplying by 10,000 or moving the decimal point 4 places to the right.

1 sq m = 1.0000. sq cm = 10,000 sq cm

To move 2 units of area to the left means dividing by 10,000 or moving the decimal point 4 places to the left.

10,000 sq cm = 1.0000. sq m = 1 sq m

Changing Area Units by Moving the Decimal Point

1. *Each* move of one unit of area to the right means moving the decimal point 2 places to the right.
2. *Each* move of one unit of area to the left means moving the decimal point 2 places to the left.

Example 14 325 sq km = ________ sq m.

Solution Square meters are 3 places to the right of square kilometers.

Thus the decimal point is moved 6 places to the right:
325 sq km = 325.000000. sq m = 325,000,000 sq m

Example 15 500 sq cm = ________ sq m.

Solution Square meters are 2 places to the left of square centimeters.

Thus the decimal point is moved 4 places to the left:
500 sq cm = .0500. sq m = 0.05 sq m

Example 16 Find the volume of the illustrated cone in kilometers and in meters.

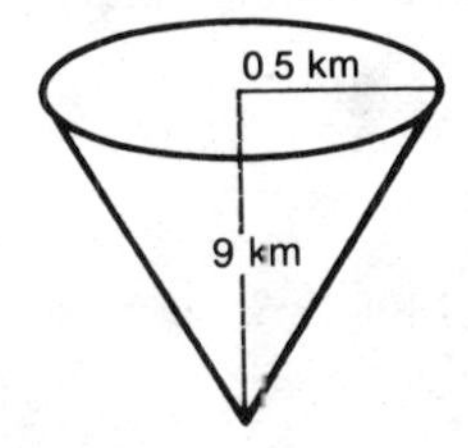

Solution *Volume in kilometers:*

$$V = \frac{1}{3} \times \pi \times r^2 \times h$$

$$= \frac{1}{\not 3} \times 3.14 \times (0.5 \text{ km})^2 \times (\overset{3}{\not 9} \text{ km})$$

$$= 3.14 \times (0.25 \text{ sq km}) \times (3 \text{ km})$$

$$= 2.355 \text{ cubic kilometers}$$

Volume in meters:

Convert the radius and height of cone to meters.
0.5 km = 500 m and 9 km = 9,000 m

$$V = \frac{1}{3} \times \pi \times r^2 \times h$$

$$= \frac{1}{3} \times \pi \times (500 \text{ m})^2 \times (\overset{3\,000}{\not{9,000}} \text{ m})$$

$$= 3.14 \times (250{,}000 \text{ sq m}) \times (3{,}000 \text{ m})$$

$$= 2{,}355{,}000{,}000 \text{ cubic meters}$$

We can also use the method of moving the decimal point to convert larger units of volume to smaller units of volume, and vice versa. Observe:

1 m = 10 dm
1 m × 1 m × 1 m = 10 dm × 10 dm × 10 dm
1 cu m = 1000 cu dm

To move from one unit of volume to the *next smaller* unit of volume means multiplying by 1,000 or moving the decimal point 3 places to the right.

1 cu m = 1.000. cu dm = 1,000 cu dm

To move from one unit of volume to the *next larger* unit of volume means dividing by 1,000 or moving the decimal point 3 places to the left.

1,000 cu dm = 1.000. cu m = 1 cu m

Observe once more:

$$\begin{aligned} 1 \text{ m} &= 100 \text{ cm} \\ 1\text{m} \times 1\text{m} \times 1 \text{ m} &= 100 \text{ cm} \times 100 \text{ cm} \times 100 \text{ cm} \\ 1 \text{ cu m} &= 1{,}000{,}000 \text{ cu cm} \end{aligned}$$

To move 2 units of volume to the right means multiplying by 1,000,000 or moving the decimal point 6 places to the right.

1 cu m = 1.000000. cu cm = 1,000,000 cu cm

To move 2 units of volume to the left means dividing by 1,000,000 or moving the decimal point 6 places to the left.

1,000,000 cu cm = 1.000000. cu m = 1 cu m

Changing Volume Units by Moving the Decimal Point

1. *Each* move of one unit of volume to the right means moving the decimal point 3 places to the right.
2. *Each* move of one unit of volume to the left means moving the decimal point 3 places to the left.

Example 17 2.355 cu km = ________________ cu m.

Solution Cubic meters are 3 places to the right of cubic kilometers.

Thus the decimal point is moved 9 places to the right.
2.355 cu km = 2.355000000. cu m
= 2,355,000,000 cu m

Example 18 6.2 cu mm = ________________ cu dm.

Solution Cubic decimeters are 2 places to the left of cubic millimeters.

Thus the decimal point is moved 6 places to the left.
6.2 cu mm = 0.000006.2 cu dm
= 0.0000062 cu dm

6.3 EXERCISES

1. Fill in the blanks.

   (a) If a kilometer is 1,000 meters, then a kilowatt is ________ watts.

   (b) A decagon has ________ sides.

   (c) A decade is ________ years.

In problems 2–25 convert to the units shown.

2. 3.5 g = ____________ mg
3. 5,000 m = ____________ km
4. 3.6 kl = ____________ l
5. 65 cm = ____________ dm
6. 9.4 kl = ____________ ml
7. 42.3 dag = ____________ kg
8. 12 ml = ____________ l
9. 15 dg = ____________ dag
10. 600 m = ____________ hm
11. 0.304 kl = ____________ cl
12. 24.6 mg = ____________ hg
13. 250 cm = ____________ dam
14. 25 hg = ____________ dg
15. 6.5 km = ____________ m
16. 25.8 dal = ____________ cl
17. 2,500 cm = ____________ m
18. 0.95 dag = ____________ mg
19. 3,450 kg = ____________ g
20. 0.286 dal = ____________ cl
21. 32,455 mm = ____________ hm
22. 12 sq cm = ____________ sq m
23. 6.5 sq km = ____________ sq m
24. 10.6 cu mm = ____________ cu cm
25. 0.151 cu m = ____________ cu dm

26. Find the perimeter of the given figure in meters and in kilometers.

____

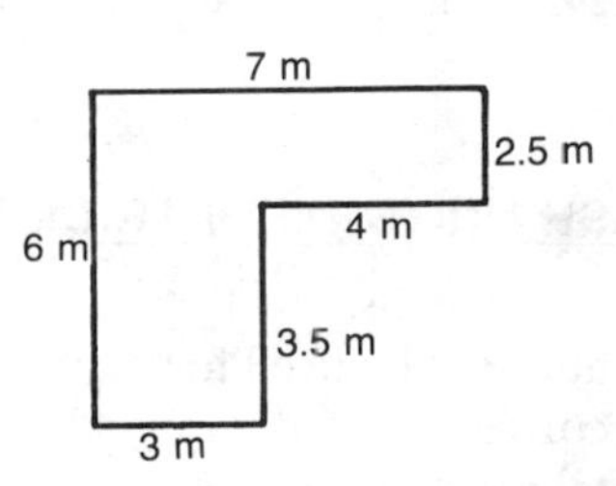

27. Find the area of the given triangle in square centimeters and in square meters.

____

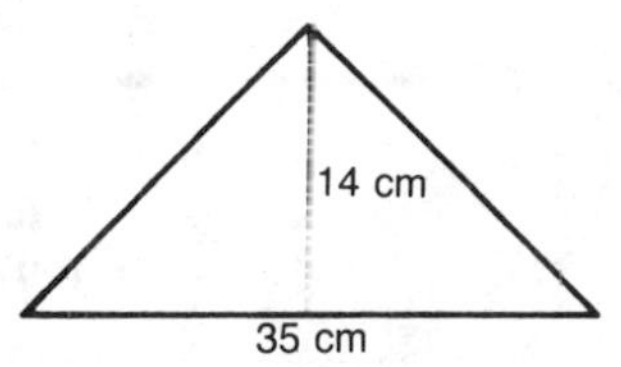

28. Find the area of the following rectangle in square meters and in square millimeters.

____

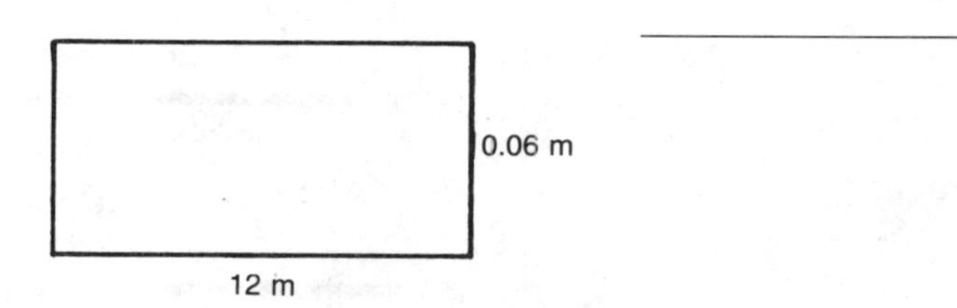

29. Find the volume of the given rectangular solid in cubic kilometers and in cubic meters.

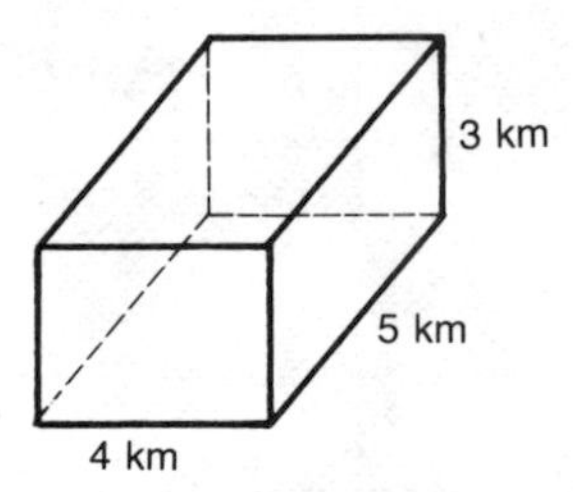

30. Find the volume of the given cylinder in cubic centimeters and in cubic meters.

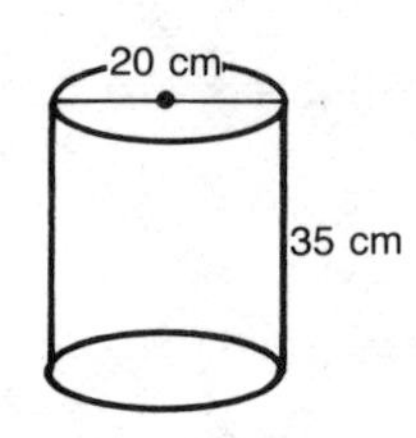

ANSWERS

1. (a) 1,000 **(b)** 10 **(c)** 10 **2.** 3,500 **3.** 5 **4.** 3,600 **5.** 6.5 **6.** 9,400,000 **7.** 0.423 **8.** 0.012 **9.** 0.15 **10.** 6 **11.** 30,400 **12.** 0.000246 **13.** 0.250 **14.** 25,000 **15.** 6,500 **16.** 25,800 **17.** 25 **18.** 9,500 **19.** 3,450,000 **20.** 286 **21.** 0.32455 **22.** 0.0012 **23.** 6,500,000 **24.** 0.0106 **25.** 151 **26.** 26 m; .026km **27.** 245 sq cm; 0.0245 sq m **28.** 0.72 sq m; 720,000 sq mm **29.** 60 cu km; 60,000,000,000 cu m **30.** 43,960 cu cm; 0.043960 cu m

6.4 CONVERTING ENGLISH UNITS TO METRIC UNITS, AND VICE VERSA

Because the English system of measure continues to be used in the United States and will be used for many years it is necessary and important for us to learn how to convert from the English system to the metric system and vice versa. The following tables will help us in making the conversions. *The conversions are approximate.*

Table 6.7 **English-Metric Conversion Table — Length**

1 inch = 2.54 centimeters	1 centimeter = 0.394 inch
1 foot = 0.305 meter	1 meter = 39.37 inches
1 yard = 0.914 meter	= 3.28 feet
1 mile = 1.61 kilometers	= 1.09 yards
	1 kilometer = 0.621 mile

Table 6.8 **English-Metric Conversion Table — Weight**

1 ounce = 28.35 grams	1 gram = 0.035 ounce
1 pound = 454 grams	1 kilogram = 2.20 pounds
= 0.454 kg	

When more than one conversion method is applicable, answers may differ due to rounding off in the tables.

Table 6.9 **English-Metric Conversion Table — Liquid Capacity**

1 pint = 0.474 liter	1 liter = 2.11 pints
1 quart = 0.946 liter	= 1.06 quarts
1 gallon = 3.785 liters	= 0.264 gallon
	= 33.8 ounces

Examples 1–4 illustrate the conversion of English units to metric units. We will begin the process by setting up a proportion.

Example 1 20 pounds = ________ grams.

Solution Let x be the number of grams equivalent to 20 pounds. Since 1 lb = 454 grams, we can set up a proportion.

$$\frac{20 \text{ lb}}{x \text{ grams}} = \frac{1 \text{ lb}}{454 \text{ grams}}$$

$$x = 20 \times 454$$

$$x = 9{,}080 \text{ grams}$$

Example 2 10 quarts = ________ liters.

Solution Let x be the number of liters equivalent to 10 quarts. Since 1 qt = 0.946 liter, we can set up a proportion:

$$\frac{10 \text{ quarts}}{x \text{ liters}} = \frac{1 \text{ quart}}{0.946 \text{ liter}}$$

$$x = 10 \times 0.946$$
$$= 9.46 \text{ liters}$$

Example 3 5 yards = ________ meters.

Solution Let x be the number of meters equivalent to 0.5 yards. Since 1 yard = 0.914 meter, we can set up a proportion.

$$\frac{0.5 \text{ yard}}{x \text{ meters}} = \frac{1 \text{ yard}}{0.914 \text{ meter}}$$

$$x = 0.5 \times 0.914$$
$$= 0.457 \text{ meter}$$

Example 4 50 inches = ________ millimeters.

Solution Let x be the number of centimeters equivalent to 50 inches.

step 1: Since 1 inch = 2.54 cm, we can set up a proportion.

$$\frac{50 \text{ inches}}{x \text{ cm}} = \frac{1 \text{ inch}}{2.54 \text{ cm}}$$

$$x = 50 \times 2.54$$

$$x = 127 \text{ cm}$$

step 2: 127 cm = 127.0. mm = 1,270 mm

Examples 5–8 illustrate the conversion of metric units to English units. We will also begin this process by setting up a proportion.

Example 5 50 kg = ________ pounds.

Solution Let x be the number of pounds equivalent to 50 kg. Since 1 kg = 2.20 lb, we can set up a proportion.

$$\frac{50 \text{ kg}}{x \text{ lb}} = \frac{1 \text{ kg}}{2.20 \text{ lb}}$$

$$x = 110 \text{ lb}$$

Example 6 15 kilometers = ________ miles.

Solution Let x be the number of miles equivalent to 15 km. Since 1 km = 0.621 mi, we can set up a proportion.

$$\frac{15 \text{ km}}{x \text{ mi}} = \frac{1 \text{ km}}{0.621 \text{ mi}}$$

$$\begin{aligned} x &= 15 \times 0.621 \\ &= 9.315 \text{ mi} \end{aligned}$$

Example 7 0.75 liter = ________ pints.

Solution Let x be the number of pints equivalent to 0.75 liter. Since 1 liter = 2.11 pints, we can set up a proportion.

$$\frac{0.75 \text{ liter}}{x \text{ pints}} = \frac{1 \text{ liter}}{2.11 \text{ pints}}$$

$$x = 0.75 \times 2.11$$

$$x = 1.5825 \text{ pints}$$

Example 8 10 kilograms = ________ ounces.

Solution Let x be the number of pounds equivalent to 10 kg.

step 1: Since 1 kg = 2.20 lb, we can set up a proportion.

$$\frac{10 \text{ kg}}{x \text{ lb}} = \frac{1 \text{ kg}}{2.20 \text{ lb}}$$

$$x = 10 \times 2.20$$

$$x = 22 \text{ lbs}$$

step 2: Since

$$\begin{aligned} 1 \text{ lb} &= 16 \text{ oz} \\ 22 \text{ lb} &= 22 \times 16 \text{ oz} \\ &= 352 \text{ oz} \end{aligned}$$

Temperature Conversion

To change from Celsius to Fahrenheit use the following formula:

$$F = \frac{9}{5}C + 32 \text{ where}$$

F is the number of Fahrenheit degrees
C is the number of Celsius degrees

Example 9 20°C = ________ °F.

Solution $F = \frac{9}{5}(20) + 32$ (multiply $\frac{9}{5}$ (20) first)

$= 36 + 32$

$= 68°F$

Example 10 12°C = ________ °F.

Solution $F = \frac{9}{5}(12) + 32$

$= \frac{108}{5} + 32$

$= 21.6 + 32$

$= 53.6°F$

$\doteq 54°F$ (rounded off to the nearest degree)

> To change from Fahrenheit to Celsius use the following formula:
>
> $$C = \frac{5}{9}(F - 32) \text{ where}$$
>
> F is the number of Fahrenheit degrees
> C is the number of Celsius degrees

Example 11 32°F = ________ °C.

$C = \frac{5}{9}(32 - 32)$ (subtract numbers within parentheses first)

$= \frac{5}{9}(0)$

$= 0°C$

Example 12 60°F = ________ °C.

Solution $C = \frac{5}{9}(60 - 32)$

$= \frac{5}{9}(28)$

$= \frac{140}{9}$

$= 15.55°C$

$\doteq 16°C$ (rounded off to the nearest degree)

6.4 EXERCISES

In problems 1–24 convert the given values to the indicated units.

1. 12 oz = ________________ g

2. 26.2 mi = ________________ km

3. 8.5 l = ________________ gal

4. 180 lb ________________ kg

5. 20 ft = ______________ m

6. 1040 g = ______________ lb

7. 215 cm = ______________ in

8. 135.2 oz = ______________ l

9. 12 m = ______________ in

10. 16.5 kg = ______________ oz

11. 8 qt = ______________ l

12. 100 oz = ______________ g

13. 12 dm = ______________ ft

14. 3 kg = ______________ lb

15. 25 l = ______________ pt

16. 40 km = ______________ mi

17. 30 gal = ______________ l

18. 2 lb = ______________ g

19. 35 pt = ______________ l

20. 120 in = ______________ mm

21. 32 m = ______________ yd

22. 20.5 l = ______________ qt

23. 3 mi = ______________ m

24. 150 cm = ______________ ft

25. Convert the following Fahrenheit temperatures to Celsius.

(a) 212°F ______________ **(b)** 86°F ______________

(c) 68°F ______________ **(d)** 41°F ______________

26. Convert the following Fahrenheit temperatures to Celsius. (Round off to the nearest degree.)

(a) 72°F ______________ **(b)** 46°F ______________

(c) 100°F ______________ **(d)** 52°F ______________

27. Convert the following Celsius temperatures to Fahrenheit.

(a) 75°C ______________ **(b)** 20°C ______________

(c) 5°C ______________ **(d)** 30°C ______________

28. Convert the following Celsius temperatures to Fahrenheit. (Round off to the nearest degree.)

(a) 32°C ______________ **(b)** 14°C ______________

(c) 9°C ______________ **(d)** 21°C ______________

29. The highest weather temperature ever recorded is 134°F. Convert this temperature to Celsius. (Round off to the nearest degree.)

____

30. Joseph is 107 centimeters tall, Rosa's height is 110 centimeters and Eric's height is 115 centimeters. What is their total height in feet? (Round off to the nearest hundredth.)

____

31. The tallest known California redwood tree is 112.413 meters tall. How tall is this tree in yards? (Round off to the nearest hundredth.)

____

32. Sidney's mother purchased a bottle of 25 vitamin C tablets. Each tablet contains 500 milligrams. How many milligrams are in this bottle? How many grams?

____

33. An American-made car has a 120-inch 304.8 wheelbase and a European car has a 250-centimeter wheelbase. What is the wheelbase of the American-made car in centimeters?

____

What is the wheelbase of the European car in inches? ______________

34. A speed limit sign reads 80 kilometers per hour. What is this speed in miles per hour?

____

35. The approximate distance of the earth from the sun is 93,000,000 miles. What is this distance in kilometers?

____

ANSWERS

1. 340.2 **2.** 42.182 **3.** 2.244 **4.** 81.72 **5.** 6.1 **6.** 2.288 **7.** 84.71 **8.** 4 **9.** 472.44 **10.** 580.8 **11.** 7.568 **12.** 2,835 **13.** 3.936 **14.** 6.6 **15.** 52.75 **16.** 24.84 **17.** 113.55 **18.** 908 **19.** 16.59 **20.** 3,048 **21.** 34.88 **22.** 21.73 **23.** 4,830 **24.** 4.925 **25. (a)** 100°C **(b)** 30°C **(c)** 20°C **(d)** 5°C **26. (a)** 22°C **(b)** 8°C **(c)** 38°C **(d)** 11°C **27. (a)** 167°F . **(b)** 68°F **(c)** 41°F **(d)** 86°F **28. (a)** 90°F **(b)** 57°F **(c)** 48°F **(d)** 70°F **29.** 57°C **30.** 10.89 **31.** 122.53 **32.** 12,500 mg; 12.5 g **33.** 304.8 cm; 98.5 in **34.** 49.68 mi/h **35.** 149,730,000 km

FOCUS ON PROBLEM SOLVING

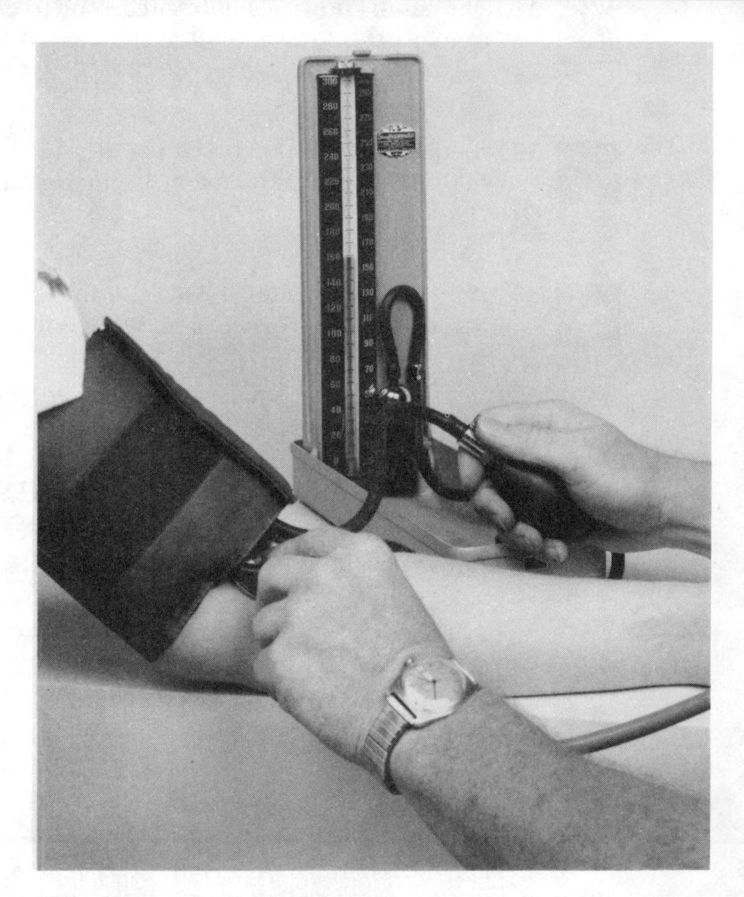

Having your blood pressure taken is an important part of any medical check-up.

In order for a doctor or nurse to administer dosages of medicine, he or she must understand the system used for weighing and measuring the drugs. The use of the metric system in administering drugs is becoming prevalent, rapidly replacing a system known as the apothecaries system.

Of the prefixes used in the metric system, only three are used with frequency in a hospital situation; kilo, centi, and milli. For example, the sphygmomanometer is an instrument used to measure blood pressure. The scale on this instrument is marked in millimeters. This is also true of other very precise instruments.

The volume of a liter is defined to be the amount contained in a cube 1 decimeter (dm) in length. Since 10 cm = 1 dm, 1,000 cubic centimeters (cc) = 1 cubic decimeter, and 1000 cc would have the capacity of 1 liter. Also, 1 liter = 1,000 ml; thus if there are 1000 ml in a liter and 1,000 cc in a liter, then 1 cc would have the capacity of 1 ml. This relationship is important because drugs are usually ordered in relation to their volume, but are measured according to the cc capacity of the instrument in which they are given. Thus a syringe and medicine glass are marked according to the accompanying picture.

A typical day in the life of a nurse may require her to solve some problems in the metric system. If the nurse is required to administer $2\frac{1}{2}$ ml of a drug, by syringe, it would be necessary for her to know that 1 cc = 1 ml. Thus 2 cc = 2 ml would be 1 full syringe (2-cc capacity); and since $\frac{1}{2}$ cc = $\frac{1}{2}$ ml, the syringe would then be filled to the $\frac{1}{2}$-cc mark.

Also the nurse is asked to compute the amount of a drug for a patient who weighs 160 pounds. Five milligrams is to be given for every pound of body weight. The nurse must first convert from pounds to kilograms, as follows.

$$1 \text{ lb} = 0.454 \text{ kg}$$
$$160 \text{ lb} = 72.64 \text{ kg}$$

To determine the amount of drug to administer she must multiply the number of kilograms of body weight by 5.

$$\begin{aligned}\text{Amount of Drug} &= 5(72.64)\\ &= 363.2 \text{ milligrams}\end{aligned}$$

BUILDING YOUR MATH VOCABULARY

TERM	DEFINITION	EXAMPLE
Celsius Scale, C	This is the metric scale for measuring temperature.	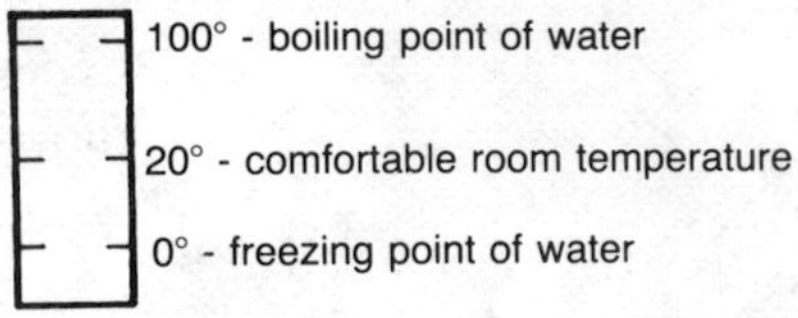 100° - boiling point of water 20° - comfortable room temperature 0° - freezing point of water
English System of Measure	Basically this is the system of measurement used in the United States.	Units: inch, foot, yard, mile; pint, quart, gallon, ounce, pound, ton; second, minute, hour, day, week
Fahrenheit Scale, F	This is the English scale for measuring temperature.	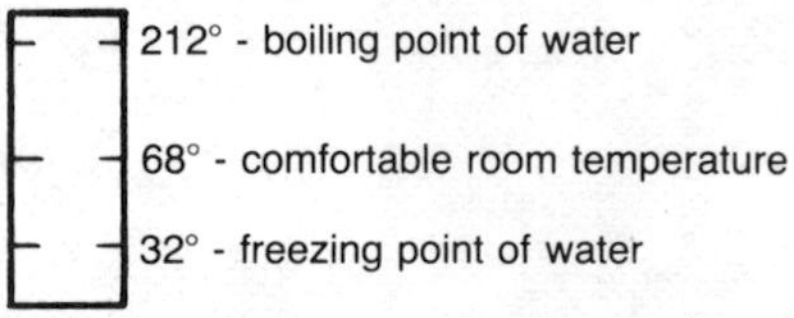 212° - boiling point of water 68° - comfortable room temperature 32° - freezing point of water
Gram, g	The standard metric unit of weight. A gram is about the weight of an ordinary paper clip.	1 gram = 0.035 ounce 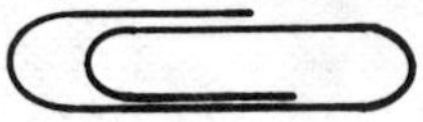
Liter, l	The standard metric unit of liquid volume. A liter is a little more than a quart.	1 liter = 1.06 quart 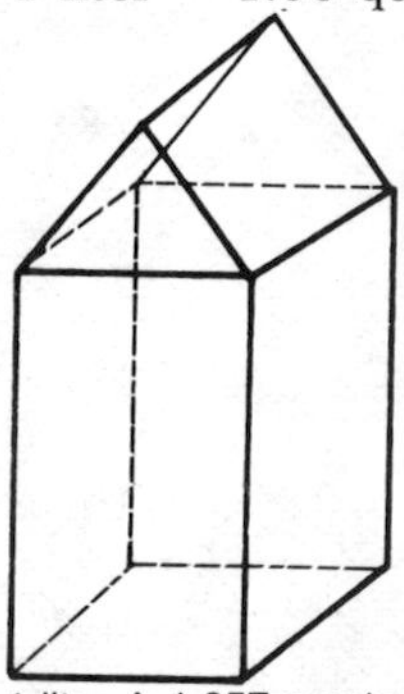 1 liter $\doteq$ 1.057 quarts
Meter, m	The standard metric unit of length. A meter is slightly longer than a yard.	1 meter = 3.28 feet  1 meter $\doteq$ 1.094 yards $\doteq$ 39.37 inches
Metric System	The international system of measurement used by all major countries of the world with the exception of the United States.	Units: meter, gram, liter

6

Chapter Review

In problems 1–20 convert to the units shown.

1. 72 in = ______ ft

2. 2.5 gal = ______ qt

3. 2.6 mi = ______ ft

4. $2\frac{1}{4}$ tons = ______ lb

5. $5\frac{1}{2}$ pt = ______ fl oz

6. 5 tbsp = ______ tsp

7. 19 qt = ______ gal

8. 1,460 da = ______ yr

9. 88 tbsp = ______ cups

10. 65 wk = ______ yr

11. 5 pt = ______ qt

12. 80 oz = ______ lb

13. 7 gal = ______ pt

14. 3 cups = ______ tsp

15. $2\frac{1}{2}$ da = ______ min

16. $1\frac{1}{4}$ mi = ______ yd

17. 4.2 yd = ______ in

18. 11.2 fl oz = ______ cups

19. 52 mo = ______ yr

20. 504 h = ______ wk

In problems 21–30 simplify the given measurement numbers.

21. 130 min ______

22. 60 wk ______

23. 12 yd 8 ft ______

24. 3 gal 7 qt ______

25. 4 tons 5,120 lb ______

26. 12 h 210 min ______

27. 6 yd 9 ft 18 in ______

28. 18 h 320 min 86 s ______

29. 8 gal 5 qt 7 pt ______

30. 5 yr 50 wk 20 days ______

In problems 31–40 perform the indicated operations and simplify.

31.
$$\begin{array}{rrr} & \text{4 tons} & \text{1650 lb} \\ + & \text{5 tons} & \text{1040 lb} \\ \hline \end{array}$$

32.
$$\begin{array}{rrrr} & \text{4 gal} & \text{3 qt} & \text{1 pt} \\ & \text{10 gal} & \text{2 qt} & \text{1 pt} \\ + & \text{6 gal} & \text{3 qt} & \text{1 pt} \\ \hline \end{array}$$

33.
$$\begin{array}{rrr} & \text{35 min} & \text{14 s} \\ - & \text{22 min} & \text{50 s} \\ \hline \end{array}$$

34.
$$\begin{array}{rrrr} & \text{6 yd} & \text{1 ft} & \text{4 in} \\ - & \text{3 yd} & \text{2 ft} & \text{9 in} \\ \hline \end{array}$$

35.
$$\begin{array}{rrrr} & \text{19 yr} & & \text{3 da} \\ - & \text{12 yr} & \text{16 wk} & \text{8 da} \\ \hline \end{array}$$

36.
$$\begin{array}{rr} \text{12 lb} & \text{14 oz} \\ & \times\ 7 \\ \hline \end{array}$$

37.
$$\begin{array}{rrr} \text{2 h} & \text{45 min} & \text{32 s} \\ & & \times\ 8 \\ \hline \end{array}$$

38. $3\overline{)12 \text{ gal } 3 \text{ qt}}$

39. $7\overline{)12 \text{ ft } 3 \text{ in}}$

40. $5\overline{)6 \text{ gal } 2 \text{ qt } 1 \text{ pt}}$

41. Convert the following Fahrenheit temperatures to Celsius. (Round off to the nearest degree.)

(a) 74°F ____________ **(b)** 50°F ____________

(c) 65°F ____________ **(d)** 98°F ____________

42. Convert the following Celsius temperatures to Fahrenheit. (Round off to the nearest degree.)

(a) 10°C ____________ **(b)** 16°C ____________

(c) 0°C ____________ **(d)** 38°C ____________

In problems 43–74 convert to the units shown.

43. 3 qt = ____________ l

44. 100 yd = ____________ m

45. 250 mg = ____________ g

46. 500 km = ____________ mi

47. 50.5 cm = ____________ mm

48. 200 l = ____________ hl

49. 25.5 dag = ____________ g

50. 100 m = ____________ yd

51. 1,000 mm = ____________ in

52. 15 gal = ____________ kl

53. 200 cl = ____________ l

54. 3.5 km = ____________ dm

55. 20 in = ____________ mm

56. 4l = ____________ ml

57. 8 sq mm = ____________ sq dm

58. 2.62 cu m = ____________ cu cm

59. 3.25 kg = ____________ lb

60. 61.5 dal = ____________ dl

61. 742 cg = ____________ g

62. 4 kg = ____________ oz

63. 100 g = ____________ lb

64. 12 dal = ____________ kl

65. 12l = ____________ qt

66. 25.4 cm = ____________ dam

67. 60.2 oz = ____________ g

68. 21 kl = ____________ pt

69. 45 hm = ____________ cm

70. 1500 g = ____________ kg

71. 200 lb = ____________ kg

72. 0.45 kl = ____________ l

73. 15 sq dam = ____________ sq cm

74. 5.4 cu dm = ____________ cu m

75. The distance from Chicago to Montreal is approximately 850 miles. Express this distance in kilometers.

____

76. The net weight of a can of soup is 10.75 ounces. Express this weight in grams.

____

77. How many liters are needed to fill a gas tank with a capacity of 16 gallons?

____

78. How many 30-centimeter by 30-centimeter tiles are needed to cover the floor of a 3.6 meter by 3.6 meter room?

____

79. A canned ham weighs 1285 grams. How many kilograms does it weigh?

____

80. A prescription requires 24 doses of medicine, each dose containing 35 milliliters. How many centiliters should the pharmacist pour into a bottle?

____

81. A wine bottle holds 1.89 liters. How many centiliters are in this bottle?

____

82. The moon is approximately 240,000 miles away from the earth. Express this distance in kilometers.

____

83. Allison can run the 100-meter dash in 9.8 s. What is her speed in miles per hour? (Round off to tenths.)

____

84. The dimensions of a cylindrical can of vegetables is 4.25 inches in height and 3 inches in diameter. What is the volume of the can in cubic centimeters? (Round off to hundredths.)

____

85. A seamstress needs 20 inches of material that costs $5.76 per yard. How much does she pay for the material?

____

ANSWERS

1. 6 **2.** 10 **3.** 13,728 **4.** 4,500 **5.** 88 **6.** 15 **7.** $4\frac{3}{4}$ **8.** 4 **9.** $5\frac{1}{2}$ **10.** $1\frac{1}{4}$ **11.** $2\frac{1}{2}$ **12.** 5 **13.** 56 **14.** 144 **15.** 3,600 **16.** 2,200 **17.** 151.2 **18.** 1.4 **19.** $4\frac{1}{3}$ **20.** 3 **21.** 2 h 10 min **22.** 1 yr 8 wk **23.** 14 yd 2 ft **24.** 4 gal 3 qt **25.** 6 tons 1,120 lb **26.** 15 h 30 min **27.** 9 yd 1 ft 6 in **28.** 23 h 21 min 26 s **29.** 10 gal 1 pt **30.** 6 yr 6 da **31.** 10 tons 690 lb **32.** 22 gal 1 qt 1 pt **33.** 12 min 24 s **34.** 2 yd 1 ft 7 in **35.** 6 yr 35 wk 2 da **36.** 90 lb 2 oz **37.** 22 h 4 min 16 s **38.** 4 gal 1 qt **39.** 1 ft 9 in **40.** 1 gal 1 qt $\frac{3}{5}$ pt **41.** **(a)** 23°C **(b)** 10°C **(c)** 18°C **(d)** 37°C **42.** **(a)** 50°F **(b)** 61°F **(c)** 32°F **(d)** 100°F **43.** 2.838 l **44.** 91.4m **45.** 0.25 g **46.** 310.5 mi **47.** 505 mm **48.** 2 hl **49.** 255 g **50.** 109 yd **51.** 39.37 in **52.** 0.056775 kl **53.** 2 l **54.** 35,000 dm **55.** 508 mm **56.** 4,000 ml **57.** 0.0008 sq dm **58.** 2,620,000 cu cm **59.** 7.15 lb **60.** 6,150 dl **61.** 7.42 g **62.** 140.8 oz **63.** 0.22 lb **64.** 0.12 kl **65.** 12.72 qt **66.** 0.0254 dam **67.** 1,706.67 g **68.** 44,310 pt **69.** 450,000 cm **70.** 1.5 kg **71.** 90.8 kg **72.** 450 l **73.** 15,000,000 sq cm **74.** 0.0054 cu m **75.** 1,368.5 km **76.** 304.7625 g **77.** 60.56 l **78.** 144 tiles **79.** 1.285 kg **80.** 84 cl **81.** 189 cl **82.** 386,400 km **83.** 22.8 mph **84.** 492.04 cu cm **85.** $3.20

Warmup Test

1. Convert each of the following to the indicated units.

(a) 120 in = __________ ft (b) 3.75 gal = __________ qt

(c) $3\frac{1}{8}$ lb = __________ oz (d) 14 pt = __________ gal

(e) 2,112 yd = __________ mi (f) $2\frac{1}{6}$ h = __________ s

2. Simplify each of the following.

(a) 185 min __________ (b) 29 in __________

(c) 15 lb 50 oz __________ (d) 10 gal 10 qt __________

(e) 6 h 92 min 73 s __________ (f) 12 yd 11 ft 14 in __________

3. Perform the indicated operations and simplify.

(a)
```
    3 tons   895 lb 12 oz
    2 tons  1672 lb 11 oz
+   1 ton   1823 lb 13 oz
```

(b)
```
    5 gal
 −  3 gal 2 qt 1 pt
```

(c)
```
3 h 22 min 14 s
            × 6
```

(d) $4\overline{)3 \text{ yd } 2 \text{ ft } 6 \text{ in}}$

4. Convert each of the following to the indicated units:

(a) 15 g = __________ mg (b) 0.25 kl = __________ l

(c) 30.25 l = __________ dal (d) 300 cm = __________ dm

(e) 5,000 ml = __________ kl (f) 35.7 dg = __________ hg

5. Convert each of the following to the indicated units:

(a) 26 gal = __________ l (b) 50 g = __________ lb

(c) 80 kg = __________ lb (d) 30 km = __________ mi

(e) 25 ft = __________ m (f) 35.5 kl = __________ qt

6. Convert the following Fahrenheit temperatures to Celsius. (Round off to the nearest degree.)

(a) 82°F __________ (b) 95°F __________ (c) 212°F __________

7. Convert the following Celsius temperatures to Fahrenheit. (Round off to the nearest degree.)

(a) 15°C __________ (b) 32°C __________ (c) 12°C __________

8. A bottle of aspirin contains 100 tablets each 325 milligrams. How many grams of aspirin are contained in the bottle?

____

9. An intercollegiate diving competition takes place on the 10-meter high diving board (distance from water surface to diving board). How high is the diving board above the surface of the water in feet?

____

10. A road sign on the Trans-Canada Highway reads "Speed Limit 95 km" per hour. What is the speed limit in miles per hour? (Round off to nearest mile.)

____

11. A customer in a grocery store bought four 2-liter bottles of soda. How many kiloliters of soda did he buy? How many quarts of soda did he buy?

____

12. A saddle maker needs 15 inches of leather strapping that costs $8.52 per yard. How much does he pay for the leather strapping?

____

Challenge Test

1. Convert each of the following to the indicated units.

(a) 96 in = ______ ft **(b)** 4.25 gal = ______ qt

(c) $5\frac{1}{4}$ lb = ______ oz **(d)** 18 pt = ______ gal

(e) 2,816 yd = ______ mi **(f)** $3\frac{2}{3}$ h = ______ s

2. Simplify each of the following:

(a) 210 min ______ **(b)** 75 in ______

(c) 75 in ______ **(d)** 12 gal 15 qt ______

(e) 8 h 110 min 88 s ______ **(f)** 15 yd 10 ft 26 in ______

3. Perform the indicated operations and simplify.

(a)
```
   15 yd  2 ft  11 in
    6 yd  2 ft   9 in
+   7 yd  1 ft  11 in
---------------------
```

(b)
```
   7 h          30 s
 − 4 h  26 min  45 s
--------------------
```

(c)
```
5 gal 3 qt 1 pt
            × 6
---------------
```

(d) $5\overline{)2 \text{ gal } 3 \text{ qt } 1 \text{ pt}}$

4. Convert each of the following to the indicated units:

(a) 26 l = ______ m **(b)** 46.72 g = ______ dag

(c) 70,000 ml = ______ kl **(d)** 0.36 kg = ______ g

(e) 650 cm = ______ dm **(f)** 75.26 dm = ______ km

5. Convert each of the following to the indicated units:

(a) 30 gal = ______ l **(b)** 62 kg = ______ lb

(c) 38 ft = ______ m **(d)** 20.5 lb = ______ g

(e) 70 km = ______ mi **(f)** 42.5 kl = ______ qt

6. Convert the following Fahrenheit temperatures to Celsius. (Round off to the nearest degree.)

(a) 63°F ______ **(b)** 36°F ______ **(c)** 122°F ______

7. Convert the following Celsius temperatures to Fahrenheit. (Round off to the nearest degree.)

(a) 5°C ______ **(b)** 21°C ______ **(c)** 13°C ______

8. A bottle of vitamin C tablets contains 100 tablets, each 500 milligrams. How many grams of vitamin C are contained in the bottle?

______

9. In track events there is a 5,000-meter foot race. How many miles is this race? (Round off to the nearest tenth of a mile.)

____

10. A road sign on the Trans-Canada highway reads "Speed Limit 65 km" per hour. What is the speed limit in miles per hour? (Round off to the nearest mile.)

____

11. Each of the Ulz Farms' milking cows averages 14 gallons of milk per week. How many liters of milk does each cow give? How many kiloliters of milk does each cow give?

____

12. A sailmaker needs 27 inches of binding that costs $9.48 per yard. How much does she pay for the binding?

____

An Algebra Warmup

CHAPTER OUTLINE

7.1 BASIC DEFINITIONS

This section deals with definitions of numbers and magnitudes of numbers. Because the concepts of **negative number** and **literal number** represent a large imaginative step in the thinking process, each topic will receive special attention.

The Negative Number

How much is $5 - 7$? In arithmetic you learned that this is an impossible chore; for how can one start with 5 objects and take away 7 objects? In algebra, with the help of the number line in Figure 7.1 we will arrive at the answer to $5 - 7$. Consider the following pattern:

$5 - 1 = 4$	4 is located one place to the left of 5 on the number line.
$5 - 2 = 3$	3 is located two places to the left of 5 on the number line.
$5 - 3 = 2$	2 is located three places to the left of 5 on the number line.
. . .	
$5 - 5 = 0$	0 is located five places to the left of 5 on the number line.
. . .	
$5 - 7 =$	The answer is located seven places to the left of 5 on the number line.

However, we are now at a point on our number line without a name.

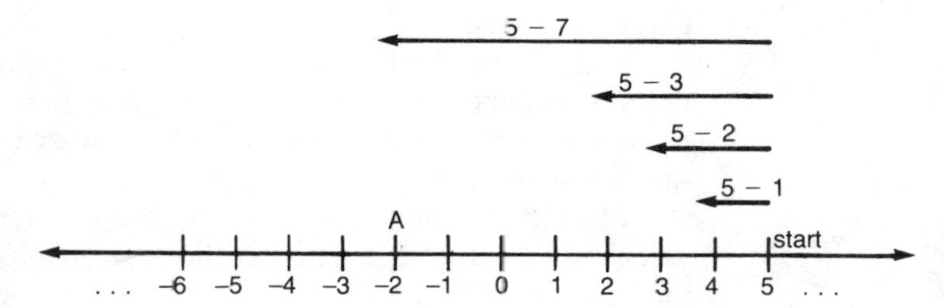

Figure 7.1

This difficulty is easily solved by naming successive unit distances left of zero (0); -1, "negative one"; -2, "negative two"; -3, "negative three"; and so on. We are naming points to the left of zero, on the number line by numbers called **negative integers.** In general, numbers to the left of zero (numbers < 0) on the number line. are called **negative numbers.** Thus, -0.76, -14.09, -5, -763.1, and -0.2 are all negative numbers. Numbers to the right of 0 on the number line, are called **positive numbers** (numbers > 0). Thus, $+0.09$, -2.76, $+445.18$, $+5.25$, and $+78.193$ are all positive numbers; $+1$, $+2$, $+3$, . . . are called **positive integers.** As a writing convenience, we drop the "+" sign preceding a positive number. For example, $+7$ becomes 7, $+49$ becomes 49, and $+99.98$ becomes 99.98.

In this context, "+" and "−" are simply direction signs telling us whether a number is located to the left (−) or to the right (+) of 0 on the number line. Zero itself is neither positive nor negative. This special number is frequently called the **origin** for it is the beginning point of our number line. Indeed, such a name is suitable for the important role that 0 plays in mathematics.

Example 1 (a) Locate $+3$ (or 3) on a number line.
(b) Locate -4 on the same number line.

Solution

−4 −3 −2 −1 0 1 2 3

When you add $-1, -2, -3, \ldots$ to the collection of whole numbers, you obtain the collection of **integers:**

$$\ldots, -3, -2, -1, 0, 1, 2, 3, \ldots$$

Many applied questions can be answered with the aid of integers. How do you express a loss of \$5.00 from misplacing your wallet? Why, −\$5,00. A temperature of 10 degrees below zero Fahrenheit may be written −10°F. The last play run by the Miami Dolphins resulted in −3 yards. Clearly, a loss of 3 yards. You are broke and you borrow \$25.00 from a friend; your financial status is −\$25.00. The first use of positive and negative numbers dates back to the bookkeeping efforts of ancient Greeks, Chinese, and Hindus more than 2,000 years ago. Chinese merchants wrote positive numbers in black ink and negative numbers in red ink in their accounts. From this ancient practice comes the slang expression "being in the red," meaning being in debt.

The Literal Number

To find 36 percent, P, of the base number 80 you multiply $P = 0.36 \times 80 = 48$. For a rectangle with a width of 5 inches and a length of 7 inches, the perimeter, P, is $P =$ (2 × 5 inches) + (2 × 7 inches) = 24 inches. You determine the area, A, of a circle with four-inch radius, as follows: $A = \pi \times (4\text{ in})^2$ or 16π sq in. Having "formulas" for solving applied problems is achieved through algebra.

In algebra we write mathematical statements, called **literal expressions,** in which letters are used to represent numbers. For example, $P = RB$, $P = 2W + 2L$, and $A = \pi R^2$ are literal expressions. Should a letter, such as P, appear in two or more different literal expressions, you distinguish the meaning by the context of the expression itself.

Each letter in a literal expression is called a **literal number.** There are four literal numbers in $I = PRT$, and there are three literal numbers in $P = 2W + 2L$. However, $A = \pi r^2$ has only two literal numbers, R and A, because π is a fixed value (approximately 3.14).

What literal number should be used to represent a collection of arithmetic values? Using the first letter in the topic name is a common memory device in mathematics. Thus, we use A for area, v for velocity, W for work, and I for interest. Should you be dealing with the circumference of any circle, the letter C is the common literal number. Can you think of the literal number that is commonly used to denote time?

If you are using the literal number, capital B, you must continue to use B in all of the related work that follows for b and B are two distinct literal numbers. In short, there are 26 lowercase literal numbers and 26 uppercase literal numbers.

Being able to portray all possible time values by the single literal number, t, is a powerful symbolic achievement. Being able to work and think in terms of literal numbers is a goal of algebra.

A symbol that denotes a fixed value is called a **constant.** Thus 6, 7.74, and π are all constants. The amount of money in your wallet at any particular moment is a constant. By contrast, a symbol denoting a quantity that may take on more than one value is called a **variable.** In mathematics the literal numbers u, v, w, x, y, and z are common names for variables. If x denotes the price of gasoline for 1979, 1980, 1981, 1982, 1983, x is a variable. If y denotes the temperature in your yard during a day in November, y is a variable.

We write mathematical statements in terms of constants and variables (or literal numbers). For a particular statement, a variable may assume one true value, or two true values, and so on. If twice a number equals 36 and we let x be the unknown, this produces the statement $2x = 36$. Can you figure out the value of x in this statement? If some number less 82 equals 18 and we let y be the unknown, this produces the statement $y - 82 = 18$. Can you figure out the value of y in this statement?

Example 2 (a) "Thirty-six plus an unknown number, x, equals 75" may be expressed $36 + x = 75$.

(b) "Some number, y, divided by 5 produces the quotient thirteen" may be expressed $\frac{y}{5} = 13$ or $y \div 5 = 13$.

Operation Symbols

The basic operation symbols for arithmetic are carried over into algebra with exactly the same meaning. However, there are symbols for multiplication and division (shown in the table below) that are more commonly used in algebra:

Table 7.1

Operation	Symbols	Examples
1. Addition	$+$	$9 + 7, x + 5, x + y$
2. Subtraction	$-$	$8 - 6, 5 - y, x - y$
3. Multiplication	$\times$	9×5
	· Raised dot	$9 \cdot 5, a \cdot b$
	( ) Parentheses	$a(b), (a)b, (a)(b)$
	Juxtaposition	$ab, ba, 5a$
4. Division	$\div$	$x \div y$
	— Division or fraction bar	$\frac{10}{5}, \frac{x}{y}, \frac{x}{5}$

Note: When you use juxtaposition to express the product of an integer and a literal number, such as 5 and a, write the integer first for clarity:

$5 \times a = 5a$, not $a5$

Absolute Value

On the number line, how far is the point associated with -5 from the origin? The answer is five units away or simply 5. On the other hand the number 17 is seventeen units away from the origin. When we speak of how far any real number is from the origin, 0, without respect to direction, we are speaking of its **absolute value.** This is an important concept and is defined next for emphasis.

> The **absolute value** or magnitude of any number a, denoted $|a|$, is the distance from the origin, 0, to the point on the number line associated with a. See Figure 7.2.

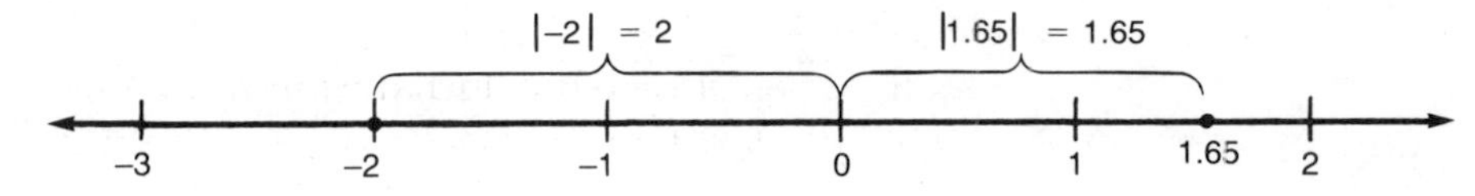

Figure 7.2

Example 3 (a) $|-8| = 8$, (b) $|15| = 15$, (c) $|-\frac{2}{3}| = \frac{2}{3}$, (d) $|\frac{11}{5}| = \frac{11}{5}$, (e) $|-2| = 2$, (f) $|1.65| = 1.65$

Of course $|0| = 0$. With this special case in mind, it is true that *the absolute value of any number is always positive or zero.* Table 7.2 presents a review of important symbols.

Example 4 Evaluate $-|-10|$.

Solution First concentrate on the absolute value itself, $|-10| = 10$. Thus the answer is -10.

Table 7.2

Symbol	Meaning
$a < b$	a is less than b
$a \leqslant b$	a is less than or equal to b
$a > b$	a is greater than b
$a \geqslant b$	a is greater than or equal to b
$a = b$	a equals b
$a \neq b$	a is not equal to b
$\lvert a \rvert$	The absolute value of a

Example 5 Write: x is less than or equal to 10.

Solution $x \leqslant 10$

Example 6 Write y is greater than -5.

Solution $y > -5$

7.1 EXERCISES

In problems 1–4 refer to the number line below

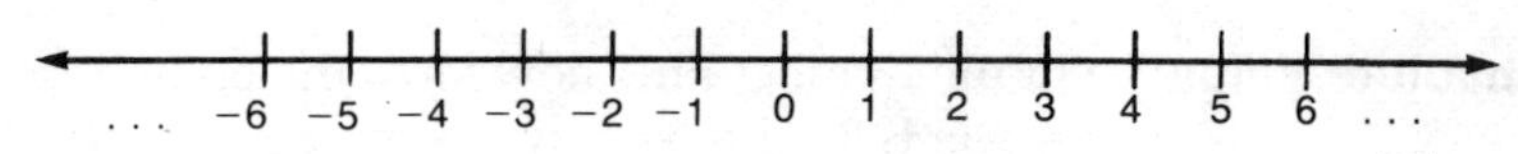

1. Name the point 6 units to the left of -1. ____ −7

2. Name the point 10 units to the left of -3. ____ −13

3. Name the point 5 units to the right of -1. ____ 4

4. Name the point 8 units to the right of -4. ____ 4

In problems 5–15 answer true or false.

5. If $-3 < -1$, then -3 must be to the left of -1 on the number line. ____ true

6. If $-7 > -10$, then -7 must be to the left of -10 on the number line. ____ false

7. $0 > -1$ ____ true

8. $10 < -10$ ____ false

9. The collection of negative integers is as follows: $-1, -2, -3, \ldots$. ____ true

10. The number 0 is called the "origin" of the real number line. ____ true

11. The number 0 is a positive number. ____ false

12. $|19| = -19$ false **13.** $|0| = 0$ true

14. $|-3| = 3$ true **15.** $-|-4| = -4$ true

In problems 16–21 express the written statement symbolically.

16. Write: t is less than or equal to 15. 1≤15

17. Write: x is less than 0. X<0

18. Write: y is greater than 19. y>19

19. Write: y is greater than or equal to 19. y≥19

20. Write: x is less than or equal to -2. X≤-2

21. Write: x is less than or equal to -100. X≤-100

In problems 22–27 express the given absolute value by a number.

22. $|-15|$ 15 **23.** $|29|$ 29

24. $|0|$ 0 **25.** $-|-6|$ -6

26. $-|-5|$ -5 **27.** $-|4|$ -4

28. Express the product of $c \times d$ in two other ways: cd c·d (c)(d)

29. Express the product of $g \times h$ in two other ways: gh g·H (g)(H)

30. Express the product of -6 and a in three ways: -6a (-6)(a)

31. Express the product of 12 and b in three ways: 12·b 126 (12)(b)

32. How far is -7 from the origin? How far is 4 from the origin? How far is -7 from 4? 7 units to left 4 to right 7 to left

33. How far is -11 from the origin? How far is 6 from the origin? How far is -11 from 6? 17 units to left

---

ANSWERS

1. -7 **2.** -13 **3.** 4 **4.** 4 **5.** true **6.** false **7.** true **8.** false **9.** true **10.** true **11.** false **12.** false **13.** true **14.** true **15.** true **16.** $t \leq 15$ **17.** $x < 0$ **18.** $y > 19$ **19.** $y \geq 19$ **20.** $x \leq -2$ **21.** $x \leq -100$ **22.** 15 **23.** 29 **24.** 0 **25.** -6 **26.** -5 **27.** -4 **28.** cd; $c \cdot d$; $(c)(d)$; $c(d)$; $(c)d$ **29.** gh; $g \cdot h$; $(g)(h)$; $g(h)$; $(g)h$ **30.** $-6a$; $-6 \cdot a$; $(-6)(a)$; $-6(a)$; $(-6)a$ (*Note:* If you write a−6, it is wrong.) **31.** $12b$; $12 \cdot b$; $(12)(b)$; $12(b)$; $(12)b$ (*Note:* If you write $b12$, it is wrong.) **32.** 7 units to the left; 4 units to the right; 11 units to the left **33.** 11 units to the left; 6 units to the right; 17 units to the left

7.2 ADDITION OF SIGNED NUMBERS

On a certain day in Duluth, Minnesota, the temperature read 12°F. During the night the temperature dropped 55°F. What was the temperature? You must add -55 and 12. What is the answer? On a first-down play, the Cleveland Browns' quarterback was "sacked" for a loss of 10 yards, but on the next two plays Cleveland gained 15 yards and 7 yards, respectively. What have they accomplished after three plays? Have they made a first down (10 yards)? These questions will be answered in this section.

A **signed number** is a number that has a "+" sign or "−" sign associated with it. Recall, if the sign is omitted, a "+" sign is understood. Hence, $+8 = 8$ and $+19 = 19$. The procedures used to perform the four basic operations (addition, subtraction, multiplication, and division) with signed numbers are shown in boxes for emphasis. To justify and understand these procedures, we rely upon the number line.

We represent signed numbers as follows: Begin at the origin, 0, and imagine every positive number as a movement to the right and every negative number as a movement to the left.

Example 1 Represent $+5$ on a number line.

Solution

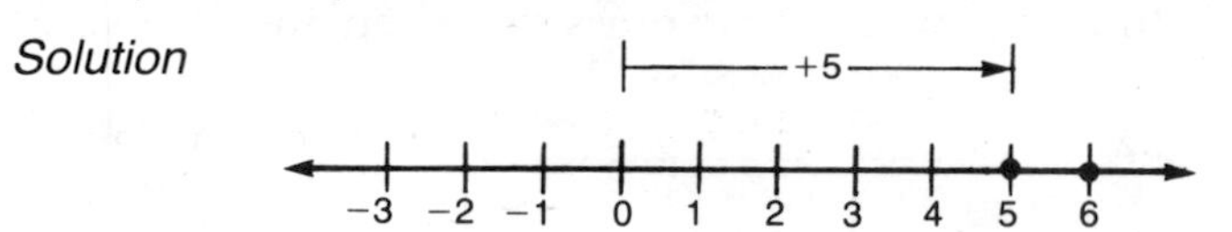

Example 2 Represent -8 on a number line.

Solution

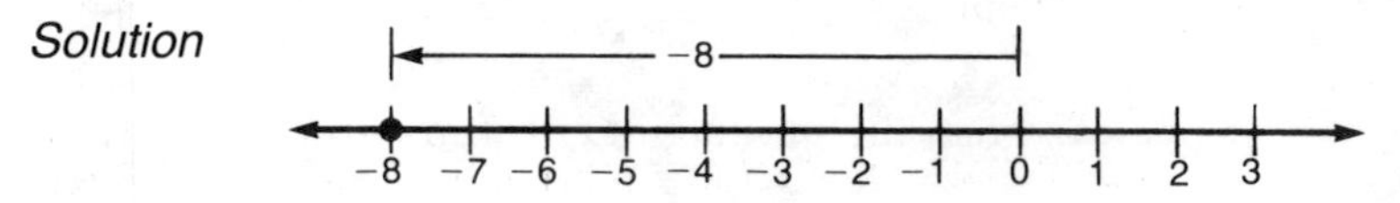

> *To add two numbers having like signs,* add their absolute values. The sign of the sum is the common sign of the numbers.

Example 3 (a) $(+5) + (+4) = +(|+5| + |+4|) = +(5 + 4) = 9$

(b) $(-6) + (-5) = -(|-6| + |-5|) = -(6 + 5) = -11$

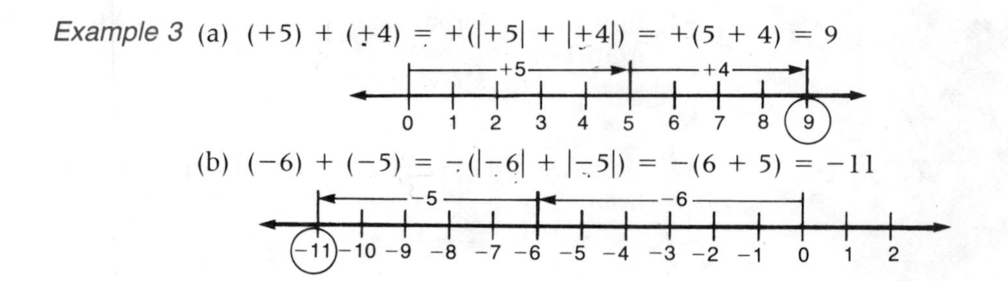

Example 4 (a) $13 + (+27) = +(|13| + |+27|) = +(13 + 27) = 40$

(b) $(-70) + (-60) = -(|-70| + |-60|) = -(70 + 60) = -130$

> *To add numbers having unlike signs,* subtract the smaller absolute value from the larger absolute value. The sign of the sum is the sign of the number having the larger absolute value.

Example 5 (a) $(+4) + (-11) = -(|-11| - |+4|) = -(11 - 4) = -7$

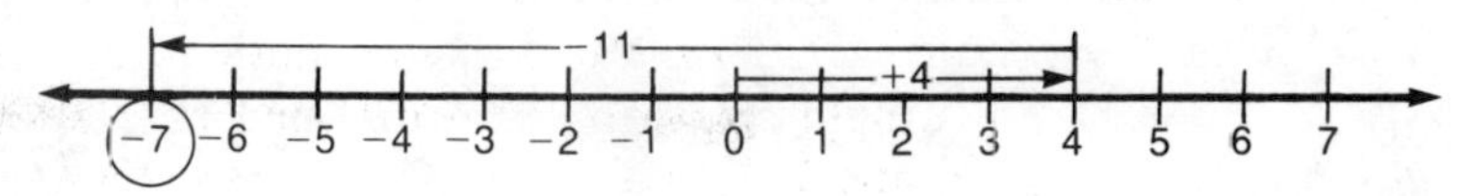

(b) $(+8) + (-3) = (|+8| - |-3|) = +(8 - 3) = +5$, or 5

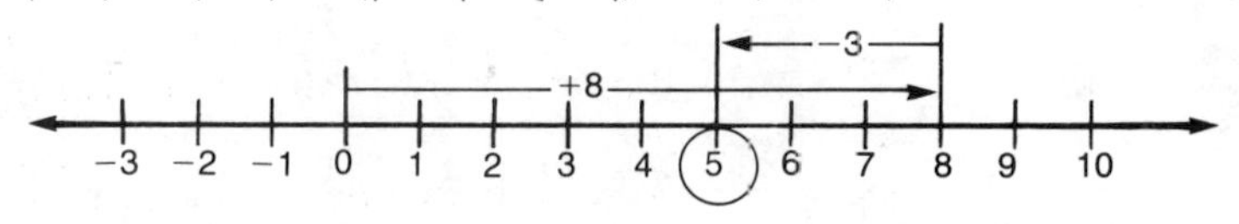

Example 6 On a certain first-down play, the Cleveland Browns' quarterback was "sacked" for a loss of 10 yards, but on the next two plays Cleveland gained 15 yards and 7 yards, respectively.

(a) What have they accomplished after three plays?
(b) Have they made a first down (gained 10 yards)?

Solution (a) Begin with a loss of 10 yards, -10. To this add $+15$ yards. The result is $(-10) + (+15) = +5$, or 5. To this sum add $+7$. The answer is $+12$, or 12.

(b) Yes, they have made a first down because 12 yards $\geqslant$ 10 yards.

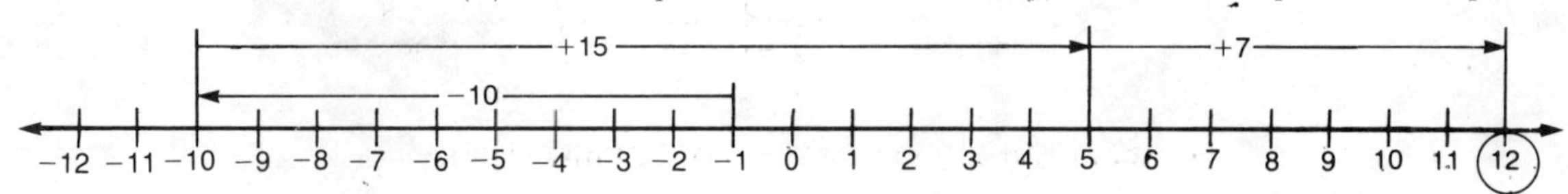

Example 7 Do the following additions of signed numbers.

(a) $(-55) + 12 =$ (b) $63 + (-79) =$
(c) $128 + (-84) =$ (d) $(-256) + 137 =$

Solution (a) $(-55) + 12 = -43$ (b) $63 + (-79) = -16$
(c) $128 + (-84) = 44$ (d) $(-256) + 137 = -119$

Example 8 Do the following addition of signed numbers:

(a) $-0.5 + 3.7$ (b) $\frac{1}{5} + \left(-\frac{1}{3}\right)$

Solution (a) $-0.5 + 3.7 = 2.2$

(b) $\frac{1}{5} + \left(-\frac{1}{3}\right) = \frac{3}{15} + \left(-\frac{5}{15}\right) = -\frac{2}{15}$

Example 9 Add: $(-3) + 2 + (-5) + (-6) + 4$.

Solution You may add left to right.

$$(-3) + 2 + (-5) + (-6) + 4 =$$
$$-1 + (-5) + (-6) + 4 =$$
$$-6 + (-6) + 4 =$$
$$-12 + 4 = -8$$

Example 10 Add $-5.66 + (-2.11) + (7.04) + (-0.63)$.

Solution

$$-5.66 + (-2.11) + (7.04) + (-0.63) =$$
$$-7.77 + (7.04) + (-0.63) =$$
$$-0.73 + (-0.63) = -1.36$$

Example 11 $|(-10) + 8| =$

Solution $|(-10) + 8| = |-2| = 2$

7.2 EXERCISES

In problems 1–54 perform the indicated additions.

1. $15 + 8 =$ ______

2. $12 + 14 =$ ______

3. $8 + (-3) =$ ______

4. $10 + (-6) =$ ______

5. $16 + (-13) =$ ______

6. $18 + (-11) =$ ______

7. $27 + (-19) =$ ______

8. $32 + (-16) =$ ______

9. $8 + (-12) =$ ______

10. $9 + (-15) =$ ______

11. $(-8) + 4 =$ ______

12. $(-42) + 36 =$ ______

13. $8 + (-17) =$ ______

14. $(-12) + (-21) =$ ______

15. $(-14) + (-9) =$ ______

16. $(-31) + (-16) =$ ______

17. $(-24) + (-26) =$ ______

18. $(-6) + (-41) =$ ______

19. $(-5) + 5 =$ ______

20. $3 + (-3) =$ ______

21. $\frac{3}{4} + \frac{5}{8} =$ ______

22. $\frac{6}{25} + \frac{9}{20} =$ ______

23. $\left(-\frac{17}{12}\right) + \frac{5}{8} =$ ______

24. $\frac{-14}{15} + \frac{7}{10} =$ ______

25. $\left(-\frac{5}{12}\right) + \left(-\frac{7}{18}\right) =$ ______

26. $\left(-2\frac{1}{2}\right) + \left(-4\frac{1}{3}\right) =$ ______

27. $\frac{3}{4} + \left(-\frac{1}{2}\right) =$ ______

28. $\left(-\frac{1}{6}\right) + \frac{7}{8} =$ ______

29. $7.5 + 6.3 =$ ______

30. $12.4 + 8.6 =$ ______

31. $(-15.2) + 24.3 =$ ______

32. $17.65 + (-9.8) =$ ______

33. $(-23.05) + (-15.82) =$ ______

34. $(-17.9) + (-18.21) =$ ______

35. $(-24.03) + 16.75 =$ ______

36. $12.3 + (-19.1) =$ ____________

37. $(-8) + 6 + 7 + (-8) =$ ____________

38. $(-18) + 0 + 13 + 7 =$ ____________

39. $6 + (-5) + (-8) + 4 =$ ____________

40. $20 + (-5) + 4 + (-9) =$ ____________

41. $7 + (-10) + (-5) + 3 + (-2) =$ ____________

42. $(-9) + 12 + (-5) + 18 + (-3) =$ ____________

43. $\left(-\frac{1}{2}\right) + \frac{1}{3} + \frac{1}{6} =$ ____________

44. $2\frac{1}{2} + \left(-\frac{3}{5}\right) + \left(-\frac{7}{10}\right) =$ ____________

45. $\left(-\frac{2}{3}\right) + \frac{3}{4} + \left(-\frac{5}{6}\right) =$ ____________

46. $\frac{3}{10} + \left(-\frac{7}{12}\right) + \left(-\frac{1}{15}\right) =$ ____________

47. $(-9.5) + (-8.6) + (-12.3) + 17 =$ ____________

48. $(-8.3) + 6.4 + (-10.8) + 15.2 =$ ____________

49. $-15.85 + 6.02 + 1.01 + (-7.05) =$ ____________

50. $-12.86 + (-15.43) + 36.62 + (-2.81) =$ ____________

51. $|(-19) + (-2)| =$ ____________

52. $|(-23) + (47)| =$ ____________

53. $|15 + 18| =$ ____________

54. $|16 + (-19)| =$ ____________

55. The temperature at 6 A.M. was $-22°F$; then it went up $18°F$ by noon. Find the temperature at noon.

____

56. An airplane follows its flight pattern at 32,000 ft. Because of heavy air traffic at this height it descends 6,000 ft. What is the airplane's new altitude?

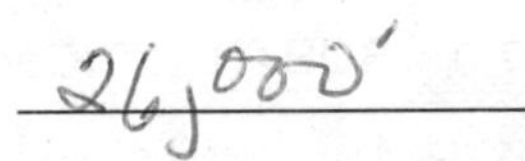

57. The temperature at 7 A.M. was $-8°F$. During the day it went up by $32°F$. What was the high temperature for the day?

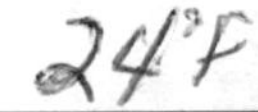

58. A man owes $110 on his Master Card. He makes a payment of $65. What amount does he still owe? (Write your answer as a negative number.)

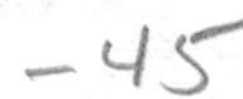

59. A bookstore begins with 4 copies of a certain textbook. If the store receives a shipment of 30 more texts and sells 32 of these texts, how many are left?

60. Nancy owes \$50 and pays back \$30. What is her financial status? (Write your answer as a negative number.)

____

61. Your favorite pro football team gains 8 yards on its first play, loses 6 yards on its second play, and then loses another 7 yards on its third play. What is the net result after these three plays?

____

62. An elevator begins at the 5th floor, rises 20 floors, and then descends 22 floors. What floor is the elevator on now?

____

63. Each share of Anaconda Stock began the day at \$25.75, rose \$1.75 in value and then descended by \$3.25 by the end of the trading day. What is the value of each share at the end of the day?

____

64. On a particular day in December, in Boston, the temperature starts at 25°F, drops 19°F during the day due to an offshore wind, and then rises 10°F at the end of the day. What is the temperature at this time?

____

ANSWERS

1. 23 **2.** 26 **3.** 5 **4.** 4 **5.** 3 **6.** 7 **7.** 8 **8.** 16 **9.** −4 **10.** −6 **11.** −4 **12.** −6 **13.** −9 **14.** −33 **15.** −23 **16.** −47 **17.** −50 **18.** −47 **19.** 0 **20.** 0 **21.** $1\frac{3}{8}$ **22.** $\frac{69}{100}$ **23.** $-\frac{19}{24}$ **24.** $-\frac{7}{30}$ **25.** $-\frac{29}{36}$ **26.** $-6\frac{5}{6}$ **27.** $\frac{1}{4}$ **28.** $\frac{17}{24}$ **29.** 13.8 **30.** 21 **31.** 9.1 **32.** 7.85 **33.** −38.87 **34.** −36.11 **35.** −7.28 **36.** −6.8 **37.** −3 **38.** 2 **39.** −3 **40.** 10 **41.** −7 **42.** 13 **43.** 0 **44.** $1\frac{1}{5}$ **45.** $-\frac{3}{4}$ **46.** $-\frac{7}{20}$ **47.** −13.4 **48.** 2.5 **49.** −15.87 **50.** 5.52 **51.** 21 **52.** 24 **53.** 33 **54.** 3 **55.** −4°F **56.** 26,000 ft **57.** 24°F **58.** −\$45 **59.** 2 **60.** −\$20 **61.** −5 yards **62.** 3rd **63.** \$24.25 **64.** 16°F

7.3 SUBTRACTION OF SIGNED NUMBERS

We know how to subtract a smaller positive number from a larger positive number:

$$10 - 6 = 4$$

In this section we will learn how to subtract a larger positive number from a smaller positive number:

$$7 - 12 = ?$$

and how to subtract signed numbers in general:

$$\begin{aligned} 8 - (-5) &= ? \\ -6 - (-4) &= ? \\ -5 - (-10) &= ? \\ -9 - 15 &= ? \end{aligned}$$

> **Definition of Subtraction**
>
> Let a and b represent any real numbers; then $a - b = a + (-b)$.

This definition leads to the following procedure.

> *To subtract one signed number from another:*
>
> 1. Change the operation symbol from subtraction to addition and change the sign of the subtrahend.
> 2. Follow the procedures for adding signed numbers.

Example 1 $7 - 12 =$

Solution $7 - 12 = 7 + (-12) = -5$ We changed subtraction to addition. We changed the sign of $+12$ to -12.

Example 2 $8 - (-5) =$

Solution $8 - (-5) = 8 + (+5) = 13$ We changed subtraction to addition. We changed the sign of -5 to $+5$.

Example 3 $-6 - (-4) =$

Solution $-6 - (-4) = -6 + (+4) = -2$

Example 4 $-5 - (-10) =$

Solution $-5 - (-10) = -5 + (+10) = 5$

Example 5 $-9 - 15 =$

Solution $-9 - 15 = -9 + (-15) = -24$

Example 6 $\$263.78 - \$160.52 =$

Solution $\$263.78 - \$160.52 = \$263.78 + (-\$160.52) = \$103.26$

Subtracting signed numbers sometimes causes difficulty. To avoid difficulty it is important to realize that "+" and "−" play dual roles in algebra.

+ means: (1) a number located to the right of 0, or (2) addition
− means: (1) a number located to the left or 0, or (2) subtraction

Consider $5 - 3$. You may interpret the sign, "−", as meaning negative three, that is, attach the "−" to the number and think of the problem as an addition problem. Now your problem becomes the addition

$$(+5) + (-3) = +2, \text{ or } 2$$

Or you may interpret the sign "−" as meaning subtraction. Now your problem becomes the subtraction

$$(+5) - (+3) = +2, \text{ or } 2$$

Of course, the same answer is realized using either interpretation. *The key idea is that you use only one interpretation of the minus sign, "−".* Select the interpretation that you find convenient. Below we have diagrammed the two correct thought processes for $a - b$.

$$a - b \begin{cases} (+a) \text{ subtract } (+b) \\ \text{OR} \\ (+a) \text{ add } (-b) \end{cases}$$

Don't write $(+7) + (-4)$ as $7 + -4$; it leads to confusion. In the initial stages, you should always use parentheses to clarify your thinking.

Example 7 Solve $8 - 18$ in two ways, in accordance with the preceding discussion.

Solution 1 $(+8)$ add $(-18) = (+8) + (-18) = -10$
Solution 2 $(+8)$ subtract $(+18) = (+8) - (+18) = -10$

Although $8 - 6 + 7 + 5 - 4 - 9$ may be interpreted as a series of additions and subtractions, it is often more convenient to look upon this problem as a single addition of signed numbers in the following manner: $8 + (-6) + 7 + 5 + (-4) + (-9)$. If we begin by adding positive numbers, $8 + 7 + 5 = 20$, and negative numbers, $-6 + (-4) + (-9) = -19$, we obtain $20 - 19$ or $20 + (-19) = 1$.

Example 8 Evaluate $-19 + 7 - 14 - 12 + 5$.

Solution $-19 + 7 + (-14) + (-12) + 5$

If we add all the positive numbers, $7 + 5$, we obtain $+12$.
If we add all the negative numbers, $-19 + (-14) + (-12)$, we obtain -45.
Adding these answers gives us $+12 + (-45) = -33$ as our answer.

Example 9 Add

$$\begin{array}{r} \$14.81 \\ -\$19.22 \\ \$27.58 \\ (+)\ -\$16.23 \\ \hline \end{array}$$

Solution step 1: $\$14.81 + \$27.58 = \$42.39$
step 2: $-\$19.22 + (-\$16.23) = -\$35.45$
step 3: $\$42.39 + (-\$35.45) = \$6.94$ (answer)

Example 10 On a certain day in Duluth, Minnesota, the temperature read 12°F. During the night the temperature dropped 55°F. What was the temperature?

Solution It started at $+12°$ and dropped $+55°$. We must do the following subtraction of signed numbers: $(+12°) - (+55°) = -43°F$

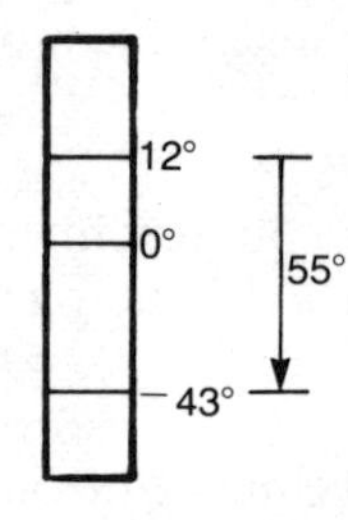

Example 11 Evaluate $|-18 - (-10)|$.

Solution $|-18 - (-10)| = |-18 + 10| = |-8| = 8$

7.3 EXERCISES

In problems 1–66 find the indicated differences.

1. $10 - 8 =$ __________
2. $12 - 6 =$ __________
3. $17 - 11 =$ __________
4. $22 - 14 =$ __________
5. $15 - (-6) =$ __________
6. $33 - (-17) =$ __________
7. $52 - (-32) =$ __________
8. $141 - (-29) =$ __________
9. $38 - (-14) =$ __________
10. $25 - (-50) =$ __________
11. $(-7) - (-5) =$ __________
12. $(-10) - (-14) =$ __________
13. $(-16) - (-17) =$ __________
14. $(-11) - (-11) =$ __________
15. $(-25) - (-30) =$ __________
16. $(-31) - (-49) =$ __________
17. $(-4) - (-2) =$ __________
18. $(-13) - (-8) =$ __________
19. $(-50) - (-25) =$ __________
20. $(-46) - (-35) =$ __________
21. $-20 - (-10) =$ __________
22. $-100 - (-50) =$ __________
23. $-6 - 10 =$ __________
24. $-11 - 15 =$ __________
25. $-28 - 22 =$ __________
26. $-19 - 19 =$ __________
27. $-43 - 15 =$ __________
28. $-52 - 60 =$ __________
29. $7 - 15 =$ __________
30. $12 - 18 =$ __________
31. $34 - 39 =$ __________
32. $28 - 40 =$ __________
33. $24 - 41 =$ __________
34. $10 - 20 =$ __________
35. $16 - (-16) =$ __________
36. $18 - (-24) =$ __________
37. $48 - 50 =$ __________
38. $26 - (-8) =$ __________
39. $22 - 42 =$ __________
40. $-15 - (-15) =$ __________
41. $-25 - (-16) =$ __________
42. $17 - (-19) =$ __________
43. $(-0.008) - (-0.06) =$ __________
44. $(-3.09) - (-4.23) =$ __________
45. $\left(\frac{3}{10}\right) - \left(-\frac{2}{5}\right) =$ __________
46. $\left(-2\frac{1}{2}\right) - \frac{1}{4} =$ __________
47. $3\frac{1}{3} - 5 =$ __________
48. $\left(-\frac{5}{6}\right) - \frac{2}{9} =$ __________
49. $(-4.21) - (-4.21) =$ __________
50. $6.75 - 8.42 =$ __________

51. $\left(-1\frac{3}{4}\right) - \frac{2}{3} =$ ________

52. $2.08 - (-1.004) =$ ________

53. $(-15) - (-6) + 8 =$ ________

54. $16 - (-4) - 20 =$ ________

55. $12 - 18 - 6 =$ ________

56. $-8 - 7 - 10 =$ -25

57. $(-18) - (-32) + (-4) =$ 10

58. $15 + (-10) - 12 =$ -7

59. $\left(-\frac{1}{2}\right) + \frac{1}{3} - \left(-\frac{1}{4}\right) =$ ________

60. $\frac{2}{5} - \left(-1\frac{1}{3}\right) + 2 =$ ________

61. $5.03 - (-2.21) - (-4.6) =$ ________

62. $-8.01 - 3.23 - (12.065) =$ ________

63. $|0 - 17| =$ 0

64. $|-25 - (-34)| =$ ________

65. $|92 - 100| =$ 8

66. $|16 - (-20)| =$ 36

67. Death Valley is 282 feet below sea level. Mount Whitney is 14,495 feet above sea level. What is the difference in altitudes between Mount Whitney and Death Valley?

____

68. The Patriots lost 8 yards on first down, gained 12 yards on second down and lost 7 yards on third down. What was the net yardage for three downs?

____

69. The temperature on Mount Washington at noon was −6°F. During the afternoon the temperature dropped 26°F. What was the temperature?

____

70. A store manager spent $450 in advertising for a pre-Christmas sale. His morning sales totaled $375. Express his financial condition as a signed number.

____

71. The depth of a reservoir was 30 feet. During a summer drought the depth dropped 8 feet. The depth increased by 5 feet during the winter snow months. What is the present depth of the reservoir?

____

72. A salesperson used his charge card during a business trip. His hotel bill was $250, his meals $112 and travel expenses $320. At the end of the month he made a payment of $225. Express the balance on his charge card as a negative number.

____

ANSWERS

1. 2 **2.** 6 **3.** 6 **4.** 8 **5.** 21 **6.** 50 **7.** 84 **8.** 170 **9.** 52 **10.** 75 **11.** −2 **12.** 4 **13.** 1 **14.** 0 **15.** 5 **16.** 18 **17.** −2 **18.** −5 **19.** −25 **20.** −11 **21.** −10 **22.** −50 **23.** −16 **24.** −26 **25.** −50 **26.** −38 **27.** −58 **28.** −112 **29.** −8 **30.** −6 **31.** −5 **32.** −12 **33.** −17 **34.** −10 **35.** 32 **36.** 42 **37.** −2 **38.** 34 **39.** −20 **40.** 0 **41.** −9 **42.** 36 **43.** 0.052 **44.** 1.14 **45.** $\frac{7}{10}$ **46.** $-2\frac{3}{4}$ **47.** $-1\frac{2}{3}$ **48.** $-1\frac{1}{18}$ **49.** 0 **50.** −1.67 **51.** $-2\frac{5}{12}$ **52.** 3.084 **53.** −1 **54.** 0 **55.** −12 **56.** −25 **57.** 10 **58.** −7 **59.** $\frac{1}{12}$ **60.** $3\frac{11}{15}$ **61.** 11.84 **62.** −23.305 **63.** 17 **64.** 9 **65.** 8 **66.** 36 **67.** 14,777 feet **68.** −3 yards **69.** −32°F **70.** −$75 **71.** 27 feet **72.** −$457

7.4 MULTIPLICATION OF SIGNED NUMBERS

In arithmetic, multiplication is considered a method for doing repeated addition of the same number. For example,

$$3 \times 4 = 4 + 4 + 4 = 12 \quad \text{and} \quad 5 \times 2 = 2 + 2 + 2 + 2 + 2 = 10$$

We may extend this concept of multiplication to the product of a positive number times a negative number.

$$(-3)(4) = (4)(-3) = (-3) + (-3) + (-3) + (-3) = -12 \text{ (addition of signed numbers)}$$

$$(-1)(5) = (5)(-1) = (-1) + (-1) + (-1) + (-1) + (-1) = -5$$

Now lets generalize this thinking to any positive integer, a, times negative one.

$$(-1)(a) = (a)(-1) = \underbrace{(-1) + (-1) + (-1) + \ldots (-1)}_{\text{"a" addends of } -1} = -a; \text{ (addition of signed numbers)}$$

Observe, multiplying a positive integer, a, by negative one, -1, always yields negative a, or $-a$. Thus $(-1)(9) = -9$, $(-1)(100) = -100$, and, by agreement, $(-1)(2.78) = -2.78$. It is now only reasonable to wonder whether negative one, -1, times negative a, $-a$, will produce positive a. Mathematicians can prove that this conjecture is true. We will simply accept the conjecture as reasonable. The result is that multiplying any signed number by negative one, -1, always changes the number from positive to negative, or vice versa. For example, $(-1)(17) = -17$, $(-1)(-17) = 17$, $(-1)(12) = -12$, and $(-1)(-12) = 12$. With this motivation, we proceed to state the following procedure for multiplying signed numbers.

> *To multiply two signed numbers,* multiply their absolute values. The sign of the product will be "+" if the two numbers have like signs and "−" if the two numbers have unlike signs.

Example 1 $(+2)(+6) =$

Solution $(+2)(+6) = +(|+2| \cdot |+6|) = +(2 \cdot 6) = +12$

Example 2 $(+5)(-3) =$

Solution $(+5)(-3) = -(|+5| \cdot |-3|) = -(5 \cdot 3) = -15$

Example 3 $(-7)(+4) =$

Solution $(-7)(+4) = -\,(|-7| \cdot |+4|) = -(7 \cdot 4) = -28$

Example 4 $(-5)(-3) =$

Solution $(-5)(-3) = +\,(|-5| \cdot |-3|) = +(5 \cdot 3) = +15$

Example 5 $(-7)(-4) =$

Solution $(-7)(-4) = +\,(|-7| \cdot |-4|) = +(7 \cdot 4) = +28$

Example 6 If you owe six people \$5.00 each, what is your debt expressed as a negative number?

Solution The \$5.00 debt will be denoted −\$5.00
Thus you owe (6) (−\$5.00) = −\$30.00

Example 7 $(-5) \cdot |-12| =$

Solution $(-5) \cdot |-12| = (-5)(12) = -60$

Usually we do not put the + sign in front of a positive number.

Example 8 $(13)(-5) =$

Solution $(13)(-5) = -65$

Example 9 $(-14)(-2)(10) =$

Solution
$$\begin{aligned}(-14)(-2)(10) &= \\ (28)(10) &= \\ 280 &\end{aligned}$$

7.4 EXERCISES

In problems 1–60 find the indicated products.

1. $(+8)(+5) =$ +40 ________ **2.** $(+9)(+10) =$ + 90 ________

3. $(+12)(+10) =$ +120 ________ **4.** $(+16)(+12) =$ ________

5. $(9)(11) =$ ________ **6.** $(20)(15) =$ ________

7. $(25)(8) =$ ________ **8.** $(16)(9) =$ ________

9. $(-5)(+6) =$ −30 ________ **10.** $(+12)(-12) =$ −144 ________

11. $(+15)(-20) =$ ________ **12.** $(-14)(+8) =$ ________

13. $(-22)(4) =$ ________ **14.** $(-15)(15) =$ ________

15. $(30)(-11) =$ ______ 16. $(16)(-25) =$ ______

17. $(-9)(-4) =$ ______ 18. $(-18)(-10) =$ ______

19. $(-44)(-5) =$ ______ 20. $(-32)(-10) =$ ______

21. $(-25)(-16) =$ ______ 22. $(-19)(-9) =$ ______

23. $(-15)(-8) =$ ______ 24. $(-13)(-13) =$ ______

25. $(6)(-12) =$ ______ 26. $(-14)(-8) =$ ______

27. $(-45)(4) =$ ______ 28. $(-7)(5) =$ ______

29. $(-5)(0) =$ ______ 30. $(-17)(-10) =$ ______

31. $(-4)(-15) =$ ______ 32. $(26)(-20) =$ ______

33. $\left(\frac{3}{4}\right)\left(-\frac{8}{9}\right) =$ ______ 34. $(-0.004)(-3.1) =$ ______

35. $(5.001)(-.02) =$ ______ 36. $\left(\frac{1}{6}\right)\left(-\frac{3}{5}\right) =$ ______

37. $\left(-\frac{4}{11}\right)\left(-\frac{22}{48}\right) =$ ______ 38. $(-2.6)(+4.15) =$ ______

39. $(-8.25)(-3.5) =$ ______ 40. $\left(-\frac{6}{7}\right)\left(-\frac{24}{30}\right) =$ ______

41. $\left(-2\frac{1}{2}\right)\left(\frac{3}{10}\right) =$ ______ 42. $\left(-3\frac{1}{3}\right)\left(-2\frac{2}{5}\right) =$ ______

43. $(-6)(-7)(-5) =$ ______ 44. $(-3)(-2)(1)(7) =$ ______

45. $(-8)(5)(0)(2) =$ ______ 46. $(10)(5)(2)(-1) =$ ______

47. $(-4)(-5)(-2)(-10) =$ ______ 48. $(12)(5)(-2)(-3) =$ ______

49. $\left(-\frac{1}{2}\right)\left(\frac{2}{5}\right)\left(\frac{10}{7}\right) =$ ______ 50. $\left(2\frac{1}{2}\right)\left(3\frac{1}{3}\right)\left(2\frac{2}{5}\right) =$ ______

51. $\left(\frac{6}{5}\right)\left(-\frac{7}{12}\right)\left(-\frac{10}{21}\right) =$ ______ 52. $\left(-3\frac{1}{4}\right)\left(-\frac{5}{16}\right)\left(-\frac{32}{39}\right) =$ ______

53. $|(-5)(4)| =$ ______ 54. $|(-6)(-8)| =$ ______

55. $|(9)(-10)| =$ ______ 56. $|(-12)(-7)| =$ ______

57. $|-5| \cdot 3 =$ ______ 58. $(-2) \cdot |-4| =$ ______

59. $(-1)\ |8| =$ ______ 60. $10 \cdot |6| =$ ______

61. A football team loses 5 yards each on three successive plays. What is their yardage for the three plays?

______

62. A storage tank is leaking water into the ground at 7 gallons per minute. Find the loss of water from the tank after 25 minutes. (Express as a signed number.)

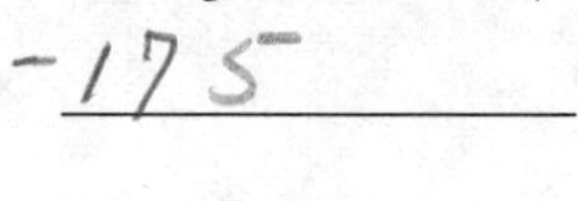

______

63. A business borrows money to buy 12 microcomputers at a cost of \$699 each. How much did the company go in debt? (Express as a signed number.)

ANSWERS

1. 40 **2.** 90 **3.** 120 **4.** 192 **5.** 99 **6.** 300 **7.** 200 **8.** 144 **9.** −30
10. −144 **11.** −300 **12.** −112 **13.** −88 **14.** −225 **15.** −330 **16.** −400
17. 36 **18.** 180 **19.** 220 **20.** 320 **21.** 400 **22.** 171 **23.** 120 **24.** 169
25. −72 **26.** 112 **27.** −180 **28.** −35 **29.** 0 **30.** 170 **31.** 60
32. −520 **33.** $-\frac{2}{3}$ **34.** 0.0124 **35.** −0.10002 **36.** $-\frac{1}{10}$ **37.** $\frac{1}{6}$ **38.** −10.79
39. 28.875 **40.** $\frac{24}{35}$ **41.** $-\frac{3}{4}$ **42.** 8 **43.** −210 **44.** 42 **45.** 0 **46.** −100
47. 400 **48.** 360 **49.** $-\frac{2}{7}$ **50.** 20 **51.** $\frac{1}{3}$ **52.** $-\frac{5}{6}$ **53.** 20 **54.** 48
55. 90 **56.** 84 **57.** 15 **58.** −8 **59.** −8 **60.** 60 **61.** −15
62. −175 gallons **63.** −\$8,388

7.5 DIVISION OF SIGNED NUMBERS

Using the following definition of division, the procedure for dividing signed numbers can be shown to be similar to that for multiplication.

> **Definition of Division**
>
> If a, b, and c are any real numbers, $b \neq 0$, if $\frac{a}{b} = c$, then $a = bc$.

This definition leads to the following.

> *To divide two signed numbers,* divide their absolute values. The sign of the quotient is "+" if the two numbers have like signs and "−" if the two numbers have unlike signs.

Example 1 $\dfrac{+30}{+5} =$

Solution $\dfrac{+30}{+5} = +6$ because $+30 = (+5)(+6)$

Example 2 $\dfrac{-20}{+4} =$

Solution $\dfrac{-20}{+4} = -5$ because $-20 = (+4)(-5)$

Example 3 $\dfrac{-32}{-8} =$

Solution $\dfrac{-32}{-8} = +4$ because $-32 = (-8)(+4)$

Example 4 $\frac{+27}{-3} =$

Solution $\frac{+27}{-3} = -9$ because $+27 = (-3)(-9)$

Example 5 (a) $\frac{-10}{5} = -2$ (b) $\frac{10}{-5} = -2$ (c) $\frac{+10}{+5} = -2$

All three answers are identical.

Example 6 $\frac{|-20|}{-4} =$

Solution $\frac{|-20|}{-4} = \frac{20}{-4} = -5$

Example 7 $\left(-\frac{3}{5}\right) \div (2) =$

Solution $\left(-\frac{3}{5}\right) \div (2) = \left(-\frac{3}{5}\right) \times \left(\frac{1}{2}\right) = -\frac{3}{10}$

Example 8 $\left(-\frac{2}{7}\right) \div \left(-\frac{3}{14}\right) =$

Solution $\left(-\frac{2}{7}\right) \div \left(-\frac{3}{14}\right) = \left(-\frac{2}{\not 7_1}\right) \times \left(-\frac{\not 14^2}{3}\right) = \frac{4}{3} = 1\frac{1}{3}$

Example 9 $(33.264) \div (-0.28) =$

Solution To begin, the sign of the answer must be negative.
$(33.264) \div (-0.28) = -118.38$

7.5 EXERCISES

In problems 1–55 find the indicated quotients.

1. $\frac{15}{5} =$ ________ **2.** $\frac{27}{3} =$ ________ **3.** $\frac{36}{12} =$ ________

4. $\frac{60}{15} =$ ________ **5.** $\frac{100}{20} =$ ________ **6.** $\frac{140}{70} =$ ________

7. $\frac{-25}{5} =$ ________ **8.** $\frac{-39}{13} =$ ________ **9.** $\frac{-24}{6} =$ ________

10. $\frac{-72}{9} =$ ________ **11.** $\frac{-52}{13} =$ ________ **12.** $\frac{-75}{15} =$ ________

13. $\frac{84}{-12} =$ ________ **14.** $\frac{96}{-16} =$ ________ **15.** $\frac{120}{-24} =$ ________

16. $\frac{66}{-33} =$ ________ **17.** $\frac{56}{-14} =$ ________ **18.** $\frac{88}{-11} =$ ________

19. $\frac{-48}{-16} =$ ________ **20.** $\frac{-36}{-9} =$ ________ **21.** $\frac{-150}{-10} =$ ________

22. $\frac{-84}{-21} =$ ________ **23.** $\frac{-108}{-12} =$ ________ **24.** $\frac{-58}{-29} =$ ________

25. $\frac{-125}{25} =$ ________ **26.** $\frac{-60}{-30} =$ ________ **27.** $\frac{78}{-6} =$ ________

28. $\frac{-64}{-8} =$ ________ **29.** $\frac{105}{-21} =$ ________ **30.** $\frac{132}{12} =$ ________

31. $\dfrac{-136}{-17}$ = ______ 32. $\dfrac{225}{-15}$ = ______ 33. $\dfrac{-240}{30}$ = ______

34. $\dfrac{156}{-13}$ = ______ 35. $\dfrac{-250}{-25}$ = ______ 36. $\dfrac{-196}{14}$ = ______

37. $\left(-\frac{3}{4}\right) \div \left(\frac{5}{8}\right)$ = ______ 38. $\left(-2\frac{1}{2}\right) \div \left(-\frac{5}{4}\right)$ = ______

39. $(-2.046) \div (0.03)$ = ______ 40. $(33.976) \div (6.2)$ = ______

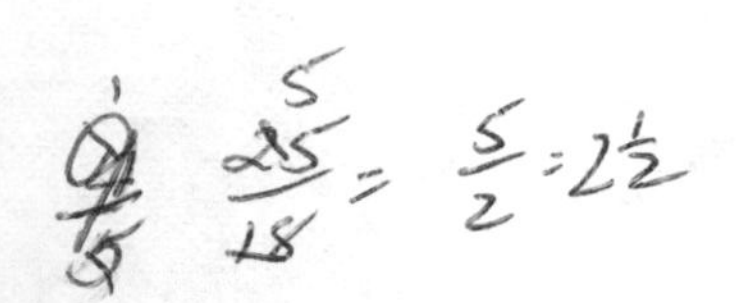

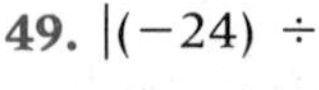

41. $\left(\frac{7}{12}\right) \div \left(-5\frac{1}{3}\right)$ = ______ 42. $\left(-\frac{9}{14}\right) \div \left(\frac{3}{7}\right)$ = ______

43. $(40.32) \div (-5.04)$ = ______ 44. $(-82.82) \div (-10.1)$ = ______

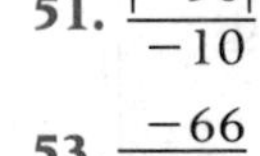

45. $(-9.18) \div (-0.204)$ = ______ 46. $(-78.75) \div (0.3)$ = ______

47. $\left(1\frac{4}{5}\right) \div \left(-\frac{18}{25}\right)$ = ______ 48. $\left(-2\frac{3}{4}\right) \div (-2)$ = ______

49. $|(-24) \div 8|$ = ______ 50. $|(-16) \div (-16)|$ = ______

51. $\dfrac{|-50|}{-10}$ = ______ 52. $\dfrac{36}{|-18|}$ = ______

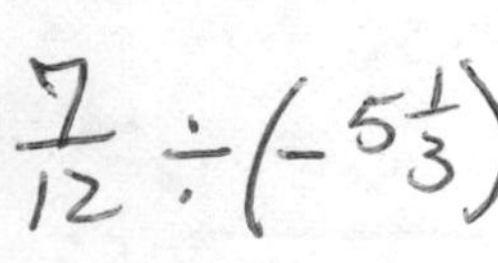

53. $\dfrac{-66}{|11|}$ = ______ 54. $\dfrac{|-72|}{-9}$ = ______

55. $\dfrac{|128|}{-32}$ = ______

ANSWERS

1. 3 2. 9 3. 3 4. 4 5. 5 6. 2 7. −5 8. −3 9. −4 10. −8
11. −4 12. −5 13. −7 14. −6 15. −5 16. −2 17. −4 18. −8
19. 3 20. 4 21. 15 22. 4 23. 9 24. 2 25. −5 26. 2 27. −13
28. 8 29. −5 30. 11 31. 8 32. −15 33. −8 34. −12 35. 10
36. −14 37. $-1\frac{1}{5}$ 38. 2 39. −68.2 40. 5.48 41. $-\frac{7}{64}$ 42. $-1\frac{1}{2}$
43. −8 44. 8.2 45. 45 46. −262.5 47. $-2\frac{1}{2}$ 48. $1\frac{3}{8}$ 49. 3 50. 1
51. −5 52. 2 53. −6 54. −8 55. −4

7.6 EXPONENTS

When two or more numbers are multiplied together, each number is called a **factor** or a divisor of the product. The product 3 × 5 has 3 and 5 as factors; the product 2 × 5 × 7 has 2, 5, and 7 as factors; the product $7x$ has 7 and x as factors; the product 3 × 3 × 3 × 3 has 3 as its only factor.

Example 1 List all of the factors of 12 greater than 1.

Solution The factors of 12 greater than 1 are 2, 3, 4, 6, and 12. (Of course, these numbers are not all simultaneously factors of 12.)

When you multiply the same factor again and again, such as in 3 × 3 × 3 × 3, the following shorthand notation is convenient: write the repeating factor, 3, and place a 4 (the number of times 3 is used as a factor) above and to the right of 3—thus, 3^4. The 3 is the base and the 4 is the exponent. Similarly, 5 × 5 × 5 × 5 × 5 × 5 × 5 has the factor 5 repeated 7 times. This may be abbreviated 5^7. The 5 is the base and 7 is the exponent.

If a is any number, and n is a natural number (1, 2, 3, . . .), then:

exponent

$$\text{base} \rightarrow a^n = \underbrace{a \times a \times a \times \cdots \times a}_{\text{there are } n \text{ factors of } a}$$

Here n is called the **exponent** to the **base** a. The number n tells us how many times to write down the base, a, with multiplication understood. a^n is called a **power** and may be read "a to the nth power." Hence, 2^7 is read "two to the seventh power," and 3^5 is read "three to the fifth power."

Example 2 $6^4 = 6 \times 6 \times 6 \times 6 = 1{,}296$

Example 3 $2^6 = 2 \times 2 \times 2 \times 2 \times 2 \times 2 = 64$

Example 4 $7^3 = 7 \times 7 \times 7 = 343$

Example 5 $b^5 = b \times b \times b \times b \times b$

Example 6 (a) $5^1 = 5$, (b) $a^1 = a$. The exponent, 1, is understood.

Example 7

$10^4 = 10 \times 10 \times 10 \times 10 = 10{,}000$
$10^3 = 10 \times 10 \times 10 = 1{,}000$; this is read "10 cubed"
$10^2 = 10 \times 10 = 100$; this is read "10 squared"
$10^1 = 10$

When raising 10 to a positive power, the number of zeros following the digit 1 in the answer is the same as the exponent; for example,

$$10^{\boxed{2}} = 1\boxed{00} \qquad 10^{\boxed{3}} = 1\boxed{000} \qquad 10^{\boxed{4}} = 1\boxed{0{,}000}$$

Example 8 $\left(\frac{1}{2}\right)^4 = \frac{1}{2} \times \frac{1}{2} \times \frac{1}{2} \times \frac{1}{2} = \frac{1}{16}$

The base is $\frac{1}{2}$; the exponent is 4.

Example 9 (a) $(-2)^4 = (-2)(-2)(-2)(-2) = +16$

The base is -2; the exponent is 4.

$-2^4 = -(2 \times 2 \times 2 \times 2) = -16$

-2^4 means the value of 2^4 is negated.

Hence $(-2)^4 \neq -2^4$

(b) $(-3)^2 = (-3)(-3) = +9$

$-3^2 = -(3 \times 3) = -9$

Hence $(-3)^2 \neq -3^2$

Example 10 (a) $(5x)^2$ means $(5x)(5x)$. If $x = 2$, then $(5x)^2 = (10)(10) = 100$.

(b) $5x^2$ means $5(x)(x)$. If $x = 2$, then $5x^2 = 5(2)(2) = 20$.

Definition: If a is any number ($a \neq 0$), then $a^0 = 1$.

Example 11 (a) $7^0 = 1$ (b) $15^0 = 1$
(c) $(-2)^0 = 1$ (d) $x^0 = 1$

Multiplication of Exponential Numbers

Consider the following products:

$6^4 \times 6^0 = 6^4 \times 1 = 6^4$

$4^3 \times 4^2 = (4 \times 4 \times 4) \times (4 \times 4) = 4^5$ Observe that $3 + 2 = 5$

$7^2 \times 7^4 = (7 \times 7) \times (7 \times 7 \times 7 \times 7) = 7^6$ Observe that $2 + 4 = 6$

If you multiply $a^5 \cdot a^6$, the first factor contributes 5 factors of a to the product and the second factor contributes 6 factors of a to the product, so the product has a total of 11 factors of a and the answer must be a^{11}, read "a to the eleventh power."

To multiply exponential numbers having the same base, add the exponents: $a^m \cdot a^n = a^{m+n}$.

Example 12 $2^2 \cdot 2^3 = 2^{2+3} = 2^5$

Example 13 $3^5 \cdot 3^4 = 3^{5+4} = 3^9$

Example 14 $5 \cdot 5^6 = 5^{1+6} = 5^7$

Example 15 $4^3 \cdot 4^4 \cdot 4^5 = 4^{3+4+5} = 4^{12}$

Example 16 $a^3 \cdot a^5 = a^{3+5} = a^8$

Example 17 $2^3 \times 3^2$ The rule does not apply when the bases are different. $2^3 = 8$; $3^2 = 9$. Thus $2^3 \times 3^2 = 8 \times 9 = 72$

Division of Exponential Numbers

Consider the following:

$$\frac{6^4}{6^2} = \frac{\overset{1}{\cancel{6}} \times \overset{1}{\cancel{6}} \times 6 \times 6}{\underset{1}{\cancel{6}} \times \underset{1}{\cancel{6}}} = 6 \times 6 = 6^2 \quad \text{Observe that } 4 - 2 = 2$$

$$\frac{10^7}{10^4} = \frac{\overset{1}{\cancel{10}} \times \overset{1}{\cancel{10}} \times \overset{1}{\cancel{10}} \times \overset{1}{\cancel{10}} \times 10 \times 10 \times 10}{\underset{1}{\cancel{10}} \times \underset{1}{\cancel{10}} \times \underset{1}{\cancel{10}} \times \underset{1}{\cancel{10}}}$$

$$= 10 \times 10 \times 10 = 10^3 \quad \text{Observe that } 7 - 4 = 3$$

$$\frac{5^3}{5^3} = \frac{\overset{1}{\cancel{5}} \times \overset{1}{\cancel{5}} \times \overset{1}{\cancel{5}}}{\underset{1}{\cancel{5}} \times \underset{1}{\cancel{5}} \times \underset{1}{\cancel{5}}} = 1$$

7-4

Consider further:

$$\frac{4^3}{4^5} = \frac{\not{4}^1 \times \not{4}^1 \times \not{4}^1}{\not{4}_1 \times \not{4}_1 \times \not{4}_1 \times 4 \times 4} = \frac{1}{4 \times 4} = \frac{1}{4^2}$$ Observe that $5 - 3 = 2$

$$\frac{3^2}{3^6} = \frac{\not{3}^1 \times \not{3}^1}{\not{3}_1 \times \not{3}_1 \times 3 \times 3 \times 3 \times 3} = \frac{1}{3 \times 3 \times 3 \times 3} = \frac{1}{3^4}$$

Observe that $6 - 2 = 4$

In the first group of examples we subtracted the exponent of the denominator from that of the numerator and in the second group of examples we reversed the process. To eliminate the problem of trying to decide how to subtract the exponents we introduce **negative exponents.**

> If a is any number $a \neq 0$, then
> $$a^{-m} = \frac{1}{a^m}$$
> where m is a natural number.

Example 18 $2^{-3} = \frac{1}{2^3} = \frac{1}{8}$

Example 19 $5^{-2} = \frac{1}{5^2} = \frac{1}{25}$

Example 20 $10^{-1} = \frac{1}{10^1} = \frac{1}{10}$

Example 21 $x^{-5} = \frac{1}{x^5}$

Note: A negative exponent *does not* change the sign of the number.

We can now give a rule for the division of exponential numbers.

> When dividing exponential numbers with the same base, you may determine the exponent of your answer by subtracting the exponent of the denominator from the exponent of the numerator; that is,
> $$\frac{a^m}{a^n} = a^{m-n} \qquad a \neq 0$$
> where m and n are integers.

Example 22 $\frac{3^8}{3^4} = 3^{8-4} = 3^4 = 81$

Example 23 $\frac{5^6}{5} = 5^{6-1} = 5^5 = 3125$

Example 24 $\frac{2^8}{2^2} = 2^{8-2} = 2^6 = 64$

Example 25 $\frac{3^5}{3^7} = 3^{5-7} = 3^{-2} = \frac{1}{3^2} = \frac{1}{9}$

Example 26 $\frac{4^3}{4^6} = 4^{3-6} = 4^{-3} = \frac{1}{4^3} = \frac{1}{64}$

Example 27 $\frac{x^{-5}}{x^{-2}} = x^{-5-(-2)} = x^{-3} = \frac{1}{x^3}$

Example 28 $\frac{a^{-3}}{a^{-8}} = a^{-3-(-8)} = a^5$

Example 29 $\frac{10^4}{5^2}$ The rule cannot be applied when the bases are different.

Finding the Power of an Exponential Number

Using our ability to multiply with like bases consider the following:

$(4^2)^3 = 4^2 \times 4^2 \times 4^2 = 4^6$

$(10^5)^4 = 10^5 \times 10^5 \times 10^5 \times 10^5 = 10^{20}$

> To find the power of an exponential number multiply the exponents
>
> $$(a^m)^n = a^{mn}$$
>
> where m and n are integers.

Example 30 $(2^3)^4 = 2^{3 \cdot 4} = 2^{12}$

Example 31 $(4^5)^2 = 4^{5 \cdot 2} = 4^{10}$

Example 32 $(x^3)^7 = x^{3 \cdot 7} = x^{21}$

Example 33 $(7^8)^0 = 7^{8 \cdot 0} = 7^0 = 1$

Example 34 If $x = 2$, $4x^3$ means $4 \cdot 2^3 = 4 \cdot 8 = 32$

Finding the Power of a Product of Numbers

Using the rule for multiplying exponential numbers with like bases, consider the following:

$(3x)^2 = (3x)\,(3x) = 3 \cdot 3 \cdot x \cdot x = 9x^2$

$(xy)^3 = (xy)(xy)(xy) = x \cdot x \cdot x \cdot y \cdot y \cdot y = x^3y^3$

> To find the power of a product of numbers,
>
> $$(ab)^m = a^m\, b^m$$
>
> where m is an integer.

Example 35 $(2xy)^3 = 2^3x^3y^3 = 8x^3y^3$

Example 36 $(ab)^4 = a^4b^4$

Example 37 $(3a^3b^2)^2 = 3^2(a^3)^2(b^2)^2$

$= 9a^6b^4$

Example 38 $(-5a^{-2}b^3)^3 = (-5)^3(a^{-2})^3(b^3)^3$

$= -125a^{-6}b^9$

$= \dfrac{-125b^9}{a^6}$

Example 39 Does $(4x)^3 = 4x^3$?

Solution $(4x)^3 = 4^3\,x^3 = 64x^3$

$64x^3 \neq 4x^3$

Finding the Power of a Quotient of Numbers

Consider the following.

$$\left(\frac{2}{3}\right)^2 = \frac{2}{3} \times \frac{2}{3} = \frac{2^2}{3^2} = \frac{4}{9}$$

$$\left(\frac{4}{5}\right)^3 = \frac{4}{5} \times \frac{4}{5} \times \frac{4}{5} = \frac{4^3}{5^3} = \frac{64}{125}$$

> If a and b are any numbers, $b \neq 0$, and m is an integer, then
> $$\left(\frac{a}{b}\right)^m = \frac{a^m}{b^m}$$

Example 40 $\left(\frac{1}{2}\right)^4 = \frac{1^4}{2^4} = \frac{1}{16}$

Example 41 $\left(\frac{3}{4}\right)^3 = \frac{3^3}{4^3} = \frac{27}{64}$

Example 42 $\left(\frac{2}{3}\right)^{-2} = \dfrac{2^{-2}}{3^{-2}} = \dfrac{\frac{1}{2^2}}{\frac{1}{3^2}} = \dfrac{\frac{1}{4}}{\frac{1}{9}}$

$= \frac{1}{4} \div \frac{1}{9}$

$= \frac{1}{4} \times \frac{9}{1} = \frac{9}{4} = 2\frac{1}{4}$

Example 43 $\left(\dfrac{x^2}{y}\right)^3 = \dfrac{(x^2)^3}{y^3} = \dfrac{x^6}{y^3}$

7.6 EXERCISES

In problems 1–14 answer true or false.

1. $5^0 = 1$ __________

2. 10^3 has a base of 10. __________

3. $2^3 \cdot 2^4 = 4^7$ ______

4. a^{15} has 15 for an exponent. ______

5. 7^3 is read "seven squared." ______

6. $3^2 \cdot 2^3 = 6^5$ ______

7. $(3^5)^3 = 3^8$ ______

8. $4^{-2} = \frac{1}{8}$ ______

9. $\frac{7^5}{7^{10}} = 7^{-5}$ ______

10. "Nine to the 6th power" is expressed as 6^9. ______

11. $(3x)^2 = 3x^2$ ______

12. $-3^4 = -81$ ______

13. $\frac{6^3}{3^2} = 2$ ______

14. $a^{13} \cdot a^{12} = a^{25}$ ______

In problems 15–24 write each of the given quantities in exponential form.

15. *bbbb* ______ 16. *aaaa* ______

17. *c* cubed ______ 18. *x* squared ______

19. 8*abab* ______ 20. 5*mmm* ______

21. *abbabb* ______ 22. (5)(5)(6)(6)(6) ______

23. (7)(2)(7)(2) ______ 24. $(3x)(3x)(3x)(3x)$ ______

In problems 25–44 evaluate the given quantities.

25. 3^4 ______ 26. 5^0 ______

27. 4^{-2} ______ 28. 8^{-1} ______

29. 5^3 ______ 30. 5^{-3} ______

31. 10^{-2} ______ 32. 2^{-5} ______

33. 8^0 ______ 34. 2^7 ______

35. $(0.6)^2$ ______ 36. $(0.4)^3$ ______

37. $\left(\frac{1}{2}\right)^3$ ______ 38. $\left(\frac{3}{4}\right)^2$ ______

39. $\left(\frac{1}{10}\right)^{-2}$ ______ 40. $\left(\frac{2}{5}\right)^{-3}$ ______

41. $(1.1)^3$ ______ 42. $\left(\frac{1}{3}\right)^2$ ______

43. 10^{-4} ______ 44. $\left(\frac{3}{2}\right)^{-3}$ ______

In problems 45–100, using rules of exponents, evaluate the given expressions. Leave your answer in exponential form with positive exponents.

45. $3^4 \cdot 3^2 =$ ____________ **46.** $5^2 \cdot 5^6 =$ ____________

47. $10 \cdot 10^4 =$ ____________ **48.** $7^4 \cdot 7^3 =$ ____________

49. $(-2)^5 \cdot (-2)^3 =$ ____________ **50.** $x^5 \cdot x^4 =$ ____________

51. $(x^4)^3 =$ ____________ **52.** $(a^5)^2 =$ ____________

53. $(2^4)^5 =$ ____________ **54.** $(7^6)^3 =$ ____________

55. $(3^3)^3 =$ ____________ **56.** $(4^3)^2 =$ ____________

57. $(10^2)^{-1} =$ ____________ **58.** $(2^{-3})^2 =$ ____________

59. $(4^{-2})^{-2} =$ ____________ **60.** $(5^{-1})^{-5} =$ ____________

61. $(8^{-1})^4 =$ ____________ **62.** $(3^5)^{-3} =$ ____________

63. $\frac{4^7}{4^2} =$ ____________ **64.** $\frac{10^5}{10^4} =$ ____________

65. $\frac{6^{11}}{6^7} =$ ____________ **66.** $\frac{2^8}{2^6} =$ ____________

67. $\frac{x^7}{x^4} =$ ____________ **68.** $\frac{x^2}{x^7} =$ ____________

69. $\frac{2^4}{2^6} =$ ____________ **70.** $\frac{10^3}{10^{10}} =$ ____________

71. $\frac{8^5}{8^9} =$ ____________ **72.** $\frac{4^2}{4^5} =$ ____________

73. $\frac{4^{-2}}{4^3} =$ ____________ **74.** $\frac{7^{-3}}{7^5} =$ ____________

75. $\frac{12^8}{12^{-4}} =$ ____________ **76.** $\frac{3^{-5}}{3^{-8}} =$ ____________

77. $\frac{x^{-2}}{x^{-3}} =$ ____________ **78.** $\frac{2^{-4}}{2^{-2}} =$ ____________

79. $\frac{5^{-1}}{5^{-4}} =$ ____________ **80.** $\frac{a^{-6}}{a^{-8}} =$ ____________

81. $5^7 \cdot 5^8 =$ ____________ **82.** $6^4 \cdot 6^{-3} =$ ____________

83. $x^{-4} \cdot x^{-5} =$ ____________ **84.** $a^{-6} \cdot a =$ ____________

85. $\frac{5^4}{5^{-2}} =$ ____________ **86.** $\frac{7^{-3}}{7^{-6}} =$ ____________

87. $\frac{7^2}{7^{11}} =$ ____________ **88.** $\frac{x^4}{x^{10}} =$ ____________

89. $(3^{-2})^{-4} =$ ____________ **90.** $(x^2)^{-1} =$ ____________

91. $(4x)^2 =$ ____________ **92.** $(3x^2y^3)^4 =$ ____________

93. $(5x^{-1})^{-2} =$ ____________ **94.** $(x^{-2}y^{-3})^{-3} =$ ____________

95. $(-xy^2)^4 =$ ____________ **96.** $(2^2xy^{-1})^2 =$ ____________

97. $\left(\frac{3}{5}\right)^3 =$ __________ **98.** $\left(\frac{2}{7}\right)^4 =$ __________

99. $\left(\frac{3x}{y^2}\right)^2 =$ __________ **100.** $\left(\frac{x^{-1}}{x^{-2}}\right)^3 =$ __________

In problems 101–110 find a numerical value for the given expression.

101. $\left(\frac{1}{2}\right)^2 =$ __________ **102.** $\left(\frac{2}{5}\right)^3 =$ __________

103. $6^2 =$ __________ **104.** $\left(\frac{4}{3}\right)^4 =$ __________

105. $3x^3$ if $x = 2$ __________ **106.** $(4x)^2$ if $x = \frac{1}{4}$ __________

107. $-4x^3$ if $x = -1$ __________ **108.** $-\frac{1}{2}x^2$ if $x = 8$ __________

109. $-(6x)^2$ if $x = \frac{2}{3}$ __________ **110.** $-\left(\frac{1}{2}x\right)^3$ if $x = 10$ __________

ANSWERS

1. true **2.** true **3.** false **4.** true **5.** false **6.** false **7.** false **8.** false
9. true **10.** false **11.** false **12.** true **13.** false **14.** true **15.** b^4 **16.** a^4
17. c^3 **18.** x^2 **19.** $8a^2b^2$ **20.** $5m^3$ **21.** a^2b^4 **22.** $5^2 \cdot 6^3$ **23.** $7^2 \cdot 2^2$
24. 3^4x^4 **25.** 81 **26.** 1 **27.** $\frac{1}{16}$ **28.** $\frac{1}{8}$ **29.** 125 **30.** $\frac{1}{125}$ **31.** $\frac{1}{100}$ **32.** $\frac{1}{32}$
33. 1 **34.** 128 **35.** 0.36 **36.** 0.064 **37.** $\frac{1}{8}$ **38.** $\frac{9}{16}$ **39.** 100
40. $15\frac{5}{8}$ **41.** 1.331 **42.** $\frac{1}{9}$ **43.** $\frac{1}{10{,}000}$ **44.** $\frac{8}{27}$ **45.** 3^6 **46.** 5^8 **47.** 10^5
48. 7^7 **49.** $(-2)^8$ **50.** x^9 **51.** x^{12} **52.** a^{10} **53.** 2^{20} **54.** 7^{18} **55.** 3^9
56. 4^6 **57.** $\frac{1}{10^2}$ **58.** $\frac{1}{2^6}$ **59.** 4^4 **60.** 5^5 **61.** $\frac{1}{8^4}$ **62.** $\frac{1}{3^{15}}$ **63.** 4^5 **64.** 10
65. 6^4 **66.** 2^2 **67.** x^3 **68.** $\frac{1}{x^5}$ **69.** $\frac{1}{2^2}$ **70.** $\frac{1}{10^7}$ **71.** $\frac{1}{8^4}$ **72.** $\frac{1}{4^3}$ **73.** $\frac{1}{4^5}$
74. $\frac{1}{7^8}$ **75.** 12^{12} **76.** 3^3 **77.** x **78.** $\frac{1}{2^2}$ **79.** 5^3 **80.** a^2 **81.** 5^{15} **82.** 6
83. $\frac{1}{x^9}$ **84.** $\frac{1}{a^5}$ **85.** 5^6 **86.** 7^3 **87.** $\frac{1}{7^9}$ **88.** $\frac{1}{x^6}$ **89.** 3^8 **90.** $\frac{1}{x^2}$ **91.** 4^2x^2
92. $3^4x^8y^{12}$ **93.** $\frac{x^2}{5^2}$ **94.** x^6y^9 **95.** x^4y^8 **96.** $\frac{2^4x^2}{y^2}$ **97.** $\frac{3^3}{5^3}$ **98.** $\frac{2^4}{7^4}$
99. $\frac{3^2x^2}{y^4}$ **100.** x^3 **101.** $\frac{1}{4}$ **102.** $\frac{8}{125}$ **103.** 36 **104.** $3\frac{13}{81}$ **105.** 24 **106.** 1
107. 4 **108.** -32 **109.** -16 **110.** -125

7.7 ONE-STEP EQUATIONS

An equation is a statement that two quantities are equal. The following are examples of equations: $2 + 7 = 9$, $6a + 3a = 11a - 2a$, and $10^2 - 64 = 36$. However, if you look close, you will see that both sides of each equation are identical. These equations are called *identities.*

Example 1 Write two equations that are identities.

Solution

$$a + 5 + 1 = a + 6$$
$$(7 + 8)x = 15x$$

Although it is true that identities are correct equations and of some importance in advanced mathematics, it is also true that they are of minor value in algebra. Identities are true for all values of the variable. Equations that are true only for some values of the variable ("some" in mathematics means at least one) are more important, and it is these equations that we will investigate. They are called **conditional**

equations or **open sentences.** For example, $8x = 32$ is a conditional equation. By inspection, we can see that $x = 4$ is the only value of the variable, x, that will make this equation true. That is, $x = 4$ "satisfies the equation." This value, $x = 4$, is called the **solution** or the **root** of the equation $8x = 32$. The equation $y - 5 = 8$, has $y = 13$ as a solution; and the equation $\frac{y}{3} = 6$ has $y = 18$ as a solution. That is, 13 is the number that "satisfies" the first equation and 18 is the number that "satisfies" the second equation. To check the first equation observe that $13 - 5 = 8$; and to check the second equation observe that $\frac{18}{3} = 6$.

> A *solution*, or *root*, of an equation is a number that, when used in place of the variable, makes the equation true. That is, a solution satisfies the equation.

To solve an equation means to determine what value(s), if any, of your variable will satisfy the equation. An equation is said to be "solved" when you can write it in the form x, meaning one x, equals some real number. We can determine with relative ease that $n = 11$ is the solution of the equation $n + 12 = 23$. We can also see that $x = 5$ is the solution of the equation $5x - 3 = 22$. However, the solution of the equation $\frac{5x}{3} + 17 = 27$ is not as easily determined. The following rules help us to solve equations.

> **Rules for Solving Equations**
>
> These rules are methods that are used to isolate the variable on one side of the equation.
>
> Rule 1. Adding the same number to each side of an equation produces a correct equation.
> Rule 2. Subtracting the same number from each side of an equation produces a correct equation.
> Rule 3. Multiplying each side of an equation by the same number produces a correct equation.
> Rule 4. Dividing each side of an equation by the same nonzero number produces a correct equation.

In this section we will concentrate on equations that require only one step in their solution. Solving these equations will be easy if you follow the rules for solving equations and carefully read the next examples.

Example 1 Solve the equation $y - 8 = 17$.

Solution It is understood that you are to solve for y.

$$y - 8 = 17$$
$$y - 8 \boxed{+8} = 17 \boxed{+8;} \text{ add } +8 \text{ to each side (rule 1)}$$
$$y + 0 = 25$$
$$y = 25; \text{ so "one } y \text{ equals 25"}$$

Example 2 Solve the equation $x + 15 = 35$.

Solution It is understood that you are to solve for x.

$$x + 15 = 35$$
$$x + 15 \boxed{-\ 15} = 35 \boxed{-\ 15}; \text{ subtract 15 from each side (rule 2)}$$
$$x + 0 = 20$$
$$x = 20$$

Example 3 Solve the equation $\frac{x}{5} = 7$.

Solution It is understood that you are to solve for x.

$$\frac{x}{5} = 7$$
$$\boxed{(\cancel{5})}\left(\frac{x}{\cancel{5}}\right) = \boxed{(5)}(7); \text{ multiply each side by 5 (rule 3)}$$
$$x = 35$$

Example 4 Solve the equation $11w = 132$.

Solution It is understood that you are to solve for w.

$$11w = 132$$
$$\frac{\cancel{11}w}{\boxed{\cancel{11}}} = \frac{132}{\boxed{11}}; \text{ divide each side by 11 (rule 4)}$$
$$w = 12$$

You "check" the solution or root of your equation by replacing the variable in the equation (everywhere it appears) with your solution. If your solution is correct, working out the arithmetic on each side of the equation independently (do the left side and do the right side) will yield the same numerical value.

Example 5 (a) Check the solution to Example 1.
(b) Check the solution to Example 3.

Checks:

(a) Replace y by 25

$$25 - 8 \stackrel{?}{=} 17$$
$$17 = 17$$

(b) Replace x by 35

$$\frac{35}{5} \stackrel{?}{=} 7$$
$$7 = 7$$

The preceding example solutions are correct.a12

Example 6 The statement $x + 1 = x + 2$ is interesting in that there is no value of x that will make it true. *It has no solution.* There is no number to which you can either add 1 or add 2 and obtain the same answer. This sort of "equation" will not be discussed further because it is really not an equation.

7.7 EXERCISES

Solve each of these equations.

1. $4n = 20$ ______________________

2. $x + \$25.32 = \$75.18$ ______________

3. $\frac{n}{7} = 7$ ______________________

4. $90r = 7.2$ ______________________

5. $x - 3 = 9$ ______ 6. $(0.08)b = 7.2$ ______

7. $z + 5 = 10$ ______ 8. $r + 7 = 0$ ______

9. $t - 6 = 4.2$ ______ 10. $-25 + k = 25$ ______

11. $\frac{1}{3}y = 4$ ______ 12. $\frac{3}{2}y = 12$ ______

13. $n + 2.4 = 9.6$ ______ 14. $0.26x = 4.68$ ______

15. $15x = 240$ ______ 16. $\frac{x}{0.9} = 30$ ______

17. $1.2x = 3.6$ ______ 18. $\frac{4}{5}y = 16$ ______

19. $t + 5 = 5$ ______ 20. $0.37t = 7.03$ ______

21. $\frac{x}{6} = 2.5$ ______ 22. $\frac{k}{0.75} = 60$ ______

23. $0.7y = 49$ ______ 24. $80s = 54.4$ ______

25. $62x = 5{,}394$ ______ 26. $\frac{x}{91} = 6{,}825$ ______

27. $y + \$75.86 = \$110.00$ ______ 28. $w + \$62.57 = \$83.40$ ______

29. $\frac{8}{7}x = 16$ ______ 30. $\frac{y}{0.72} = 54$ ______

ANSWERS

1. $n = 5$ **2.** $x = \$49.86$ **3.** $n = 49$ **4.** $r = 0.08$ **5.** $x = 12$ **6.** $b = 90$
7. $z = 5$ **8.** $r = -7$ **9.** $t = 10.2$ **10.** $k = 50$ **11.** $y = 12$ **12.** $y = 8$
13. $n = 7.2$ **14.** $x = 18$ **15.** $x = 16$ **16.** $x = 27$ **17.** $x = 3$ **18.** $y = 20$ **19.** $t = 0$
20. $t = 19$ **21.** $x = 15$ **22.** $k = 45$ **23.** $y = 70$ **24.** $s = 0.68$ **25.** $x = 87$
26. $x = 621{,}075$ **27.** $y = \$34.14$ **28.** $w = \$20.83$ **29.** $x = 14$ **30.** $y = 38.88$

7.8 MULTISTEP EQUATIONS

The next examples illustrate equations that require two or more steps to complete their solution. Each equation will have exactly one root; this will be true of all equations in this chapter.

Example 1 Solve $5n - 3 = 17$.

Solution

$$5n - 3 = 17$$

$$5n - 3 \boxed{+ 3} = 17 \boxed{+ 3}; \text{ we add 3 to both sides (rule 1)}$$

$$5n = 20$$

$$\frac{\overset{1}{\cancel{5}}n}{\boxed{\cancel{5}_1}} = \frac{20}{\boxed{5}}; \text{ we divided both sides by 5 (rule 4)}$$

$$n = 4$$

Check

$$5(4) - 3 \stackrel{?}{=} 17$$

$$20 - 3 \stackrel{?}{=} 17$$

$$17 = 17$$

Example 2 Solve $\frac{1}{5}x + 3 = 10$.

Solution

$$\frac{1}{5}x + 3 \boxed{-\ 3} = 10 \boxed{-\ 3;}$$ we subtracted 3 from each side (rule 2)

$$\frac{1}{5}x = 7$$

$$\boxed{\cancel{5}^{1}}\left(\frac{1}{\cancel{5}_{1}}x\right) = \boxed{5} \cdot 7;$$ we multiplied each side by 5 (rule 3)

$$x = 35$$

Terms such as 7 and 9, $5x$ and $8x$, and $4ab$ and $9ab$ are called *like terms* or *similar terms.* Two terms are alike when their literal portion is identical. Terms such as $4a$ and $17b$ are *unlike terms.* Also, $5x$ and $3ab$ are unlike terms.

You may add like terms. For example, $5x + 3x = 8x$, $17y + (-19y) = -2y$, $3a + (-5a) + 16a = 14a$, and $8a^2 + 14a^2 = 22a^2$. Whenever like terms appear on the same side of an equation we "combine" them. To **combine** like terms means to add them. This will be illustrated in the next examples.

Example 3 Solve $6n + 8 - 4n = 17$.

Solution

$$6n + 8 - 4n = 17$$

$$2n + 8 = 17;$$ we combined (added) the n's

$$2n + 8 \boxed{-\ 8} = 17 \boxed{-\ 8;}$$ we subtracted 8 from each side (rule 2)

$$2n = 9$$

$$\frac{\cancel{2}^{1}n}{\boxed{\cancel{2}}_{1}} = \frac{9}{\boxed{2}};$$ we divided by 2 (rule 4)

$$n = \frac{9}{2} = 4\frac{1}{2}$$

Example 4 Solve $3x + 0.7 + 2x = 1.3 + 0.9$.

Solution

$5x + 0.7 = 2.2$; combine like terms
$5x = 1.5$; we subtracted 0.7 from both sides (rule 2)
$x = 0.3$; we divided both sides by 5 (rule 4)

Check

$$(3)(0.3) + 0.7 + (2)(0.3) \stackrel{?}{=} 1.3 + 0.9$$
$$0.9 + 0.7 + 0.6 \stackrel{?}{=} 2.2$$
$$2.2 = 2.2$$

Example 5 Solve $-\frac{2x}{3} + 1 = -5$.

Solution

$-\frac{2x}{3} = -6$; add -1 to both sides (rule 1)
$x = 9$; multiply both sides by $-\frac{3}{2}$ (rule 3)

Example 6 Solve $1.01x + 2.19 = 3.20$.

Solution

$1.01x = 1.01$; add -2.19 to both sides (rule 1)
$x = 1$; divide both sides by 1.01 (rule 4)

Example 7 Solve $5x - x + 6 = -x - 14$.

Solution

$4x + 6 = -x - 14$; combine like terms
$5x + 6 = -14$; add x to both sides (rule 1)
$5x = -20$; add -6 to both sides (rule 1)
$x = -4$; divide both sides by 5 (rule 4)

By adding -5 to each side of the equation $2x + 5 = -7$, you obtain the equation $2x = -12$. By dividing each side of the equation $2x = -12$ by 2, you obtain the equation $x = -6$. All three of these equations, $2x + 5 = -7$, $2x = -12$, and $x = -6$, have the same solution set; and the solution of -6 can be seen, at a glance, by merely looking at the last of these equations. Two equations that have the same solution set are called **equivalent equations.** The step-by-step process of solving an equation, in x, simply involves turning the equation into an equivalent equation again and again until a point is reached at which the solution is staring us in the face. At this time our equivalent equation reads, "One x equals some real number."

7.8 EXERCISES

Solve each of the following equations.

1. $4x + 5 = 17$ ____________
2. $7n - 5 = 23$ ____________
3. $7x - 1 = 13$ ____________
4. $5z - 7 = -42$ ____________
5. $5 - x = 3 + x$ ____________
6. $3 - 2b = 6 + b$ ____________
7. $2 - 2x = 8 + x$ ____________
8. $3w - 2 = 6 + w$ ____________
9. $5x - 4 = 3x + 1$ ____________
10. $3x = 6x - 6$ ____________
11. $5 - 3x + 9 = 7x$ ____________
12. $3x - 5 = -4x + 23$ ____________
13. $7x - 8 + 2x = 19$ ____________
14. $5x + 3 - 2x = 18$ ____________
15. $11t - 5 = 4t - 13$ ____________
16. $2x - 5 = x + 7$ ____________
17. $12n + 6 = 18 - 6n$ ____________
18. $8x - 13x = 2x + 56$ ____________
19. $15 = +4 + 21t - 17$ ____________
20. $12x - 7x - 4x + 9 = 19$ ____________
21. $\frac{2x}{5} = 4$ ____________
22. $\frac{x}{2} - 3 = 0$ ____________
23. $\frac{y}{3} - 7 = -6$ ____________
24. $\frac{1}{5}z + 3 = 13$ ____________
25. $\frac{x}{5} - \frac{3}{2} = \frac{7}{10}$ ____________
26. $\frac{x}{3} + \frac{1}{2} = \frac{5}{6}$ ____________
27. $\frac{y}{3} + 5 + 6 = 13 + y$ ____________
28. $3x - 1\frac{7}{8} + 2x = 7\frac{1}{2}$ ____________
29. $\frac{1}{8}y + \frac{3}{8}y + 1 = 3^2$ ____________
30. $\frac{2y + 7}{3} = 9$ ____________
31. $0.7x - 0.3 + 0.2x = 2.4$ ____________
32. $6 + 4(2 - t) = 3t$ ____________

ANSWERS

1. $x = 3$ 2. $n = 4$ 3. $x = 2$ 4. $z = -7$ 5. $x = 1$ 6. $b = -1$ 7. $x = -2$ 8. $w = 4$ 9. $x = \frac{5}{2}$ 10. $x = 2$ 11. $x = \frac{7}{5}$ 12. $x = 4$ 13. $x = 3$ 14. $x = 5$ 15. $t = -1\frac{1}{7}$ 16. $x = 12$ 17. $n = \frac{2}{3}$ 18. $x = -8$ 19. $t = 1\frac{1}{3}$ 20. $x = 10$ 21. $x = 10$ 22. $x = 6$ 23. $y = 3$ 24. $z = 50$ 25. $x = 11$ 26. $x = 1$ 27. $y = -3$ 28. $x = 1\frac{7}{8}$ 29. $y = 16$ 30. $y = 10$ 31. $x = 3$ 32. $t = 2$

FOCUS ON PROBLEM SOLVING

A 35 mm lens. Note that the bottom numbers are the f-numbers.

If you look at the lens mounted on a 35 mm camera, you will see a ring that has a series of numbers marked on it:

3.5 4 5.6 8

These are the *f*-numbers or relative apertures of the lens. In an adjustable camera the *f*-number is one of the ways that the amount of light reaching the film can be controlled (the other being the length of time it takes the light to reach the film, or *shutter speed*). The smaller the *f*-number the greater the amount of light that reaches the film. For the above-mentioned *f*-numbers (written *f*/3.5, *f*/4, *f*/5.6, etc.), *f*/3.5 will let in the greatest amount of light and *f*/22 will let in the least amount.

The aperture is controlled by turning the ring on the lens; the size of the aperture or circular opening, a ring of overlapping leaves, varies. As we move through the *f*-numbers the diameters of the opening change, thus changing the area of the opening. As the diameter of the opening doubles, the area of the opening increases four times.

For example, if the diameter of a circle is 2 millimeters,

$$\begin{aligned}\text{Area} &= 3.14 \times (1 \text{ mm})^2\\ &= 3.14 \times 1 \text{ sq. mm}\\ &= 3.14 \text{ sq mm}\end{aligned}$$

If, however, the diameter of a circle is 4 millimeters,

$$\begin{aligned}\text{Area} &= 3.14 \times (2 \text{ mm})^2\\ &= 3.14 \times 4 \text{ sq. mm}\\ &= 12.56 \text{ sq mm}\end{aligned}$$

Thus, as we move through the *f*-numbers, which contain intermediate settings, from smallest to largest, f /4 will let in $\frac{1}{2}$ as much light as *f* /3.5, *f* /5.6 will let in $\frac{1}{2}$ as much light as *f* /4, f /8 will let in $\frac{1}{2}$ as much light as *f* /5.6, and so forth.

How do we find the relative aperture or *f*-numbers of a lens? It is determined by two variables; the focal length of the lens—the distance measured from the center of the lens to the image it forms on film—and the diameter of the lens. Thus the formula or equations:

$$f\text{-number} = \frac{\text{focal length of lens}}{\text{diameter of lens}}$$

If the focal length of a lens is 300 mm and its diameter is 75 mm, the f-number can be found as follows:

$$f\text{-number} = \frac{300 \text{ mm}}{75 \text{ mm}} = 4, \text{ written f /4}$$

If the focal length of a lens is 500 mm and its diameter is 62.5 mm, the *f*-number is:

$$f\text{-number} = \frac{500 \text{ mm}}{62.5 \text{ mm}} = 8, \text{ written f /8}$$

If the *f*-number of a lens is 2 and the focal length of the lens is 55 mm, the diameter of the lens is:

$$2 = \frac{55 \text{ mm}}{\text{diameter of the lens}}$$

$$2 \times \text{diameter of the lens} = 55 \text{ mm}$$
$$\text{diameter of the lens} = 27.5 \text{ mm}$$

BUILDING YOUR MATH VOCABULARY

TERM	DEFINITION	EXAMPLE
Absolute Value	The distance of a number from the origin.	$\lvert -3 \rvert = 3$ $\lvert 0 \rvert = 0$ $\lvert 3 \rvert = 3$
Base	A number being multiplied by itself in an exponential statement.	$5^4 = 625$; 5 is the base
Constant	A symbol denoting a fixed value.	$7, -10, \pi$
Equation	A statement that two quantities are equal.	$3x + 5 = x + 17$
Equivalent Equations	Equations that have the same solution set.	$7x + 6 = x - 12$ and $6x = -18$ } solution: $x = -3$
Exponent	An exponent tells how many times another number, called the base, is to be used as a factor.	$5^4 = 625$; 4 is the exponent
Factor	Another name for a divisor.	1, 5, 7 and 35 are factors of 35
Integers	The numbers $\ldots -3, -2, -1, 0, 1, 2, 3, \ldots$	-6 is a negative integer 3 is a positive integer
Literal Number	A letter (variable) used to represent some number.	In $V = \frac{1}{3}\pi r^2$, V and r are literal numbers
Negative Number	A number to the left of zero on the number line (number < 0).	$-\frac{1}{2}$ and -4 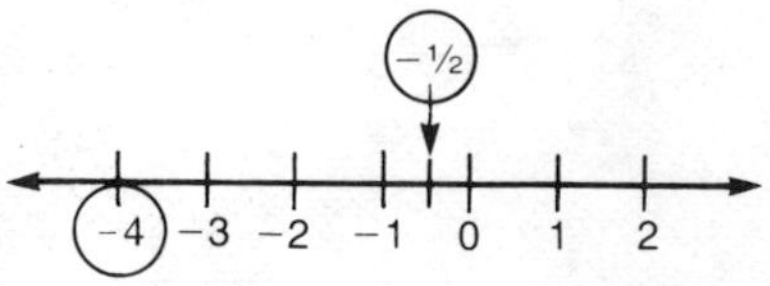
Origin	The starting point, or zero point, on the number line.	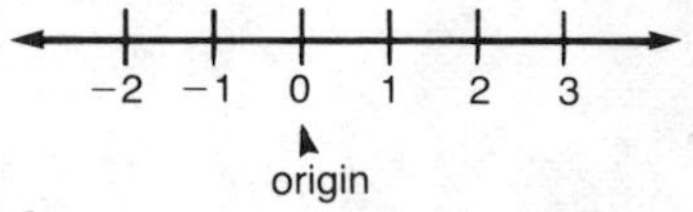
Positive Number	A number to the right of zero on the number line (number > 0).	$\frac{1}{4}$ and 3 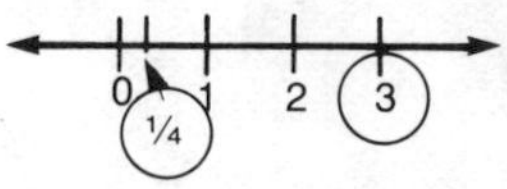
Signed Number	A number that has a "+" sign or a "−" sign associated with it.	$-10, +8$
Solution or Root	A number that satisfies an equation.	6 is a solution or root of the equation $x + 5 = 11$
Variable	A symbol denoting a quantity that can change value.	In $V = \frac{1}{3}\pi r^2$, V and r are variables

7

Chapter Review

In problems 1–6 answer true or false.

1. $-10 > -2$ __________

2. $a \cdot b = (a)(b)$ __________

3. $|12| = 12$ __________

4. $|-12| = -12$ __________

5. $-|-12| = -12$ __________

6. $(-5)(-4) = -20$ __________

In problems 7–57 perform the indicated operations.

7. $(+10) + (-15) =$ __________

8. $(-16) + (-12) =$ __________

9. $(+12) - (-5) =$ __________

10. $-10 - 6 =$ __________

11. $2\frac{3}{4} + \left(-7\frac{1}{2}\right) =$ __________

12. $-\frac{5}{6} + \frac{4}{9} =$ __________

13. $\left(-5\frac{1}{5}\right) + \left(-3\frac{3}{4}\right) =$ __________

14. $25 + 10 - 30 + 5 =$ __________

15. $-9 + 14 - 21 + 9 =$ __________

16. $-15 + 27 + 3 - 5 - 10 =$ __________

17. $30 - 40 + 20 - 30 + 6 =$ __________

18. $12.2 + 14.5 - 7.7 + 21.9 - 4.1 =$ __________

19. $10.4 + 13.6 - 9.7 + 20.3 =$ __________

20. $\frac{3}{5} - \frac{1}{2} + \frac{1}{4} + \frac{2}{5} =$ __________

21. $15 - (-8) + 6 - (-3) - 17 =$ __________

22. $-22 - (+15) - (-16) - (-8) + 2 =$ __________

23. $+15 - 6 - (-12) - 4 + (-8) + 2 =$ __________

24. $15.2 - (+6.5) - (-3.8) - 12.5 =$ __________

25. $(+5)(+9) =$ __________

26. $(-55) \div (-5) =$ __________

27. $(-3)(-5.4) =$ __________

28. $(-45)(12) =$ __________

29. $(+88) \div (+8) =$ __________

30. $(-78) \div 6 =$ __________

31. $(+0.2) \div (-0.04) =$ __________

32. $(-6)(19) =$ __________

33. $(0.025) \div (-0.5) =$ __________

34. $(-20)(-6) =$ ______

35. $(0.452)(-0.03) =$ ______

36. $(-30)\left(-2\frac{1}{5}\right) =$ ______

37. $\left(4\frac{1}{2}\right) \div \left(-2\frac{2}{3}\right) =$ ______

38. $(-35)(0.05) =$ ______

39. $(-35) \div (-0.05) =$ ______

40. $(3.5)(-0.5) =$ ______

41. $\left(-\frac{9}{10}\right)\left(-\frac{15}{4}\right) =$ ______

42. $(+1.35) \div (+0.09) =$ ______

43. $\left(6\frac{3}{4}\right) \div \left(-2\frac{5}{8}\right) =$ ______

44. $(-5)(-4)(-3) =$ ______

45. $(-15)(+12)(-10) =$ ______

46. $(+8)(+6)(0)(-5) =$ ______

47. $(-17)(-24)(-1)(-2) =$ ______

48. $(-4)(-5)(+20)(-1) =$ ______

49. $|(-8)(12)| =$ ______

50. $|25 \div (-5)| =$ ______

51. $(-10)|-15| =$ ______

52. $\left|\frac{-14}{7}\right| =$ ______

53. $|(16)(3)| =$ ______

54. $|(-4)(2)(0)| =$ ______

55. $|-77 \div (-11)| =$ ______

56. $|0 \div (-6)| =$ ______

57. $|(-12)(-8)(-1)| =$ ______

In problems 58–64 answer true or false.

58. $5^2 \cdot 2^3 = 10^5$ ______

59. $(a^4)^5 = a^9$ ______

60. $6^{-2} = \frac{1}{36}$ ______

61. $8^0 = 0$ ________________

62. $\frac{x^7}{x^3} = x^4$ ________________

63. $5^3 \cdot 5^6 = 5^9$ ________________

64. $(-7)^2 = -7^2$ ________________

In problems 65–84 perform the indicated operation and leave your answer in exponential form with positive exponents.

65. $x^3 \cdot x^9 =$ ________________

66. $(3^7)^4 =$ ________________

67. $\frac{x^{12}}{x^{10}} =$ ________________

68. $\frac{x^6}{x^{-4}} =$ ________________

69. $(-4^2)^3 =$ ________________

70. $\frac{-4^5}{4^5} =$ ________________

71. $\frac{a^3}{a^9} =$ ________________

72. $(5x)^2 =$ ________________

73. $a^4 \cdot a^3 \cdot a^6 =$ ________________

74. $x^4 \cdot x^2 \cdot x^{-6} =$ ________________

75. $[2^6]^{-2} =$ ________________

76. $x^{-3} \cdot x^4 =$ ________________

77. $\frac{x^{-6}}{x^{-10}} =$ ________________

78. $x^{-2} \cdot x^{-3} \cdot x^{-5} =$ ________________

79. $\frac{x^{-5}}{x^4} =$ ________________

80. $(2xy^2)^4 =$ ________________

81. $\left(\frac{3x}{2y^2}\right)^3 =$ ________________

82. $(-x^{-2}y^3)^2 =$ ________________

83. $\left(\frac{2^{-1}}{3^{-1}}\right)^2 =$ ________________

84. $\frac{(x^4y)^3}{x^8y^2} =$ ________________

In problems 85–100 solve the given equations.

85. $10x = 250 =$ ________________

86. $6 + y = -8$ ________________

87. $\frac{a}{15} = -4$ ________________

88. $3.5x = 28$ ______

89. $6b - 3 = 4b + 9$ ______

90. $\frac{8x}{9} = 0$ ______

91. $5c - 8 + 2c = 6c + 5 + 2c$ ______

92. $\frac{y}{4} + \frac{2y}{4} + 2 = y + 3$ ______

93. $\frac{x}{2} + \frac{x}{6} - \frac{1}{3} = 3 - \frac{x}{3}$ ______

94. $3x + 2x - 8x + 10 = -2$ ______

95. $\frac{5 - 3a}{2} + 7 = -a + \frac{1}{2}$ ______

96. $\frac{2n}{5} + \frac{1}{3} = -n - 2$ ______

97. $15 - 5a + 3 = 6a + 7$ ______

98. $4.2 - x + 3.6 = 5.4 - 0.2x$ ______

99. $\frac{4}{7}x - 2 = \frac{1}{2}x + 4$ ______

100. $\frac{3a}{10} - \frac{2}{5} + \frac{a}{10} = \frac{1}{2} - \frac{7a}{10} + \frac{1}{5}$ ______

ANSWERS

1. false **2.** true **3.** true **4.** false **5.** true **6.** false **7.** -5 **8.** -28
9. 17 **10.** -16 **11.** $-4\frac{3}{4}$ **12.** $-\frac{7}{18}$ **13.** $-8\frac{19}{20}$ **14.** 10 **15.** -7 **16.** 0
17. -14 **18.** 36.8 **19.** 34.6 **20.** $\frac{3}{4}$ **21.** 15 **22.** -11 **23.** 11 **24.** 0
25. 45 **26.** 11 **27.** 16.2 **28.** -540 **29.** 11 **30.** -13 **31.** -5 **32.** -114
33. -0.05 **34.** 120 **35.** -0.01356 **36.** 66 **37.** $-1\frac{11}{16}$ **38.** -1.75 **39.** 700
40. -1.75 **41.** $3\frac{3}{8}$ **42.** 15 **43.** $-2\frac{4}{7}$ **44.** -60 **45.** 1800 **46.** 0 **47.** 816
48. -400 **49.** 96 **50.** 5 **51.** -150 **52.** 2 **53.** 48 **54.** 0 **55.** 7
56. 0 **57.** 96 **58.** false **59.** false **60.** true **61.** false **62.** true **63.** true
64. false **65.** x^{12} **66.** 3^{28} **67.** x^2 **68.** x^{10} **69.** -4^6 **70.** -1 **71.** $\frac{1}{a^6}$
72. 5^2x^2, or $25x^2$ **73.** a^{13} **74.** 1 **75.** $\frac{1}{2^{12}}$ **76.** x **77.** x^4 **78.** $\frac{1}{x^{10}}$ **79.** $\frac{1}{x^9}$
80. $2^4x^4y^8 = 16x^4y^8$ **81.** $\frac{3^3x^3}{2^3y^6} = \frac{27x^3}{8y^6}$ **82.** $\frac{y^6}{x^4}$ **83.** $\frac{3^2}{2^2}$, or $\frac{9}{4}$ **84.** x^4y
85. $x = 25$ **86.** $y = -14$ **87.** $a = -60$ **88.** $x = 8$ **89.** $b = 6$ **90.** $x = 0$
91. $c = -13$ **92.** $y = -4$ **93.** $x = 3\frac{1}{3}$ **94.** $x = 4$ **95.** $a = 18$ **96.** $n = -1\frac{2}{3}$
97. $a = 1$ **98.** $x = 3$ **99.** $x = 84$ **100.** $a = 1$

7

Warmup Test

In problems 1–6 answer true or false.

1. $2^3 \cdot 3^2 = 6^5$ __________

2. $4^0 = 1$ __________

3. $\frac{10^5}{5^2} = 2^3$ __________

4. $(-3x)^2 = -9x^2$ __________

5. $(5^2)^4 = 5^6$ __________

6. $2^2 \cdot 2^3 = 4^5$ __________

In problems 7–12 perform the indicated operations.

7. $-15 + (-9) =$ __________

8. $(-18) \div 9 =$ __________

9. $(-6)(-12) =$ __________

10. $-24 - (-15) =$ __________

11. $(-6)\,|-7| =$ __________

12. $\frac{|-25|}{5} =$ __________

13. $8 - 9 + 11 + 6 - 14 - (-2) =$ __________

14. The Miami Dolphins gained 9 yards on their first play, lost 14 yards and gained 12 yards. What is their yardage after the three plays? __________

In problems 15–20 evaluate the given expressions. Leave your answers in exponential form with positive exponents.

15. $5^3 \cdot 5^6 =$ __________

16. $(2x^2y^3)^2 =$ __________

17. $\frac{x^7}{x^9} =$ __________

18. $\left(\frac{3x}{4y}\right)^3 =$ __________

19. $\frac{3^{-2}}{3^{-5}} =$ __________

20. $\left(\frac{5x^{-1}}{y^2}\right)^{-2} =$ __________

In problems 21–23 solve each of the given equations and check your solution.

21. $6 + 5m = 3m - 8$ __________

22. $\frac{2}{3} + \frac{1}{2}x = x - \frac{5}{6}$ __________

23. $3x + 5x + 10 = 2x - 8 + 6$ __________

7

Challenge Test

In problems 1–6 answer true or false.

1. $3^5 \cdot 3^8 = 3^{13}$ ________________

2. $5^{-1} = .25$ ________________

3. $7^0 = 0$ ________________

4. $-6^2 = -36$ ________________

5. $(4^3)^2 = 4^6$ ________________

6. $8^{-1} \cdot 8^{-1} = 8^{-2}$ ________________

In problems 7–12 perform the indicated operations.

7. $(18) + (-27) =$ ________________

8. $(-42) \div (-7) =$ ________________

9. $(13)(-3) =$ ________________

10. $-15 - (-10) =$ ________________

11. $(-5)|-4| =$ ________________

12. $\dfrac{-16}{|-8|} =$ ________________

13. $7 - 3 + 13 + 5 - 4 - (-3) =$ ________________

14. A woman owns a share of stock that increased by \$6 on Monday, decreased by \$4 on Tuesday, and decreased by \$7 on Wednesday. At this time, how much had this share of stock increased or decreased?

____

In problems 15–20 evaluate the given expressions. Leave your answers in exponential form with positive exponents.

15. $5^4 \cdot 5^6 =$ ________________

16. $(3x^2y)^3 =$ ________________

17. $\dfrac{x^{10}}{x^8} =$ ________________

18. $\left(\dfrac{2x}{5y}\right)^2 =$ ________________

19. $\dfrac{4^{-2}}{4^{-3}} =$ ________________

20. $\left(\dfrac{7x^{-1}}{y}\right)^2 =$ ________________

In problems 21–23 solve each of the given equations and check your solution.

21. $2m - 3 = 21 - 6m$ ________________

22. $6x + 4 - 3x + 11 - 2 = 49$ ________________

23. $\frac{1}{3}x - \frac{2}{3} = \frac{1}{5}x + 2 - \frac{4}{5}$ ________________

Appendix I

Arithmetic Practice Problem Sets

This appendix consists of three collections (A, B, and C), a total of 1,800 arithmetic problems with answers. Each collection consists of 12 sets of 50 problems. You should be able to solve successfully at least 45 of 50 problems within any set. If you fail to achieve this goal in any set of the first collection, proceed to the similar practice problem set in the next collection of problems.

ARITHMETIC PRACTICE PROBLEMS: COLLECTION A

PRACTICE PROBLEM SET 1A: ADDITION OF WHOLE NUMBERS

Do the indicated addition.

1. $9 + 2$ **2.** $7 + 6$ **3.** $5 + 5$ **4.** $8 + 9$

5. $8 + 10$ **6.** $12 + 6$ **7.** $18 + 4$ **8.** $37 + 5$

9. $9 + 15$ **10.** $23 + 24$ **11.** $61 + 17$ **12.** $85 + 17$

13. $87 + 64$ **14.** $52 + 35$ **15.** $79 + 65$ **16.** $54 + 88$

17. $95 + 57$ **18.** $63 + 47$ **19.** $43 + 59$ **20.** $78 + 78$

21. $73 + 58$ **22.** $56 + 93$ **23.** $100 + 32$ **24.** $105 + 75$

25. $77 + 64$ **26.** $111 + 66$ **27.** $125 + 33$ **28.** $219 + 341$

29. $437 + 219$ **30.** $334 + 492$ **31.** $572 + 608$ **32.** $759 + 185$

33. $640 + 319$ **34.** $871 + 643$ **35.** $905 + 478$ **36.** $890 + 377$

37. $680 + 721$ **38.** $1{,}000 + 291$ **39.** $8{,}074 + 295$ **40.** $8{,}953 + 2{,}508$

41. 15,663 + 19,076

42. 42,702 + 29,887

43. 80,993 + 56,071

44. 62 + 57 + 1,110

45. 219 + 57 + 603

46. 170 + 9,971 + 8,573

47. 217 + 419 + 565

48. 9,074 + 658 + 29 + 4,014

49. 2,998 + 7,631 + 44,607 + 26,558

50. 37,617 + 50,044 + 163,728 + 770,906

ANSWERS

1. 11 **2.** 13 **3.** 10 **4.** 17 **5.** 18 **6.** 18 **7.** 22 **8.** 42 **9.** 24 **10.** 47 **11.** 78 **12.** 102 **13.** 151 **14.** 87 **15.** 144 **16.** 142 **17.** 152 **18.** 110 **19.** 102 **20.** 156 **21.** 131 **22.** 149 **23.** 132 **24.** 180 **25.** 141 **26.** 177 **27.** 158 **28.** 560 **29.** 656 **30.** 826 **31.** 1,180 **32.** 944 **33.** 959 **34.** 1,514 **35.** 1,383 **36.** 1,267 **37.** 1,401 **38.** 1,291 **39.** 8,369 **40.** 11,461 **41.** 34,739 **42.** 72,589 **43.** 137,064 **44.** 1,229 **45.** 879 **46.** 18,714 **47.** 1,201 **48.** 13,775 **49.** 81,794 **50.** 1,022,295

PRACTICE PROBLEM SET 2A: SUBTRACTION OF WHOLE NUMBERS

Do the indicated subtraction.

1. 9 − 3	**2.** 7 − 5	**3.** 4 − 4	**4.** 8 − 5
5. 9 − 2	**6.** 10 − 4	**7.** 13 − 5	**8.** 19 − 11
9. 51 − 17	**10.** 47 − 35	**11.** 35 − 21	**12.** 89 − 57
13. 61 − 23	**14.** 74 − 68	**15.** 55 − 26	**16.** 85 − 42
17. 58 − 49	**18.** 96 − 76	**19.** 63 − 37	**20.** 59 − 27
21. 100 − 54	**22.** 144 − 76	**23.** 291 − 107	**24.** 303 − 274
25. 909 − 536	**26.** 47 − 19	**27.** 498 − 0	**28.** 180 − 65
29. 83 − 44	**30.** 216 − 65	**31.** 455 − 101	**32.** 7,271 − 1,000
33. 500 − 468	**34.** 800 − 721	**35.** 1,760 − 808	**36.** 1,275 − 850
37. 4,922 − 1,705	**38.** 9,788 − 2,009	**39.** 7,153 − 5,143	**40.** 9,762 − 3,004
41. 1,000 − 573	**42.** 8,690 − 7,023	**43.** 6,178 − 1,999	**44.** 73,021 − 15,674

45. $\begin{array}{r} 8{,}000 \\ -\ 6{,}792 \\ \hline \end{array}$ **46.** $\begin{array}{r} 10{,}000 \\ -\ 5{,}050 \\ \hline \end{array}$ **47.** $\begin{array}{r} 8{,}125 \\ -\ 6{,}376 \\ \hline \end{array}$ **48.** $\begin{array}{r} 39{,}708 \\ -\ 12{,}691 \\ \hline \end{array}$

49. $\begin{array}{r} 100{,}000 \\ -\ 39{,}769 \\ \hline \end{array}$ **50.** $\begin{array}{r} 99{,}971 \\ -\ 74{,}098 \\ \hline \end{array}$

ANSWERS

1. 6 **2.** 2 **3.** 0 **4.** 3 **5.** 7 **6.** 6 **7.** 8 **8.** 8 **9.** 34 **10.** 12 **11.** 14 **12.** 32 **13.** 38 **14.** 6 **15.** 29 **16.** 43 **17.** 9 **18.** 20 **19.** 26 **20.** 32 **21.** 46 **22.** 68 **23.** 184 **24.** 29 **25.** 373 **26.** 28 **27.** 498 **28.** 115 **29.** 39 **30.** 151 **31.** 354 **32.** 6,271 **33.** 32 **34.** 79 **35.** 952 **36.** 425 **37.** 3,217 **38.** 7,779 **39.** 2,010 **40.** 6,758 **41.** 427 **42.** 1,667 **43.** 4,179 **44.** 57,347 **45.** 1,208 **46.** 4,950 **47.** 1,749 **48.** 27,017 **49.** 60,231 **50.** 25,873

Multiplication Table

x	1	2	3	4	5	6	7	8	9	10	11	12
1	1	2	3	4	5	6	7	8	9	10	11	12
2	2	4	6	8	10	12	14	16	18	20	22	24
3	3	6	9	12	15	18	21	24	27	30	33	36
4	4	8	12	16	20	24	28	32	36	40	44	48
5	5	10	15	20	25	30	35	40	45	50	55	60
6	6	12	18	24	30	36	42	48	54	60	66	72
7	7	14	21	28	35	42	49	56	63	70	77	84
8	8	16	24	32	40	48	56	64	72	80	88	96
9	9	18	27	36	45	54	63	72	81	90	99	108
10	10	20	30	40	50	60	70	80	90	100	110	120
11	11	22	33	44	55	66	77	88	99	110	121	132
12	12	24	36	48	60	72	84	96	108	120	132	144

How to use this table is illustrated by the following example:
To multiply 7×8, move across the 7 row and down the 8 column. The answer will be at the intersection (56)

PRACTICE PROBLEM 3A: MULTIPLICATION OF WHOLE NUMBERS

Do the indicated multiplication.

1. 7×3
2. 5×4
3. 7×7
4. 9×8
5. 0×2
6. 8×7
7. 9×9
8. 7×6
9. 10×8
10. 10×0
11. 5×8
12. 11×2
13. 15×3
14. 12×6
15. 13×10
16. 14×9
17. 17×8
18. 25×6
19. 46×7
20. 52×4
21. 67×5
22. 73×8
23. 49×5
24. 65×4
25. 19×10
26. 13×13
27. 25×11
28. 32×32

29. 76×14

30. 16×15

31. 46×22

32. 27×58

33. 50×36

34. 63×46

35. 65×29

36. 87×53

37. 44×75

38. 19×98

39. 68×47

40. 92×76

41. 106×8

42. 212×5

43. 453×27

44. 317×56

45. 600×9

46. 245×85

47. 921×68

48. 705×50

49. $7{,}214 \times 58$

50. $8{,}092 \times 96$

ANSWERS

1. 21 **2.** 20 **3.** 49 **4.** 72 **5.** 0 **6.** 56 **7.** 81 **8.** 42 **9.** 80 **10.** 0 **11.** 40 **12.** 22 **13.** 45 **14.** 72 **15.** 130 **16.** 126 **17.** 136 **18.** 150 **19.** 322 **20.** 208 **21.** 335 **22.** 584 **23.** 245 **24.** 260 **25.** 190 **26.** 169 **27.** 275 **28.** 1,024 **29.** 1,064 **30.** 240 **31.** 1,012 **32.** 1,566 **33.** 1,800 **34.** 2,898 **35.** 1,885 **36.** 4,611 **37.** 3,300 **38.** 1,862 **39.** 3,196 **40.** 6,992 **41.** 848 **42.** 1,060 **43.** 12,231 **44.** 17,752 **45.** 5,400 **46.** 20,825 **47.** 62,628 **48.** 35,250 **49.** 418,412 **50.** 776,832

PRACTICE PROBLEM SET 4A: DIVISION OF WHOLE NUMBERS

Do the indicated division.

1. $3\overline{)24}$
2. $5\overline{)20}$
3. $6\overline{)36}$
4. $6\overline{)42}$
5. $5\overline{)25}$
6. $4\overline{)28}$
7. $9\overline{)81}$
8. $7\overline{)56}$
9. $6\overline{)60}$
10. $8\overline{)40}$
11. $10\overline{)100}$
12. $12\overline{)144}$
13. $15\overline{)90}$
14. $12\overline{)84}$
15. $17\overline{)102}$
16. $18\overline{)54}$
17. $16\overline{)160}$
18. $20\overline{)100}$
19. $9\overline{)171}$
20. $11\overline{)132}$
21. $6\overline{)450}$
22. $14\overline{)210}$
23. $17\overline{)136}$
24. $15\overline{)180}$
25. $5\overline{)24}$
26. $7\overline{)51}$
27. $8\overline{)76}$
28. $4\overline{)47}$
29. $9\overline{)29}$
30. $3\overline{)102}$
31. $23\overline{)1,035}$
32. $18\overline{)918}$
33. $32\overline{)1,504}$
34. $43\overline{)3,225}$
35. $36\overline{)2,340}$
36. $71\overline{)2,059}$
37. $56\overline{)1,568}$
38. $50\overline{)4,900}$
39. $40\overline{)3,120}$
40. $57\overline{)2,052}$
41. $37\overline{)7,901}$
42. $62\overline{)4,663}$
43. $54\overline{)9,048}$
44. $81\overline{)7,775}$

45. $41\overline{)4{,}305}$ **46.** $26\overline{)5{,}460}$ **47.** $37\overline{)19{,}092}$ **48.** $65\overline{)50{,}830}$

49. $64\overline{)49{,}792}$ **50.** $83\overline{)76{,}443}$

ANSWERS

1. 8 **2.** 4 **3.** 6 **4.** 7 **5.** 5 **6.** 7 **7.** 9 **8.** 8 **9.** 10 **10.** 5
11. 10 **12.** 12 **13.** 6 **14.** 7 **15.** 6 **16.** 3 **17.** 10 **18.** 5 **19.** 19
20. 12 **21.** 75 **22.** 15 **23.** 8 **24.** 12 **25.** 4, $R = 4$ **26.** 7, $R = 2$
27. 9, $R = 4$ **28.** 11, $R = 3$ **29.** 3, $R = 2$ **30.** 34 **31.** 45 **32.** 51 **33.** 47
34. 75 **35.** 65 **36.** 29 **37.** 28 **38.** 98 **39.** 78 **40.** 36 **41.** 213, $R = 20$
42. 75, $R = 13$ **43.** 167, $R = 30$ **44.** 95, $R = 80$ **45.** 105 **46.** 210 **47.** 516
48. 782 **49.** 778 **50.** 921

PRACTICE PROBLEM SET 5A: ADDITION OF FRACTIONS

Do the indicated addition. State each answer as a proper fraction or as a mixed number. Reduce all answers to simplest form.

1. $\frac{2}{5} + \frac{1}{5} =$ ______
2. $\frac{3}{6} + \frac{2}{6} =$ ______
3. $\frac{6}{8} + \frac{1}{8} =$ ______
4. $\frac{4}{10} + \frac{3}{10} =$ ______
5. $\frac{2}{9} + \frac{7}{9} =$ ______
6. $\frac{3}{10} + \frac{5}{10} =$ ______
7. $\frac{6}{5} + \frac{7}{5} =$ ______
8. $\frac{2}{3} + \frac{1}{3} =$ ______
9. $\frac{4}{7} + \frac{5}{7} =$ ______
10. $\frac{4}{13} + \frac{7}{13} =$ ______
11. $\frac{1}{2} + \frac{1}{3} =$ ______
12. $\frac{3}{4} + \frac{1}{3} =$ ______
13. $\frac{1}{2} + \frac{2}{3} =$ ______
14. $\frac{2}{5} + \frac{3}{4} =$ ______
15. $\frac{4}{7} + \frac{2}{5} =$ ______
16. $\frac{3}{5} + \frac{1}{6} =$ ______
17. $\frac{5}{8} + \frac{2}{3} =$ ______
18. $\frac{7}{10} + \frac{4}{5} =$ ______
19. $\frac{7}{9} + \frac{1}{3} =$ ______
20. $\frac{3}{4} + \frac{5}{8} =$ ______
21. $\frac{3}{10} + \frac{9}{20} =$ ______
22. $\frac{2}{5} + \frac{9}{10} =$ ______
23. $\frac{1}{9} + \frac{1}{4} =$ ______
24. $\frac{3}{5} + \frac{7}{15} =$ ______
25. $\begin{array}{r} \frac{4}{5} \\ + \frac{3}{10} \\ \hline \end{array}$
26. $\begin{array}{r} \frac{5}{14} \\ + \frac{3}{7} \\ \hline \end{array}$
27. $\begin{array}{r} \frac{5}{6} \\ + \frac{2}{4} \\ \hline \end{array}$
28. $\begin{array}{r} \frac{1}{3} \\ + \frac{5}{8} \\ \hline \end{array}$
29. $\begin{array}{r} \frac{7}{16} \\ + \frac{3}{4} \\ \hline \end{array}$
30. $\begin{array}{r} \frac{3}{5} \\ + \frac{3}{4} \\ \hline \end{array}$
31. $\frac{5}{6} + \frac{3}{7} =$ ______
32. $\frac{5}{16} + \frac{3}{5} =$ ______
33. $\frac{5}{8} + \frac{3}{20} =$ ______
34. $\frac{4}{6} + \frac{6}{14} =$ ______
35. $\frac{1}{4} + \frac{5}{14} =$ ______
36. $\frac{5}{6} + \frac{7}{8} =$ ______
37. $\frac{3}{10} + \frac{2}{15} =$ ______
38. $\frac{5}{12} + \frac{7}{18} =$ ______
39. $\frac{17}{20} + \frac{29}{40} =$ ______
40. $\frac{5}{6} + \frac{7}{9} =$ ______
41. $\frac{3}{7} + \frac{5}{16} =$ ______
42. $\frac{5}{8} + \frac{5}{6} =$ ______
43. $\frac{1}{15} + \frac{1}{25} =$ ______
44. $\frac{9}{7} + 3 =$ ______
45. $5 + \frac{31}{8} =$ ______
46. $3 + \frac{3}{2} =$ ______
47. $\frac{5}{3} + 4 =$ ______
48. $\frac{13}{28} + \frac{5}{42} =$ ______
49. $\frac{4}{15} + \frac{6}{33} =$ ______
50. $\frac{4}{25} + \frac{37}{50} =$ ______

ANSWERS

1. $\frac{3}{5}$ **2.** $\frac{5}{6}$ **3.** $\frac{7}{8}$ **4.** $\frac{7}{10}$ **5.** 1 **6.** $\frac{4}{5}$ **7.** $2\frac{3}{5}$ **8.** 1 **9.** $1\frac{2}{7}$ **10.** $\frac{11}{13}$ **11.** $\frac{5}{6}$
12. $1\frac{1}{12}$ **13.** $1\frac{1}{6}$ **14.** $1\frac{3}{20}$ **15.** $\frac{34}{35}$ **16.** $\frac{23}{30}$ **17.** $1\frac{7}{24}$ **18.** $1\frac{1}{2}$ **19.** $1\frac{1}{9}$
20. $1\frac{3}{8}$ **21.** $\frac{3}{4}$ **22.** $1\frac{3}{10}$ **23.** $\frac{13}{36}$ **24.** $1\frac{1}{15}$ **25.** $1\frac{1}{10}$ **26.** $\frac{11}{14}$ **27.** $1\frac{1}{3}$ **28.** $\frac{23}{24}$
29. $1\frac{3}{16}$ **30.** $1\frac{7}{20}$ **31.** $1\frac{11}{42}$ **32.** $\frac{73}{80}$ **33.** $\frac{31}{40}$ **34.** $1\frac{2}{21}$ **35.** $\frac{17}{28}$ **36.** $1\frac{17}{24}$
37. $\frac{13}{30}$ **38.** $\frac{29}{36}$ **39.** $1\frac{23}{40}$ **40.** $1\frac{11}{18}$ **41.** $\frac{83}{112}$ **42.** $1\frac{11}{24}$ **43.** $\frac{8}{75}$ **44.** $4\frac{2}{7}$
45. $8\frac{7}{8}$ **46.** $4\frac{1}{2}$ **47.** $5\frac{2}{3}$ **48.** $\frac{49}{84}$ **49.** $\frac{74}{165}$ **50.** $\frac{9}{10}$

PRACTICE PROBLEM SET 6A: SUBTRACTION OF FRACTIONS

Do the indicated subtraction. State each answer as a proper fraction or as a mixed number. Reduce all answers to simplest form.

1. $\frac{5}{6} - \frac{1}{6} =$ ______
2. $\frac{4}{7} - \frac{2}{7} =$ ______
3. $\frac{8}{9} - \frac{4}{9} =$ ______
4. $\frac{9}{10} - \frac{6}{10} =$ ______
5. $\frac{3}{8} - 0 =$ ______
6. $\frac{5}{8} - \frac{3}{8} =$ ______
7. $\frac{7}{12} - \frac{4}{12} =$ ______
8. $\frac{13}{25} - \frac{7}{25} =$ ______
9. $\frac{13}{15} - \frac{5}{15} =$ ______
10. $\frac{11}{18} - \frac{9}{18} =$ ______
11. $\frac{7}{8} - \frac{3}{8} =$ ______
12. $\frac{5}{7} - \frac{5}{7} =$ ______
13. $\frac{2}{3} - \frac{1}{2} =$ ______
14. $\frac{3}{4} - \frac{1}{2} =$ ______
15. $\frac{8}{10} - \frac{4}{5} =$ ______
16. $\frac{5}{9} - \frac{1}{3} =$ ______
17. $\frac{9}{10} - \frac{3}{20} =$ ______
18. $\frac{5}{6} - \frac{5}{12} =$ ______
19. $\frac{4}{5} - \frac{3}{4} =$ ______
20. $\frac{3}{3} - 1 =$ ______
21. $\frac{3}{4} - \frac{1}{6} =$ ______
22. $\frac{5}{7} - \frac{2}{21} =$ ______
23. $\frac{4}{5} - \frac{2}{3} =$ ______
24. $\frac{3}{6} - \frac{2}{9} =$ ______
25. $\begin{array}{r} \frac{7}{10} \\ -\ \frac{3}{5} \\ \hline \end{array}$
26. $\begin{array}{r} \frac{11}{12} \\ -\ \frac{5}{8} \\ \hline \end{array}$
27. $\begin{array}{r} \frac{17}{20} \\ -\ \frac{8}{15} \\ \hline \end{array}$
28. $\begin{array}{r} \frac{7}{16} \\ -\ \frac{5}{12} \\ \hline \end{array}$
29. $\begin{array}{r} \frac{5}{6} \\ -\ \frac{2}{7} \\ \hline \end{array}$
30. $\begin{array}{r} \frac{7}{8} \\ -\ \frac{1}{5} \\ \hline \end{array}$
31. $2 - \frac{3}{4} =$ ______
32. $10 - \frac{7}{8} =$ ______
33. $6 - \frac{4}{9} =$ ______
34. $5 - \frac{2}{3} =$ ______
35. $\frac{5}{8} - \frac{2}{7} =$ ______
36. $\frac{5}{9} - \frac{7}{15} =$ ______
37. $\frac{8}{3} - \frac{3}{4} =$ ______
38. $\frac{9}{2} - \frac{3}{5} =$ ______
39. $\frac{3}{2} - \frac{5}{4} =$ ______
40. $\frac{11}{4} - \frac{4}{3} =$ ______
41. $\frac{3}{7} - \frac{1}{9} =$ ______
42. $\frac{73}{100} - \frac{9}{25} =$ ______
43. $\frac{2}{3} - \frac{6}{19} =$ ______
44. $\frac{5}{9} - \frac{5}{12} =$ ______
45. $\frac{74}{75} - \frac{4}{25} =$ ______
46. $\frac{7}{9} - \frac{3}{7} =$ ______
47. $\frac{21}{35} - \frac{5}{14} =$ ______
48. $\frac{27}{60} - \frac{3}{40} =$ ______
49. $\frac{53}{70} - \frac{11}{35} =$ ______
50. $\frac{17}{31} - \frac{4}{9} =$ ______

ANSWERS

1. $\frac{2}{3}$ **2.** $\frac{2}{7}$ **3.** $\frac{4}{9}$ **4.** $\frac{3}{10}$ **5.** $\frac{3}{8}$ **6.** $\frac{1}{4}$ **7.** $\frac{1}{4}$ **8.** $\frac{6}{25}$ **9.** $\frac{8}{15}$ **10.** $\frac{1}{9}$ **11.** $\frac{1}{2}$ **12.** 0 **13.** $\frac{1}{6}$ **14.** $\frac{1}{4}$ **15.** 0 **16.** $\frac{2}{9}$ **17.** $\frac{3}{4}$ **18.** $\frac{5}{12}$ **19.** $\frac{1}{20}$ **20.** 0 **21.** $\frac{7}{12}$ **22.** $\frac{13}{21}$ **23.** $\frac{2}{15}$ **24.** $\frac{5}{18}$ **25.** $\frac{1}{10}$ **26.** $\frac{7}{24}$ **27.** $\frac{19}{60}$ **28.** $\frac{1}{48}$ **29.** $\frac{23}{42}$ **30.** $\frac{27}{40}$ **31.** $1\frac{1}{4}$ **32.** $9\frac{1}{8}$ **33.** $5\frac{5}{9}$ **34.** $4\frac{1}{3}$ **35.** $\frac{19}{56}$ **36.** $\frac{4}{45}$ **37.** $1\frac{11}{12}$ **38.** $3\frac{9}{10}$ **39.** $\frac{1}{4}$ **40.** $1\frac{5}{12}$ **41.** $\frac{20}{63}$ **42.** $\frac{37}{100}$ **43.** $\frac{20}{57}$ **44.** $\frac{5}{36}$ **45.** $\frac{62}{75}$ **46.** $\frac{22}{63}$ **47.** $\frac{17}{70}$ **48.** $\frac{3}{8}$ **49.** $\frac{31}{70}$ **50.** $\frac{29}{279}$

PRACTICE PROBLEM SET 7A: MULTIPLICATION OF FRACTIONS

Do the indicated multiplication. State each answer as a proper fraction or as a mixed number. Reduce all answers to simplest form.

1. $\frac{1}{3} \times \frac{2}{3} =$ ______ **2.** $\frac{3}{4} \times \frac{1}{2} =$ ______ **3.** $\frac{4}{5} \times \frac{1}{3} =$ ______

4. $\frac{5}{8} \times \frac{5}{8} =$ ______ **5.** $\frac{2}{5} \times \frac{3}{6} =$ ______ **6.** $\frac{4}{7} \times \frac{7}{9} =$ ______

7. $\frac{1}{2} \times \frac{9}{2} =$ ______ **8.** $3 \times \frac{2}{7} =$ ______ **9.** $5 \times \frac{2}{3} =$ ______

10. $0 \times \frac{9}{10} =$ ______ **11.** $\frac{3}{5} \times \frac{10}{11} =$ ______ **12.** $\frac{8}{9} \times \frac{3}{8} =$ ______

13. $\frac{16}{5} \times \frac{4}{12} =$ ______ **14.** $\frac{7}{8} \times 0 =$ ______ **15.** $\frac{7}{12} \times 96 =$ ______

16. $\frac{3}{5} \times \frac{3}{10} =$ ______ **17.** $\frac{1}{7} \times \frac{7}{8} =$ ______ **18.** $\frac{11}{12} \times \frac{4}{5} =$ ______

19. $\frac{2}{5} \times \frac{4}{5} =$ ______ **20.** $\frac{15}{20} \times \frac{10}{12} =$ ______ **21.** $\frac{18}{25} \times \frac{10}{9} =$ ______

22. $\frac{10}{30} \times \frac{25}{25} =$ ______ **23.** $\frac{2}{3} \times 10 =$ ______ **24.** $\frac{3}{40} \times 100 =$ ______

25. $\frac{5}{12} \times \frac{42}{65} =$ ______ **26.** $\frac{16}{18} \times \frac{2}{3} =$ ______ **27.** $\frac{16}{22} \times \frac{55}{72} =$ ______

28. $\frac{12}{5} \times \frac{5}{9} =$ ______ **29.** $\frac{9}{2} \times \frac{17}{15} =$ ______ **30.** $\frac{7}{10} \times \frac{6}{21} =$ ______

31. $\frac{10}{11} \times \frac{10}{11} =$ ______ **32.** $\frac{3}{4} \times \frac{8}{6} =$ ______ **33.** $\frac{3}{8} \times \frac{15}{41} =$ ______

34. $\frac{16}{9} \times \frac{45}{18} =$ ______ **35.** $\frac{8}{40} \times \frac{32}{16} =$ ______ **36.** $\frac{19}{38} \times \frac{50}{50} =$ ______

37. $100 \times \frac{5}{4} =$ ______ **38.** $8 \times \frac{3}{4} =$ ______ **39.** $\frac{1}{25} \times \frac{75}{100} =$ ______

40. $\frac{1}{10} \times \frac{1}{10} =$ ______ **41.** $\frac{51}{100} \times \frac{50}{17} =$ ______ **42.** $\frac{3}{2} \times 1{,}000 =$ ______

43. $\frac{4}{5} \times \frac{7}{9} =$ ______ **44.** $\frac{10}{9} \times \frac{5}{11} =$ ______ **45.** $1{,}000 \times \frac{1}{100} =$ ______

46. $\frac{15}{8} \times \frac{16}{3} =$ ______ **47.** $\frac{15}{7} \times 3 =$ ______ **48.** $\frac{3}{8} \times \frac{24}{5} =$ ______

49. $\frac{49}{8} \times \frac{4}{7} =$ ______ **50.** $\frac{13}{2} \times \frac{55}{13} =$ ______

ANSWERS

1. $\frac{2}{9}$ **2.** $\frac{3}{8}$ **3.** $\frac{4}{15}$ **4.** $\frac{25}{64}$ **5.** $\frac{1}{5}$ **6.** $\frac{4}{9}$ **7.** $2\frac{1}{4}$ **8.** $\frac{6}{7}$ **9.** $3\frac{1}{3}$ **10.** 0
11. $\frac{6}{11}$ **12.** $\frac{1}{3}$ **13.** $1\frac{1}{15}$ **14.** 0 **15.** 56 **16.** $\frac{9}{50}$ **17.** $\frac{1}{8}$ **18.** $\frac{11}{15}$ **19.** $\frac{8}{25}$
20. $\frac{5}{8}$ **21.** $\frac{4}{5}$ **22.** $\frac{1}{3}$ **23.** $6\frac{2}{3}$ **24.** $7\frac{1}{2}$ **25.** $\frac{7}{26}$ **26.** $\frac{16}{27}$ **27.** $\frac{5}{9}$ **28.** $1\frac{1}{3}$
29. $5\frac{1}{10}$ **30.** $\frac{1}{5}$ **31.** $\frac{100}{121}$ **32.** 1 **33.** $\frac{45}{328}$ **34.** $4\frac{4}{9}$ **35.** $\frac{2}{5}$ **36.** $\frac{1}{2}$
37. 125 **38.** 6 **39.** $\frac{3}{100}$ **40.** $\frac{1}{100}$ **41.** $1\frac{1}{2}$ **42.** 1,500 **43.** $\frac{28}{45}$ **44.** $\frac{50}{99}$
45. 10 **46.** 10 **47.** $6\frac{3}{7}$ **48.** $1\frac{4}{5}$ **49.** $3\frac{1}{2}$ **50.** $27\frac{1}{2}$

PRACTICE PROBLEM SET 8A: DIVISION OF FRACTIONS

Do the indicated division. State each answer as a proper fraction or as a mixed number. Reduce all answers to simplest form.

1. $\frac{1}{3} \div \frac{1}{2} =$ ______ **2.** $\frac{3}{5} \div \frac{2}{3} =$ ______ **3.** $\frac{1}{4} \div \frac{1}{5} =$ ______

4. $\frac{2}{3} \div \frac{2}{3} =$ ______ **5.** $\frac{4}{7} \div \frac{3}{14} =$ ______ **6.** $\frac{5}{10} \div \frac{15}{5} =$ ______

7. $\frac{8}{9} \div 1 =$ ______ **8.** $0 \div \frac{1}{2} =$ ______ **9.** $\frac{1}{2} \div 0 =$ ______

10. $\frac{2}{5} \div 2 =$ ______ **11.** $\frac{5}{6} \div 12 =$ ______ **12.** $\frac{4}{7} \div 5 =$ ______

13. $\frac{3}{11} \div 4 =$ ______ **14.** $\frac{7}{9} \div \frac{7}{9} =$ ______ **15.** $\frac{2}{9} \div \frac{16}{19} =$ ______

16. $\frac{21}{40} \div \frac{11}{15} =$ ______ **17.** $\frac{5}{6} \div \frac{1}{6} =$ ______ **18.** $\frac{12}{13} \div \frac{4}{9} =$ ______

19. $\frac{5}{12} \div \frac{15}{36} =$ ______ **20.** $\frac{9}{10} \div \frac{1}{10} =$ ______ **21.** $\frac{15}{33} \div \frac{4}{11} =$ ______

22. $\frac{1}{6} \div \frac{1}{3} =$ ______ **23.** $\frac{1}{8} \div \frac{1}{2} =$ ______ **24.** $\frac{11}{13} \div \frac{5}{26} =$ ______

25. $\frac{10}{11} \div \frac{30}{32} =$ ______ **26.** $\frac{4}{9} \div \frac{1}{9} =$ ______ **27.** $3 \div \frac{3}{2} =$ ______

28. $\frac{15}{8} \div \frac{5}{32} =$ ______ **29.** $\frac{7}{4} \div \frac{56}{10} =$ ______ **30.** $\frac{9}{2} \div \frac{27}{5} =$ ______

31. $\frac{1}{10} \div \frac{1}{10} =$ ______ **32.** $\frac{1}{10} \div 10 =$ ______ **33.** $\frac{15}{12} \div \frac{15}{18} =$ ______

34. $0 \div \frac{2}{7} =$ ______ **35.** $\frac{28}{45} \div \frac{7}{18} =$ ______ **36.** $\frac{20}{22} \div \frac{30}{44} =$ ______

37. $\frac{5}{2} \div 2 =$ ______ **38.** $\frac{7}{4} \div 4 =$ ______ **39.** $\frac{14}{3} \div \frac{21}{9} =$ ______

40. $\frac{8}{5} \div \frac{48}{7} =$ ______ **41.** $\frac{25}{30} \div \frac{5}{10} =$ ______ **42.** $\frac{5}{7} \div \frac{3}{5} =$ ______

43. $\frac{1}{9} \div \frac{2}{3} =$ ______ **44.** $\frac{15}{16} \div \frac{39}{16} =$ ______ **45.** $6 \div \frac{10}{3} =$ ______

46. $\frac{13}{2} \div \frac{26}{100} =$ ______ **47.** $\frac{3}{8} \div \frac{11}{4} =$ ______ **48.** $\frac{33}{16} \div \frac{5}{2} =$ ______

49. $\frac{38}{5} \div \frac{19}{12} =$ ______ **50.** $\frac{5}{2} \div \frac{47}{14} =$ ______

ANSWERS

1. $\frac{2}{3}$ **2.** $\frac{9}{10}$ **3.** $1\frac{1}{4}$ **4.** 1 **5.** $2\frac{2}{3}$ **6.** $\frac{1}{6}$ **7.** $\frac{8}{9}$ **8.** 0 **9.** no solution **10.** $\frac{1}{5}$

11. $\frac{5}{72}$ **12.** $\frac{4}{35}$ **13.** $\frac{3}{44}$ **14.** 1 **15.** $\frac{19}{72}$ **16.** $\frac{63}{88}$ **17.** 5 **18.** $2\frac{1}{13}$

19. 1 **20.** 9 **21.** $1\frac{1}{4}$ **22.** $\frac{1}{2}$ **23.** $\frac{1}{4}$ **24.** $4\frac{2}{5}$ **25.** $\frac{32}{33}$ **26.** 4 **27.** 2

28. 12 **29.** $\frac{5}{16}$ **30.** $\frac{5}{6}$ **31.** 1 **32.** $\frac{1}{100}$ **33.** $1\frac{1}{2}$ **34.** 0 **35.** $1\frac{3}{5}$ **36.** $1\frac{1}{3}$

37. $1\frac{1}{4}$ **38.** $\frac{7}{16}$ **39.** 2 **40.** $\frac{7}{30}$ **41.** $1\frac{2}{3}$ **42.** $1\frac{4}{21}$ **43.** $\frac{1}{6}$ **44.** $\frac{5}{13}$ **45.** $1\frac{4}{5}$

46. 25 **47.** $\frac{3}{22}$ **48.** $\frac{33}{40}$ **49.** $4\frac{4}{5}$ **50.** $\frac{35}{47}$

PRACTICE PROBLEM SET 9A: ADDITION OF DECIMALS

Do the indicated addition.

1. 4.3 + 0.7	**2.** 0.5 + 0.7	**3.** 0.6 + 1.1	**4.** 1.0 + 0.9
5. 9.2 + 3.6	**6.** 5.5 + 5.5	**7.** 7.5 + 4.3	**8.** 10.4 + 8.5
9. 7.7 + 6.5	**10.** 8.6 + 9.9	**11.** 4.7 + 3.8	**12.** 9.8 + 6.5
13. 6.1 + 0	**14.** 11.6 + 8.8	**15.** 15.7 + 29.8	**16.** 56.13 + 17.24
17. 19.05 + 10.12	**18.** 20.75 + 37.64	**19.** 72.76 + 45.38	**20.** 50.50 + 17.01
21. 104.40 + 98.78	**22.** 210.17 + 66.77	**23.** 315.44 + 78.65	**24.** 219.36 + 55.79
25. 2.1 6.0 + 0.7	**26.** 0.7 0.8 + 1.4	**27.** 5.1 0.9 + 3.2	**28.** 2.4 7.8 + 5.0
29. 101.71 + 99.18	**30.** 567.74 + 102.64	**31.** 470.83 + 209.77	**32.** 317.88 + 198.03
33. 0.077 + 0.081	**34.** 0.056 + 0.007	**35.** 0.163 + 0.095	**36.** 0.0001 + 0.0010
37. 4.56 15.981 + 6.005	**38.** 55.162 7.04 + 9.002	**39.** 101.06 207.49 + 12.017	**40.** 0.0156 2.713 + 55.66
41. 195.61 742.77 + 100.00	**42.** 500.00 700.00 + 800.00	**43.** 501.01 306.06 + 65.72	**44.** 50.05 210.77 + 998.54

45.
```
   3.2
   4.8
   5.6
+  2.1
```

46.
```
   0.02
   1.12
   0.05
+  7.25
```

47.
```
   126.21
    74.63
    21.79
+    3.74
```

48.
```
  54,763.21
+ 47,063.09
```

49.
```
  98,100.16
+ 56,763.78
```

50.
```
  387,496.43
+ 782,051.66
```

ANSWERS

1. 5 **2.** 1.2 **3.** 1.7 **4.** 1.9 **5.** 12.8 **6.** 11 **7.** 11.8 **8.** 18.9 **9.** 14.2 **10.** 18.5 **11.** 8.5 **12.** 16.3 **13.** 6.1 **14.** 20.4 **15.** 45.5 **16.** 73.37 **17.** 29.17 **18.** 58.39 **19.** 118.14 **20.** 67.51 **21.** 203.18 **22.** 276.94 **23.** 394.09 **24.** 275.15 **25.** 8.8 **26.** 2.9 **27.** 9.2 **28.** 15.2 **29.** 200.89 **30.** 670.38 **31.** 680.60 **32.** 515.91 **33.** 0.158 **34.** 0.063 **35.** 0.258 **36.** 0.0011 **37.** 26.546 **38.** 71.204 **39.** 320.567 **40.** 58.3886 **41.** 1,038.38 **42.** 2,000 **43.** 372.79 **44.** 1,259.36 **45.** 15.7 **46.** 8.44 **47.** 226.37 **48.** 101,826.30 **49.** 154,863.94 **50.** 1,169,548.09

PRACTICE PROBLEM SET 10A: SUBTRACTION OF DECIMALS

Do the indicated subtraction.

1. 0.6 − 0.2	**2.** 0.9 − 0.5	**3.** 0.8 − 0.1	**4.** 5.7 − 0.4
5. 2.9 − 0.6	**6.** 1.8 − 1.2	**7.** 4.7 − 1.0	**8.** 9.9 − 0.5
9. 6.6 − 4.2	**10.** 9.7 − 7.2	**11.** 6.1 − 0.3	**12.** 5.5 − 0.7
13. 4.8 − 0.9	**14.** 2.1 − 0.4	**15.** 9.3 − 1.5	**16.** 8.8 − 2.9
17. 9.2 − 5.6	**18.** 10.5 − 3.9	**19.** 8.1 − 5.7	**20.** 12.2 − 9.8
21. 191.73 − 62.21	**22.** 598.01 − 55.63	**23.** 707.55 − 192.73	**24.** 464.71 − 182.65
25. 201.02 − 169.76	**26.** 291.44 − 186.56	**27.** 853.72 − 761.99	**28.** 901.90 − 271.64
29. 100 − 63.71	**30.** 100 − 19.07	**31.** 476.19 − 100	**32.** 593.07 − 300
33. 19.76 − 8.09	**34.** 72.09 − 17.60	**35.** 18.13 − 4.02	**36.** 89.70 − 55.07
37. 23.4 − 0.78	**38.** 19.2 − 0.63	**39.** 55.6 − 0.07	**40.** 85.4 − 0.05
41. 19.01 − 0.563	**42.** 55 − 0.75	**43.** 37.001 − 0.06	**44.** 58.007 − 23.144
45. 0.0023 − 0.0015	**46.** 0.0097 − 0.0038	**47.** 5,630.19 − 4,058.02	**48.** 9,707.61 − 4,333.05

49. $\begin{array}{r} 46{,}046.73 \\ -\ 29{,}502.36 \\ \hline \end{array}$ **50.** $\begin{array}{r} 63{,}702.61 \\ -\ 19{,}077.66 \\ \hline \end{array}$

ANSWERS

1. 0.4 **2.** 0.4 **3.** 0.7 **4.** 5.3 **5.** 2.3 **6.** 0.6 **7.** 3.7 **8.** 9.4 **9.** 2.4
10. 2.5 **11.** 5.8 **12.** 4.8 **13.** 3.9 **14.** 1.7 **15.** 7.8 **16.** 5.9 **17.** 3.6
18. 6.6 **19.** 2.4 **20.** 2.4 **21.** 129.52 **22.** 542.38 **23.** 514.82 **24.** 282.06
25. 31.26 **26.** 104.88 **27.** 91.73 **28.** 630.26 **29.** 36.29 **30.** 80.93
31. 376.19 **32.** 293.07 **33.** 11.67 **34.** 54.49 **35.** 14.11 **36.** 34.63
37. 22.62 **38.** 18.57 **39.** 55.53 **40.** 85.35 **41.** 18.447 **42.** 54.25
43. 36.941 **44.** 34.863 **45.** 0.0008 **46.** 0.0059 **47.** 1,572.17 **48.** 5,374.56
49. 16,544.37 **50.** 44,624.95

PRACTICE PROBLEM SET 11A: MULTIPLICATION OF DECIMALS

Do the indicated multiplication.

1. $\begin{array}{r} 0.3 \\ \times\ 2 \\ \hline \end{array}$ **2.** $\begin{array}{r} 0.8 \\ \times\ 4 \\ \hline \end{array}$ **3.** $\begin{array}{r} 1.1 \\ \times\ 3 \\ \hline \end{array}$ **4.** $\begin{array}{r} 5.2 \\ \times\ 1 \\ \hline \end{array}$

5. $\begin{array}{r} 3.6 \\ \times\ 7 \\ \hline \end{array}$ **6.** $\begin{array}{r} 9.8 \\ \times\ 5 \\ \hline \end{array}$ **7.** $\begin{array}{r} 5.6 \\ \times\ 8 \\ \hline \end{array}$ **8.** $\begin{array}{r} 4.3 \\ \times\ 9 \\ \hline \end{array}$

9. $\begin{array}{r} 0.5 \\ \times\ 10 \\ \hline \end{array}$ **10.** $\begin{array}{r} 2.4 \\ \times\ 10 \\ \hline \end{array}$ **11.** $\begin{array}{r} 3.8 \\ \times\ 100 \\ \hline \end{array}$ **12.** $\begin{array}{r} 0.06 \\ \times\ 100 \\ \hline \end{array}$

13. $\begin{array}{r} 2.7 \\ \times\ 9 \\ \hline \end{array}$ **14.** $\begin{array}{r} 5.2 \\ \times\ 8 \\ \hline \end{array}$ **15.** $\begin{array}{r} 0.4 \\ \times\ 0.3 \\ \hline \end{array}$ **16.** $\begin{array}{r} 3.3 \\ \times\ 0.4 \\ \hline \end{array}$

17. $\begin{array}{r} 1.1 \\ \times\ 1.1 \\ \hline \end{array}$ **18.** $\begin{array}{r} 2.8 \\ \times\ 0.3 \\ \hline \end{array}$ **19.** $\begin{array}{r} 1.5 \\ \times\ 2.6 \\ \hline \end{array}$ **20.** $\begin{array}{r} 7.9 \\ \times\ 5.4 \\ \hline \end{array}$

21. $\begin{array}{r} 6.3 \\ \times\ 5.5 \\ \hline \end{array}$ **22.** $\begin{array}{r} 4.7 \\ \times\ 4.7 \\ \hline \end{array}$ **23.** $\begin{array}{r} 5.6 \\ \times\ 4.1 \\ \hline \end{array}$ **24.** $\begin{array}{r} 7.5 \\ \times\ 3.4 \\ \hline \end{array}$

25. $\begin{array}{r} 11.3 \\ \times\ 0.2 \\ \hline \end{array}$ **26.** $\begin{array}{r} 12.7 \\ \times\ 0.4 \\ \hline \end{array}$ **27.** $\begin{array}{r} 10.9 \\ \times\ 8.5 \\ \hline \end{array}$ **28.** $\begin{array}{r} 14.5 \\ \times\ 5.2 \\ \hline \end{array}$

29. 27.3×5.7	**30.** 35.6×9.1	**31.** 56.7×4.4	**32.** 91.6×8.7
33. 19.8×1.9	**34.** 27×0.4	**35.** 47×0.6	**36.** 58×1.7
37. 74×2.7	**38.** $19.5 \times 1{,}000$	**39.** $0.006 \times 1{,}000$	**40.** 3.104×10
41. 1.01×1.01	**42.** 2.20×2.02	**43.** 7.1×7.01	**44.** $10{,}000 \times 0.0027$
45. 3.9×2.5	**46.** 64.17×1.2	**47.** 55.09×21.3	**48.** 23.2×19.75
49. 162.17×58.36	**50.** 589.64×72.13		

ANSWERS

1. 0.6 **2.** 3.2 **3.** 3.3 **4.** 5.2 **5.** 25.2 **6.** 49 **7.** 44.8 **8.** 38.7 **9.** 5 **10.** 24 **11.** 380 **12.** 6 **13.** 24.3 **14.** 41.6 **15.** 0.12 **16.** 1.32 **17.** 1.21 **18.** 0.84 **19.** 3.9 **20.** 42.66 **21.** 34.65 **22.** 22.09 **23.** 22.96 **24.** 25.5 **25.** 2.26 **26.** 5.08 **27.** 92.65 **28.** 75.4 **29.** 155.61 **30.** 323.96 **31.** 249.48 **32.** 796.92 **33.** 37.62 **34.** 10.8 **35.** 28.2 **36.** 98.6 **37.** 199.8 **38.** 19,500 **39.** 6 **40.** 31.04 **41.** 1.0201 **42.** 4.444 **43.** 49.771 **44.** 27 **45.** 9.75 **46.** 77.004 **47.** 1,173.417 **48.** 458.2 **49.** 9,464.2412 **50.** 42,530.733

PRACTICE PROBLEM SET 12A: DIVISION OF DECIMALS

Do the indicated division.

1. $4\overline{)4.8}$ **2.** $2\overline{)8.6}$ **3.** $5\overline{)5.5}$ **4.** $7\overline{)21.7}$

5. $10\overline{)6.1}$ **6.** $10\overline{)19.8}$ **7.** $14\overline{)30.8}$ **8.** $13\overline{)66.3}$

9. $11\overline{)51.7}$ **10.** $18\overline{)129.6}$ **11.** $3\overline{)2.46}$ **12.** $7\overline{)3.71}$

13. $3\overline{)0.12}$ **14.** $8\overline{)38.4}$ **15.** $9\overline{)87.21}$ **16.** $7\overline{)35.42}$

17. $4\overline{)278.8}$ **18.** $8\overline{)33.12}$ **19.** $5\overline{)16.15}$ **20.** $9\overline{)6.93}$

21. $2.7\overline{)9.612}$ **22.** $3.8\overline{)17.898}$ **23.** $9.2\overline{)12.42}$

24. $6.1\overline{)478.24}$ **25.** $1.1\overline{)0.242}$ **26.** $5.4\overline{)25.38}$

27. $9.5\overline{)54.15}$ **28.** $3.8\overline{)15.96}$ **29.** $9.1\overline{)52.78}$

30. $5.5\overline{)41.8}$ **31.** 26.4 ÷ 2 = ________

32. 98.4 ÷ 5 = ________ **33.** 14.03 ÷ 2.3 = ________

34. 38.54 ÷ 4.7 = ________ **35.** 32.2 ÷ 0.35 = ________

36. 87.296 ÷ 2.56 = ________ **37.** 45.44 ÷ 0.71 = ________

38. 0.3626 ÷ 0.0037 = ________ **39.** 0.10752 ÷ 0.0192 = ________

40. 1,678.95 ÷ 27.3 = ________

Answers to problems 41–50 are to be rounded off to the nearest hundredth.

41. 0.5 ÷ 76 = ________ **42.** 3.9 ÷ 81 = ________

43. 15.6 ÷ 98 = ________ **44.** 76.1 ÷ 3.7 = ________

45. 100.17 ÷ 23 = ________ **46.** 46.84 ÷ 3.9 = ________

47. 0.05 ÷ 2.7 = ________ **48.** 6.08 ÷ 19 = ________

49. 4.27 ÷ 0.037 = ________ **50.** 1.672 ÷ 298.7 = ________

ANSWERS

1. 1.2 **2.** 4.3 **3.** 1.1 **4.** 3.1 **5.** 0.61 **6.** 1.98 **7.** 2.2 **8.** 5.1 **9.** 4.7 **10.** 7.2 **11.** 0.82 **12.** 0.53 **13.** 0.04 **14.** 4.8 **15.** 9.69 **16.** 5.06 **17.** 69.7 **18.** 4.14 **19.** 3.23 **20.** 0.77 **21.** 3.56 **22.** 4.71 **23.** 1.35 **24.** 78.4 **25.** 0.22 **26.** 4.7 **27.** 5.7 **28.** 4.2 **29.** 5.8 **30.** 7.6 **31.** 13.2 **32.** 19.68 **33.** 6.1 **34.** 8.2 **35.** 92 **36.** 34.1 **37.** 64 **38.** 98 **39.** 5.6 **40.** 61.5 **41.** 0.01 **42.** 0.05 **43.** 0.16 **44.** 20.57 **45.** 4.36 **46.** 12.01 **47.** 0.02 **48.** 0.32 **49.** 115.41 **50.** 0.01

ARITHMETIC PRACTICE PROBLEMS: COLLECTION B

PRACTICE PROBLEM SET 1B: ADDITION OF WHOLE NUMBERS

Do the indicated addition.

1. 6 + 7	**2.** 9 + 4	**3.** 8 + 7	**4.** 7 + 7
5. 5 + 11	**6.** 15 + 9	**7.** 19 + 6	**8.** 43 + 4
9. 15 + 10	**10.** 27 + 18	**11.** 44 + 33	**12.** 81 + 29
13. 77 + 65	**14.** 41 + 28	**15.** 88 + 37	**16.** 55 + 73
17. 91 + 66	**18.** 52 + 69	**19.** 34 + 47	**20.** 36 + 85
21. 45 + 78	**22.** 56 + 78	**23.** 25 + 19	**24.** 88 + 89
25. 98 + 66	**26.** 104 + 51	**27.** 101 + 100	**28.** 217 + 355
29. 545 + 129	**30.** 127 + 721	**31.** 493 + 672	**32.** 861 + 759
33. 365 + 482	**34.** 764 + 799	**35.** 892 + 144	**36.** 709 + 890
37. 570 + 680	**38.** 4,658 + 977	**39.** 7,063 + 367	**40.** 9,871 + 8,068
41. 22,509 + 17,044	**42.** 98,763 + 20,762	**43.** 70,994 + 63,076	**44.** 53 167 + 98

45.

$$\begin{array}{r} 244 \\ 309 \\ +\ 175 \\ \hline \end{array}$$

46.

$$\begin{array}{r} 611 \\ 88 \\ +\ 1{,}273 \\ \hline \end{array}$$

47.

$$\begin{array}{r} 553 \\ 918 \\ +\ 607 \\ \hline \end{array}$$

48.

$$\begin{array}{r} 4{,}308 \\ 991 \\ 56 \\ +\ 3{,}017 \\ \hline \end{array}$$

49.

$$\begin{array}{r} 2{,}490 \\ 22{,}763 \\ +\ 19{,}234 \\ \hline \end{array}$$

50.

$$\begin{array}{r} 50{,}193 \\ 19{,}015 \\ 4{,}388 \\ +\ 17{,}106 \\ \hline \end{array}$$

ANSWERS

1. 13 **2.** 13 **3.** 15 **4.** 14 **5.** 16 **6.** 24 **7.** 25 **8.** 47 **9.** 25 **10.** 45 **11.** 77 **12.** 110 **13.** 142 **14.** 69 **15.** 125 **16.** 128 **17.** 157 **18.** 121 **19.** 81 **20.** 121 **21.** 123 **22.** 134 **23.** 44 **24.** 177 **25.** 164 **26.** 155 **27.** 201 **28.** 572 **29.** 674 **30.** 848 **31.** 1,165 **32.** 1,620 **33.** 847 **34.** 1,563 **35.** 1,036 **36.** 1,599 **37.** 1,250 **38.** 5,635 **39.** 7,430 **40.** 17,939 **41.** 39,553 **42.** 119,525 **43.** 134,070 **44.** 318 **45.** 728 **46.** 1,972 **47.** 2,078 **48.** 8,372 **49.** 44,487 **50.** 90,702

PRACTICE PROBLEM SET 2B: SUBTRACTION OF WHOLE NUMBERS

Do the indicated subtraction.

1. $8 - 4$	**2.** $9 - 6$	**3.** $7 - 2$	**4.** $8 - 7$
5. $7 - 3$	**6.** $8 - 8$	**7.** $12 - 5$	**8.** $29 - 17$
9. $43 - 11$	**10.** $55 - 35$	**11.** $34 - 29$	**12.** $98 - 67$
13. $52 - 16$	**14.** $67 - 58$	**15.** $48 - 37$	**16.** $89 - 54$
17. $68 - 49$	**18.** $96 - 36$	**19.** $53 - 27$	**20.** $84 - 57$
21. $117 - 63$	**22.** $201 - 55$	**23.** $314 - 126$	**24.** $440 - 207$
25. $981 - 681$	**26.** $52 - 19$	**27.** $19 - 6$	**28.** $190 - 75$
29. $73 - 44$	**30.** $152 - 86$	**31.** $219 - 109$	**32.** $555 - 407$
33. $100 - 62$	**34.** $500 - 298$	**35.** $1060 - 860$	**36.** $700 - 284$
37. $5,064 - 1,072$	**38.** $9,877 - 2,005$	**39.** $6,711 - 5,432$	**40.** $2,060 - 985$
41. $1,000 - 357$	**42.** $5,192 - 3,155$	**43.** $4,566 - 1,123$	**44.** $6,718 - 3,907$
45. $7,300 - 6,781$	**46.** $8,591 - 2,748$	**47.** $5,151 - 5,101$	**48.** $54,102 - 43,172$

49. $\begin{array}{r} 10{,}000 \\ -\ 7{,}643 \\ \hline \end{array}$ **50.** $\begin{array}{r} 54{,}673 \\ -\ 48{,}764 \\ \hline \end{array}$

ANSWERS

1. 4 **2.** 3 **3.** 5 **4.** 1 **5.** 4 **6.** 0 **7.** 7 **8.** 12 **9.** 32 **10.** 20
11. 5 **12.** 31 **13.** 36 **14.** 9 **15.** 11 **16.** 35 **17.** 19 **18.** 60 **19.** 26
20. 27 **21.** 54 **22.** 146 **23.** 188 **24.** 233 **25.** 300 **26.** 33 **27.** 13
28. 115 **29.** 29 **30.** 66 **31.** 110 **32.** 148 **33.** 38 **34.** 202 **35.** 200
36. 416 **37.** 3,992 **38.** 7,872 **39.** 1,279 **40.** 1,075 **41.** 643 **42.** 2,037
43. 3,443 **44.** 2,811 **45.** 519 **46.** 5,843 **47.** 50 **48.** 10,930 **49.** 2,357
50. 5,909

PRACTICE PROBLEM SET 3B: MULTIPLICATION OF WHOLE NUMBERS

Do the indicated multiplication.

1. 5×4

2. 8×3

3. 6×6

4. 9×0

5. 7×2

6. 6×7

7. 0×4

8. 9×9

9. 10×7

10. 11×3

11. 5×9

12. 7×6

13. 18×2

14. 10×5

15. 12×7

16. 11×12

17. 25×4

18. 30×5

19. 22×5

20. 61×4

21. 53×3

22. 72×8

23. 47×6

24. 75×4

25. 85×10

26. 19×9

27. 43×7

28. 56×8

29. 71×10

30. 13×11

31. 55×12

32. 46×23

33. 40×27

34. 63×24

35. 45×76

36. 86×52

37. 88×76

38. 92×25

39. 45×50

40. 63×49

41. 205×8

42. 763×24

43. 591×65

44. 802×89

45. 400×7

46. 542×53

47. 875×66

48. 281×59

49. $6,394 \times 47$

50. $5,446 \times 83$

ANSWERS

1. 20 **2.** 24 **3.** 36 **4.** 0 **5.** 14 **6.** 42 **7.** 0 **8.** 81 **9.** 70 **10.** 33 **11.** 45 **12.** 42 **13.** 36 **14.** 50 **15.** 84 **16.** 132 **17.** 100 **18.** 150 **19.** 110 **20.** 244 **21.** 159 **22.** 576 **23.** 282 **24.** 300 **25.** 850 **26.** 171 **27.** 301 **28.** 448 **29.** 710 **30.** 143 **31.** 660 **32.** 1,058 **33.** 1,080 **34.** 1,512 **35.** 3,420 **36.** 4,472 **37.** 6,688 **38.** 2,300 **39.** 2,250 **40.** 3,087 **41.** 1,640 **42.** 18,312 **43.** 38,415 **44.** 71,378 **45.** 2,800 **46.** 28,726 **47.** 57,750 **48.** 16,579 **49.** 300,518 **50.** 452,018

PRACTICE PROBLEM SET 4B: DIVISION OF WHOLE NUMBERS

Do the indicated division.

1. $5\overline{)40}$
2. $6\overline{)36}$
3. $4\overline{)40}$
4. $3\overline{)27}$
5. $8\overline{)48}$
6. $7\overline{)49}$
7. $2\overline{)12}$
8. $9\overline{)72}$
9. $4\overline{)28}$
10. $8\overline{)80}$
11. $11\overline{)132}$
12. $15\overline{)225}$
13. $14\overline{)98}$
14. $13\overline{)65}$
15. $11\overline{)55}$
16. $10\overline{)90}$
17. $8\overline{)120}$
18. $12\overline{)144}$
19. $10\overline{)260}$
20. $9\overline{)153}$
21. $7\overline{)224}$
22. $15\overline{)105}$
23. $14\overline{)196}$
24. $16\overline{)192}$
25. $9\overline{)51}$
26. $8\overline{)42}$
27. $7\overline{)59}$
28. $9\overline{)78}$

29. $8\overline{)3{,}808}$ **30.** $5\overline{)8{,}575}$ **31.** $6\overline{)4{,}077}$ **32.** $9\overline{)5{,}697}$

33. $23\overline{)1{,}058}$ **34.** $19\overline{)1{,}064}$ **35.** $47\overline{)2{,}538}$ **36.** $81\overline{)6{,}237}$

37. $56\overline{)7{,}224}$ **38.** $95\overline{)20{,}995}$ **39.** $23\overline{)15{,}157}$ **40.** $64\overline{)44{,}928}$

41. $76\overline{)98{,}043}$ **42.** $36\overline{)99{,}761}$ **43.** $72\overline{)49{,}320}$ **44.** $86\overline{)46{,}354}$

45. $41\overline{)102{,}641}$ **46.** $15\overline{)798{,}043}$ **47.** $78\overline{)68{,}094}$ **48.** $55\overline{)49{,}720}$

49. $62\overline{)53{,}072}$ **50.** $93\overline{)57{,}381}$

ANSWERS

1. 8 **2.** 6 **3.** 10 **4.** 9 **5.** 6 **6.** 7 **7.** 6 **8.** 8 **9.** 7 **10.** 10
11. 12 **12.** 15 **13.** 7 **14.** 5 **15.** 5 **16.** 9 **17.** 15 **18.** 12 **19.** 26
20. 17 **21.** 32 **22.** 7 **23.** 14 **24.** 12 **25.** 5, $R = 6$ **26.** 5, $R = 2$
27. 8, $R = 3$ **28.** 8, $R = 6$ **29.** 476 **30.** 1715 **31.** 679, $R = 3$ **32.** 633
33. 46 **34.** 56 **35.** 54 **36.** 77 **37.** 129 **38.** 221 **39.** 659 **40.** 702
41. 1,290, $R = 3$ **42.** 2,771, $R = 5$ **43.** 685 **44.** 539 **45.** 2,503, $R = 18$
46. 53,202, $R = 13$ **47.** 873 **48.** 904 **49.** 856 **50.** 617

PRACTICE PROBLEM SET 5B: ADDITION OF FRACTIONS

Do the indicated addition. State each answer as a proper fraction or as a mixed number. Reduce all answers to simplest form.

1. $\frac{3}{7} + \frac{2}{7} =$ ______ **2.** $\frac{1}{6} + \frac{3}{6} =$ ______ **3.** $\frac{5}{9} + \frac{4}{9} =$ ______

4. $\frac{8}{12} + \frac{3}{12} =$ ______ **5.** $\frac{1}{5} + \frac{3}{5} =$ ______ **6.** $\frac{5}{10} + \frac{7}{10}$ ______

7. $\frac{4}{7} + \frac{8}{7} =$ ______ **8.** $\frac{3}{4} + \frac{1}{4} =$ ______ **9.** $\frac{8}{6} + \frac{4}{6} =$ ______

10. $\frac{9}{13} + \frac{2}{13} =$ ______ **11.** $\frac{1}{4} + \frac{1}{2} =$ ______ **12.** $\frac{2}{3} + \frac{1}{4} =$ ______

13. $\frac{3}{5} + \frac{1}{2} =$ ______ **14.** $\frac{1}{4} + \frac{2}{5} =$ ______ **15.** $\frac{3}{7} + \frac{2}{3} =$ ______

16. $\frac{1}{3} + \frac{5}{8} =$ ______ **17.** $\frac{5}{6} + \frac{1}{12} =$ ______ **18.** $\frac{3}{10} + \frac{4}{5} =$ ______

19. $\frac{4}{7} + \frac{2}{5} =$ ______ **20.** $\frac{5}{16} + \frac{7}{8} =$ ______ **21.** $\frac{3}{10} + \frac{7}{20} =$ ______

22. $\frac{3}{8} + \frac{1}{3} =$ ______ **23.** $\frac{2}{7} + \frac{2}{3} =$ ______ **24.** $\frac{7}{18} + \frac{5}{6} =$ ______

25. $\frac{2}{5} + \frac{1}{10}$ **26.** $\frac{9}{16} + \frac{1}{2}$ **27.** $\frac{7}{10} + \frac{3}{4}$

28. $\frac{6}{16} + \frac{5}{8}$ **29.** $\frac{3}{8} + \frac{7}{12}$ **30.** $\frac{4}{8} + \frac{7}{9}$

31. $\frac{29}{9} + \frac{1}{3} =$ ______ **32.** $\frac{5}{6} + \frac{34}{15} =$ ______ **33.** $\frac{9}{5} + \frac{5}{10} =$ ______

34. $\frac{10}{7} + \frac{17}{15} =$ ______ **35.** $\frac{13}{3} + \frac{17}{13} =$ ______ **36.** $\frac{5}{12} + \frac{29}{9} =$ ______

37. $\frac{67}{12} + \frac{45}{20} =$ ______ **38.** $\frac{37}{8} + \frac{52}{9} =$ ______ **39.** $\frac{27}{4} + \frac{4}{5} =$ ______

40. $\frac{3}{18} + \frac{2}{45} =$ ______ **41.** $\frac{7}{20} + \frac{3}{10} =$ ______ **42.** $\frac{9}{20} + \frac{7}{30} =$ ______

43. $\frac{7}{15} + \frac{4}{21}$ ______ **44.** $\frac{1}{5} + \frac{7}{12} =$ ______ **45.** $\frac{5}{16} + \frac{5}{12} =$ ______

46. $\frac{9}{11} + \frac{1}{2} =$ ______ **47.** $\frac{7}{10} + \frac{3}{14} =$ ______ **48.** $\frac{11}{18} + \frac{15}{42} =$ ______

49. $\frac{3}{8} + \frac{13}{20} =$ ______ **50.** $\frac{5}{16} + \frac{11}{24} =$ ______

ANSWERS

1. $\frac{5}{7}$ **2.** $\frac{2}{3}$ **3.** 1 **4.** $\frac{11}{12}$ **5.** $\frac{4}{5}$ **6.** $1\frac{1}{5}$ **7.** $1\frac{5}{7}$ **8.** 1 **9.** 2 **10.** $\frac{11}{13}$ **11.** $\frac{3}{4}$
12. $\frac{11}{12}$ **13.** $1\frac{1}{10}$ **14.** $\frac{13}{20}$ **15.** $1\frac{2}{21}$ **16.** $\frac{23}{24}$ **17.** $\frac{11}{12}$ **18.** $1\frac{1}{10}$ **19.** $\frac{34}{35}$
20. $1\frac{3}{16}$ **21.** $\frac{13}{20}$ **22.** $\frac{17}{24}$ **23.** $\frac{20}{21}$ **24.** $1\frac{2}{9}$ **25.** $\frac{1}{2}$ **26.** $1\frac{1}{16}$ **27.** $1\frac{9}{20}$
28. 1 **29.** $\frac{23}{24}$ **30.** $1\frac{5}{18}$ **31.** $3\frac{5}{9}$ **32.** $3\frac{1}{10}$ **33.** $2\frac{3}{10}$ **34.** $2\frac{59}{105}$ **35.** $5\frac{25}{39}$
36. $3\frac{23}{36}$ **37.** $7\frac{5}{6}$ **38.** $10\frac{29}{72}$ **39.** $7\frac{11}{20}$ **40.** $\frac{19}{90}$ **41.** $\frac{13}{20}$ **42.** $\frac{41}{60}$ **43.** $\frac{23}{35}$
44. $\frac{47}{60}$ **45.** $\frac{35}{48}$ **46.** $1\frac{7}{22}$ **47.** $\frac{32}{35}$ **48.** $\frac{61}{63}$ **49.** $1\frac{1}{40}$ **50.** $\frac{37}{48}$

PRACTICE PROBLEM SET 6B: SUBTRACTION OF FRACTIONS

Do the indicated subtraction. State each answer as a proper fraction or as a mixed number. Reduce all answers to simplest form.

1. $\frac{3}{4} - \frac{1}{4} =$ ______
2. $\frac{4}{5} - \frac{2}{5} =$ ______
3. $\frac{9}{10} - \frac{4}{10} =$ ______
4. $\frac{6}{7} - \frac{4}{7} =$ ______
5. $\frac{8}{11} - \frac{5}{11} =$ ______
6. $\frac{11}{12} - \frac{6}{12} =$ ______
7. $\frac{11}{15} - \frac{8}{15} =$ ______
8. $\frac{80}{100} - \frac{72}{100} =$ ______
9. $\frac{10}{14} - \frac{5}{14} =$ ______
10. $\frac{17}{18} - \frac{11}{18} =$ ______
11. $\frac{7}{8} - \frac{1}{8} =$ ______
12. $\frac{9}{11} - \frac{9}{11} =$ ______
13. $\frac{1}{2} - \frac{1}{2} =$ ______
14. $\frac{2}{5} - \frac{1}{4} =$ ______
15. $\frac{17}{20} - \frac{3}{10} =$ ______
16. $\frac{13}{15} - \frac{2}{3} =$ ______
17. $\frac{6}{7} - \frac{1}{21} =$ ______
18. $\frac{19}{30} - \frac{1}{3} =$ ______
19. $\frac{2}{3} - \frac{1}{5} =$ ______
20. $\frac{5}{6} - \frac{19}{42} =$ ______
21. $\frac{5}{8} - \frac{3}{5} =$ ______
22. $\frac{15}{18} - \frac{4}{9} =$ ______
23. $\frac{7}{12} - \frac{9}{24} =$ ______
24. $\frac{5}{9} - \frac{1}{12} =$ ______
25. $\frac{3}{4} - \frac{2}{3}$
26. $\frac{9}{10} - \frac{3}{5}$
27. $\frac{5}{7} - \frac{2}{14}$
28. $\frac{11}{16} - \frac{5}{12}$
29. $\frac{43}{51} - \frac{3}{17}$
30. $\frac{19}{40} - \frac{3}{16}$
31. $\frac{11}{6} - \frac{3}{5} =$ ______
32. $\frac{19}{7} - \frac{3}{8} =$ ______
33. $\frac{93}{20} - \frac{22}{15} =$ ______
34. $\frac{187}{50} - \frac{217}{100} =$ ______
35. $\frac{87}{100} - \frac{13}{35} =$ ______
36. $\frac{60}{11} - \frac{29}{9} =$ ______
37. $2 - \frac{2}{3} =$ ______
38. $1 - \frac{9}{10} =$ ______
39. $4 - \frac{3}{4} =$ ______
40. $1 - \frac{11}{100} =$ ______
41. $\frac{5}{12} - \frac{1}{8} =$ ______
42. $\frac{7}{16} - \frac{3}{12} =$ ______
43. $\frac{18}{21} - \frac{1}{2} =$ ______
44. $\frac{5}{22} - \frac{1}{6} =$ ______
45. $\frac{5}{7} - \frac{1}{4} =$ ______
46. $\frac{11}{15} - \frac{4}{21} =$ ______
47. $\frac{2}{9} - \frac{1}{42} =$ ______
48. $\frac{29}{60} - \frac{7}{40} =$ ______
49. $\frac{15}{16} - \frac{9}{32} =$ ______
50. $\frac{11}{24} - \frac{5}{16} =$ ______

ANSWERS

1. $\frac{1}{2}$ **2.** $\frac{2}{5}$ **3.** $\frac{1}{2}$ **4.** $\frac{2}{7}$ **5.** $\frac{3}{11}$ **6.** $\frac{5}{12}$ **7.** $\frac{1}{5}$ **8.** $\frac{2}{25}$ **9.** $\frac{5}{14}$ **10.** $\frac{1}{3}$ **11.** $\frac{3}{4}$

12. 0 **13.** 0 **14.** $\frac{3}{20}$ **15.** $\frac{11}{20}$ **16.** $\frac{1}{5}$ **17.** $\frac{17}{21}$ **18.** $\frac{3}{10}$ **19.** $\frac{7}{15}$

20. $\frac{8}{21}$ **21.** $\frac{1}{40}$ **22.** $\frac{7}{18}$ **23.** $\frac{5}{24}$ **24.** $\frac{17}{36}$ **25.** $\frac{1}{12}$ **26.** $\frac{3}{10}$ **27.** $\frac{4}{7}$ **28.** $\frac{13}{48}$

29. $\frac{2}{3}$ **30.** $\frac{23}{80}$ **31.** $1\frac{7}{30}$ **32.** $2\frac{19}{56}$ **33.** $3\frac{11}{60}$ **34.** $1\frac{57}{100}$ **35.** $\frac{349}{700}$ **36.** $2\frac{23}{99}$

37. $1\frac{1}{3}$ **38.** $\frac{1}{10}$ **39.** $3\frac{1}{4}$ **40.** $\frac{89}{100}$ **41.** $\frac{7}{24}$ **42.** $\frac{3}{16}$ **43.** $\frac{5}{14}$ **44.** $\frac{2}{33}$

45. $\frac{13}{28}$ **46.** $\frac{57}{105}$ **47.** $\frac{25}{126}$ **48.** $\frac{37}{120}$ **49.** $\frac{21}{32}$ **50.** $\frac{7}{48}$

PRACTICE PROBLEM SET 7B: MULTIPLICATION OF FRACTIONS

Do the indicated multiplication. State each answer as a proper fraction or as a mixed number. Reduce all answers to simplest form.

1. $\frac{1}{2} \times \frac{3}{4} =$ ______ **2.** $\frac{2}{5} \times \frac{1}{3} =$ ______ **3.** $\frac{3}{7} \times \frac{2}{7} =$ ______

4. $\frac{5}{8} \times \frac{2}{3} =$ ______ **5.** $\frac{9}{10} \times \frac{5}{7} =$ ______ **6.** $\frac{7}{8} \times \frac{8}{5} =$ ______

7. $4 \times \frac{3}{5} =$ ______ **8.** $7 \times \frac{5}{6} =$ ______ **9.** $\frac{5}{11} \times \frac{4}{6} =$ ______

10. $0 \times \frac{8}{9} =$ ______ **11.** $\frac{11}{16} \times \frac{4}{9} =$ ______ **12.** $\frac{3}{5} \times \frac{5}{9} =$ ______

13. $\frac{5}{6} \times 0 =$ ______ **14.** $14 \times \frac{5}{7} =$ ______ **15.** $\frac{5}{12} \times 48 =$ ______

16. $\frac{2}{3} \times \frac{5}{6} =$ ______ **17.** $\frac{1}{3} \times \frac{3}{4} =$ ______ **18.** $\frac{11}{4} \times \frac{8}{33} =$ ______

19. $\frac{7}{8} \times \frac{4}{14} =$ ______ **20.** $\frac{10}{25} \times \frac{5}{15} =$ ______ **21.** $\frac{15}{20} \times \frac{45}{30} =$ ______

22. $1 \times \frac{99}{100} =$ ______ **23.** $\frac{15}{7} \times \frac{21}{18} =$ ______ **24.** $\frac{5}{2} \times \frac{6}{10} =$ ______

25. $\frac{3}{7} \times \frac{9}{10} =$ ______ **26.** $\frac{7}{8} \times \frac{3}{14} =$ ______ **27.** $\frac{12}{9} \times \frac{15}{17} =$ ______

28. $\frac{3}{8} \times \frac{5}{16} =$ ______ **29.** $\frac{5}{14} \times \frac{7}{15} =$ ______ **30.** $\frac{3}{8} \times \frac{26}{7} =$ ______

31. $\frac{5}{12} \times \frac{24}{20} =$ ______ **32.** $\frac{16}{40} \times \frac{35}{64} =$ ______ **33.** $\frac{19}{36} \times \frac{48}{95} =$ ______

34. $\frac{18}{9} \times \frac{46}{18} =$ ______ **35.** $\frac{4}{13} \times \frac{13}{8} =$ ______ **36.** $\frac{17}{9} \times \frac{81}{34} =$ ______

37. $5 \times \frac{3}{2} =$ ______ **38.** $3 \times \frac{9}{4} =$ ______ **39.** $\frac{8}{5} \times 2 =$ ______

40. $\frac{7}{2} \times 10 =$ ______ **41.** $\frac{7}{5} \times \frac{5}{2} =$ ______ **42.** $\frac{1}{3} \times \frac{4}{3} =$ ______

43. $\frac{29}{6} \times \frac{11}{2} =$ ______ **44.** $\frac{5}{4} \times \frac{11}{6} =$ ______ **45.** $\frac{3}{4} \times \frac{16}{3} =$ ______

46. $\frac{26}{5} \times \frac{40}{13} =$ ______ **47.** $\frac{38}{7} \times \frac{7}{2} =$ ______ **48.** $\frac{55}{13} \times \frac{39}{5} =$ ______

49. $\frac{89}{12} \times \frac{24}{7} =$ ______ **50.** $\frac{21}{8} \times \frac{17}{5} =$ ______

ANSWERS

1. $\frac{3}{8}$ **2.** $\frac{2}{15}$ **3.** $\frac{6}{49}$ **4.** $\frac{5}{12}$ **5.** $\frac{9}{14}$ **6.** $1\frac{2}{5}$ **7.** $2\frac{2}{5}$ **8.** $5\frac{5}{6}$ **9.** $\frac{10}{33}$ **10.** 0

11. $\frac{11}{36}$ **12.** $\frac{1}{3}$ **13.** 0 **14.** 10 **15.** 20 **16.** $\frac{5}{9}$ **17.** $\frac{1}{4}$ **18.** $\frac{2}{3}$ **19.** $\frac{1}{4}$

20. $\frac{2}{15}$ **21.** $1\frac{1}{8}$ **22.** $\frac{99}{100}$ **23.** $2\frac{1}{2}$ **24.** $1\frac{1}{2}$ **25.** $\frac{27}{70}$ **26.** $\frac{3}{16}$ **27.** $1\frac{3}{17}$

28. $\frac{15}{128}$ **29.** $\frac{1}{6}$ **30.** $1\frac{11}{28}$ **31.** $\frac{1}{2}$ **32.** $\frac{7}{32}$ **33.** $\frac{4}{15}$ **34.** $5\frac{1}{9}$ **35.** $\frac{1}{2}$ **36.** $4\frac{1}{2}$

37. $7\frac{1}{2}$ **38.** $6\frac{3}{4}$ **39.** $3\frac{1}{5}$ **40.** 35 **41.** $3\frac{1}{2}$ **42.** $\frac{4}{9}$ **43.** $26\frac{7}{12}$ **44.** $2\frac{7}{24}$

45. 4 **46.** 16 **47.** 19 **48.** 33 **49.** $25\frac{3}{7}$ **50.** $8\frac{37}{40}$

PRACTICE PROBLEM SET 8B: DIVISION OF FRACTIONS

Do the indicated division. State each answer as a proper fraction or as a mixed number. Reduce all answers to simplest form.

1. $\frac{1}{2} \div \frac{1}{3} =$ ______	2. $\frac{3}{4} \div \frac{1}{2} =$ ______	3. $\frac{2}{3} \div \frac{5}{4} =$ ______
4. $\frac{1}{6} \div \frac{3}{4} =$ ______	5. $\frac{4}{5} \div \frac{3}{10} =$ ______	6. $\frac{7}{10} \div \frac{2}{5} =$ ______
7. $5 \div \frac{5}{6} =$ ______	8. $2 \div \frac{3}{4} =$ ______	9. $\frac{9}{10} \div 1 =$ ______
10. $\frac{15}{4} \div 2 =$ ______	11. $\frac{2}{5} \div \frac{10}{15} =$ ______	12. $\frac{7}{8} \div \frac{21}{24} =$ ______
13. $\frac{2}{3} \div \frac{1}{6} =$ ______	14. $\frac{3}{5} \div 5 =$ ______	15. $\frac{4}{5} \div \frac{1}{2} =$ ______
16. $\frac{3}{4} \div \frac{1}{4} =$ ______	17. $\frac{7}{8} \div \frac{1}{8} =$ ______	18. $\frac{5}{6} \div \frac{2}{3} =$ ______
19. $\frac{10}{21} \div \frac{7}{5} =$ ______	20. $\frac{11}{15} \div \frac{5}{22} =$ ______	21. $\frac{16}{33} \div \frac{4}{11} =$ ______
22. $\frac{1}{3} \div \frac{1}{9} =$ ______	23. $\frac{1}{2} \div \frac{1}{4} =$ ______	24. $\frac{8}{3} \div \frac{5}{12} =$ ______
25. $\frac{8}{9} \div \frac{5}{4} =$ ______	26. $\frac{9}{2} \div \frac{19}{16} =$ ______	27. $\frac{5}{12} \div \frac{5}{6} =$ ______
28. $\frac{9}{35} \div \frac{18}{14} =$ ______	29. $\frac{14}{13} \div \frac{21}{26} =$ ______	30. $\frac{15}{12} \div \frac{25}{36} =$ ______
31. $\frac{10}{11} \div \frac{40}{33} =$ ______	32. $\frac{20}{22} \div \frac{25}{22} =$ ______	33. $\frac{6}{14} \div \frac{18}{42} =$ ______
34. $0 \div \frac{2}{3} =$ ______	35. $\frac{8}{7} \div 0 =$ ______	36. $\frac{28}{45} \div \frac{14}{18} =$ ______
37. $\frac{3}{2} \div 2 =$ ______	38. $\frac{11}{4} \div 4 =$ ______	39. $\frac{15}{8} \div \frac{3}{4} =$ ______
40. $\frac{17}{5} \div \frac{4}{10} =$ ______	41. $2 \div \frac{5}{3} =$ ______	42. $5 \div \frac{4}{3} =$ ______
43. $\frac{3}{2} \div \frac{11}{8} =$ ______	44. $\frac{8}{5} \div \frac{11}{10} =$ ______	45. $\frac{17}{6} \div \frac{10}{9} =$ ______
46. $\frac{11}{4} \div \frac{13}{8} =$ ______	47. $\frac{46}{3} \div \frac{20}{9} =$ ______	48. $\frac{133}{8} \div \frac{5}{3} =$ ______
49. $\frac{46}{3} \div \frac{43}{6} =$ ______	50. $\frac{69}{16} \div \frac{19}{8} =$ ______	

ANSWERS

1. $1\frac{1}{2}$ 2. $1\frac{1}{2}$ 3. $\frac{8}{15}$ 4. $\frac{2}{9}$ 5. $2\frac{2}{3}$ 6. $1\frac{3}{4}$ 7. 6 8. $2\frac{2}{3}$ 9. $\frac{9}{10}$ 10. $1\frac{7}{8}$
11. $\frac{3}{5}$ 12. 1 13. 4 14. $\frac{3}{25}$ 15. $1\frac{3}{5}$ 16. 3 17. 7 18. $1\frac{1}{4}$ 19. $\frac{50}{147}$
20. $3\frac{17}{75}$ 21. $1\frac{1}{3}$ 22. 3 23. 2 24. $6\frac{2}{5}$ 25. $\frac{32}{45}$ 26. $3\frac{15}{19}$ 27. $\frac{1}{2}$ 28. $\frac{1}{5}$
29. $1\frac{1}{3}$ 30. $1\frac{4}{5}$ 31. $\frac{3}{4}$ 32. $\frac{4}{5}$ 33. 1 34. 0 35. no solution 36. $\frac{4}{5}$ 37. $\frac{3}{4}$
38. $\frac{11}{16}$ 39. $2\frac{1}{2}$ 40. $8\frac{1}{2}$ 41. $1\frac{1}{5}$ 42. $3\frac{3}{4}$ 43. $1\frac{1}{11}$ 44. $1\frac{5}{11}$ 45. $2\frac{11}{20}$
46. $1\frac{9}{13}$ 47. $6\frac{9}{10}$ 48. $9\frac{39}{40}$ 49. $2\frac{6}{43}$ 50. $1\frac{31}{38}$

PRACTICE PROBLEM SET 9B: ADDITION OF DECIMALS

Do the indicated addition.

1. 5.2 + 4.6 = ________
2. 9.1 + 3.4 = ________
3. 15.7 + 8.2 = ________
4. 4.4 + 9.4 = ________
5. 10.5 + 13.7 = ________
6. 12.9 + 4.8 = ________
7. 6.7 + 7.7 = ________
8. 29.5 + 15.5 = ________
9. 16.5 + 23.5 = ________
10. 11.27 + 14.33 = ________
11. 9.02 + 16.75 = ________
12. 47.70 + 16.19 = ________
13. 93.50 + 26.02 = ________
14. 50.26 + 38.44 = ________
15. 191.07 + 35.09 = ________
16. 25.94 + 37.16 = ________
17. 47.61 + 131.79 = ________
18. 89.76 + 19.55 = ________
19. 11.08 + 46.19 = ________
20. 70.58 + 104.72 = ________
21. 12.9 + 15.07 = ________
22. 33.75 + 20.6 = ________
23. 201.06 + 19.4 = ________
24. 50.05 + 210.17 = ________
25. 101.01 + 19.1 = ________
26. 5.2 + 0.3 + 9 = ________
27. 5 + 4.7 + 0.1 = ________
28. 0.56 + 0.21 = ________
29. 5.003 + 0.002 = ________
30. 15.2 + 3.5 + 10 = ________
31. 0.09 + 0.001 = ________
32. 0.27 + 0.009 = ________
33. 6 + 13 + 4.24 = ________
34. 519.62 + 18.015 = ________
35. 477.08 + 92.76 = ________
36. 607.631 + 194.207 = ________
37. 178.876 + 19.05 = ________
38. 2.946 + 0.762 + 8.001 = ________
39. 0.445 + 5.06 + 16.009 = ________
40. 29.07 + 7.706 + 18.9 = ________

41. 24.345 + 19.233 + 53.110 = ________________

42. 42.06 + 98.16 + 27.07 = ________________

43. 576.92 + 891.05 + 17.44 = ________________

44. 6.5 + 0.5 + 1.9 + 2.4 = ________________

45. 9.01 + 5 + 4.3 + 0.27 = ________________

46. 0.02 + 0.006 + 0.0007 = ________________

47. 0.03 + 0.07 + 6.004 = ________________

48. 3,276.56 + 6,704.75 = ________________

49. 9,074.62 + 8,861.11 = ________________

50. 53,173,017.02 + 47,277,193.007 + 190,000,516 = ________________

ANSWERS

1. 9.8 **2.** 12.5 **3.** 23.9 **4.** 13.8 **5.** 24.2 **6.** 17.7 **7.** 14.4 **8.** 45 **9.** 40 **10.** 25.6 **11.** 25.77 **12.** 63.89 **13.** 119.52 **14.** 88.7 **15.** 226.16 **16.** 63.1 **17.** 179.4 **18.** 109.31 **19.** 57.27 **20.** 175.3 **21.** 27.97 **22.** 54.35 **23.** 220.46 **24.** 260.22 **25.** 120.11 **26.** 14.5 **27.** 9.8 **28.** 0.77 **29.** 5.005 **30.** 28.7 **31.** 0.091 **32.** 0.279 **33.** 23.24 **34.** 537.635 **35.** 569.84 **36.** 801.838 **37.** 197.926 **38.** 11.709 **39.** 21.514 **40.** 55.676 **41.** 96.688 **42.** 167.29 **43.** 1,485.41 **44.** 11.3 **45.** 18.58 **46.** 0.0267 **47.** 6.104 **48.** 9,981.31 **49.** 17,935.73 **50.** 290,450,726.027

PRACTICE PROBLEM SET 10B: SUBTRACTION OF DECIMALS

Do the indicated subtraction.

1. 7.5 − 3.2 = ____________
2. 4.3 − 1.1 = ____________
3. 10.5 − 9.4 = ____________
4. 8.9 − 4.4 = ____________
5. 12.7 − 8.2 = ____________
6. 15.7 − 8.3 = ____________
7. 19.2 − 7.2 = ____________
8. 10 − 5.4 = ____________
9. 8 − 2.7 = ____________
10. 15 − 6.3 = ____________
11. 18 − 2.07 = ____________
12. 19.44 − 0 = ____________
13. 19.81 − 3 = ____________
14. 28.65 − 13 = ____________
15. 23.6 − 20.8 = ____________
16. 7.6 − 5.9 = ____________
17. 14.3 − 9.9 = ____________
18. 18.4 − 7.8 = ____________
19. 401.26 − 131.13 = ____________
20. 191.76 − 55.68 = ____________
21. 596.17 − 236.45 = ____________
22. 56.17 − 19.85 = ____________
23. 70.64 − 43.77 = ____________
24. 102.16 − 19.38 = ____________
25. 214.50 − 87.65 = ____________
26. 329.48 − 251.67 = ____________
27. 547.43 − 298.88 = ____________
28. 15 − 2.74 = ____________
29. 33 − 16.81 = ____________
30. 724.32 − 68 = ____________
31. 27.09 − 7.265 = ____________
32. 19.299 − 18.73 = ____________
33. 463.73 − 9.76 = ____________
34. 19.1 − 5.004 = ____________
35. 47.6 − 3.0002 = ____________
36. 625.46 − 77.509 = ____________
37. 23.5 − 0.921 = ____________
38. 65.7 − 0.84 = ____________
39. 279.37 − 47.665 = ____________
40. 3.005 − 1.296 = ____________
41. 7.507 − 3.418 = ____________
42. 390.994 − 352.165 = ____________
43. 671.098 − 583.017 = ____________
44. 1,056.9 − 473.09 = ____________
45. 0.0024 − 0.0003 = ____________
46. 0.056 − 0.029 = ____________
47. 0.00103 − 0.00065 = ____________
48. 0.00445 − 0.0008 = ____________
49. 8.001 − 2.104 = ____________
50. 6.3701 − 1.007 = ____________

ANSWERS

1. 4.3 **2.** 3.2 **3.** 1.1 **4.** 4.5 **5.** 4.5 **6.** 7.4 **7.** 12 **8.** 4.6 **9.** 5.3 **10.** 8.7 **11.** 15.93 **12.** 19.44 **13.** 16.81 **14.** 15.65 **15.** 2.8 **16.** 1.7 **17.** 4.4 **18.** 10.6 **19.** 270.13 **20.** 136.08 **21.** 359.72 **22.** 36.32 **23.** 26.87 **24.** 82.78 **25.** 126.85 **26.** 77.81 **27.** 248.55 **28.** 12.26 **29.** 16.19 **30.** 656.32 **31.** 19.825 **32.** 0.569 **33.** 453.97 **34.** 14.096 **35.** 44.5998 **36.** 547.951 **37.** 22.579 **38.** 64.86 **39.** 231.705 **40.** 1.709 **41.** 4.089 **42.** 38.829 **43.** 88.081 **44.** 583.81 **45.** 0.0021 **46.** 0.027 **47.** 0.00038 **48.** 0.00365 **49.** 5.897 **50.** 5.3631

PRACTICE PROBLEM SET 11B: MULTIPLICATION OF DECIMALS

Do the indicated multiplication.

1. $4.3 \times 5 =$ ______
2. $6.4 \times 5 =$ ______
3. $8.8 \times 5 =$ ______
4. $1.7 \times 2 =$ ______
5. $6.9 \times 8 =$ ______
6. $10.5 \times 3 =$ ______
7. $\begin{array}{r} 2.1 \\ \times\ 3 \\ \hline \end{array}$
8. $\begin{array}{r} 9.2 \\ \times\ 8 \\ \hline \end{array}$
9. $\begin{array}{r} 7.4 \\ \times\ 6 \\ \hline \end{array}$
10. $\begin{array}{r} 6.5 \\ \times\ 7.4 \\ \hline \end{array}$
11. $\begin{array}{r} 19.8 \\ \times\ 5.6 \\ \hline \end{array}$
12. $\begin{array}{r} 10.6 \\ \times\ 9.2 \\ \hline \end{array}$
13. $\begin{array}{r} 4.73 \\ \times\ 8.8 \\ \hline \end{array}$
14. $\begin{array}{r} 15.01 \\ \times\ 5.6 \\ \hline \end{array}$
15. $\begin{array}{r} 17.29 \\ \times\ 3.05 \\ \hline \end{array}$
16. $\begin{array}{r} 18.17 \\ \times\ 4.43 \\ \hline \end{array}$
17. $\begin{array}{r} 29.05 \\ \times\ 6.72 \\ \hline \end{array}$
18. $\begin{array}{r} 15.15 \\ \times\ 1.69 \\ \hline \end{array}$

19. $2.7 \times 1.6 =$ ______________
20. $3.9 \times 3.4 =$ ______________
21. $5.8 \times 9.2 =$ ______________
22. $53.1 \times 5.4 =$ ______________
23. $16.07 \times 9.1 =$ ______________
24. $65.72 \times 8.8 =$ ______________
25. $5.3 \times 10 =$ ______________
26. $9.7 \times 100 =$ ______________
27. $17.09 \times 100 =$ ______________
28. $8.4 \times 1{,}000 =$ ______________
29. $0.0027 \times 1{,}000 =$ ______________
30. $0.67 \times 1{,}000 =$ ______________
31. $0.78 \times 6.8 =$ ______________
32. $6.37 \times 0.05 =$ ______________
33. $9.39 \times 0.007 =$ ______________
34. $1.47 \times 0.09 =$ ______________
35. $100 \times 2.76 =$ ______________
36. $10 \times 0.02 =$ ______________
37. $4.29 \times 0.01 =$ ______________
38. $53.66 \times 0.01 =$ ______________

39. $0.01 \times 0.01 =$ ________

40. $5.28 \times 0.04 =$ ________

41. $6.78 \times 5.64 =$ ________

42. $56.4 \times 32.19 =$ ________

43. $0.007 \times 15 =$ ________

44. $0.008 \times 20 =$ ________

45. $3.72 \times 4.016 =$ ________

46. $5.93 \times 4.07 =$ ________

47. $17.04 \times 29.706 =$ ________

48. $49.6 \times 0.00004 =$ ________

49. $0.06 \times 4{,}956{,}155 =$ ________

50. $1.07 \times 7{,}096.011 =$ ________

ANSWERS

1. 21.5 **2.** 32 **3.** 44 **4.** 3.4 **5.** 55.2 **6.** 31.5 **7.** 6.3 **8.** 73.6 **9.** 44.4
10. 48.1 **11.** 110.88 **12.** 97.52 **13.** 41 624 **14.** 84.056 **15.** 52.7345
16. 80.4931 **17.** 195.216 **18.** 25.6035 **19.** 4.32 **20.** 13.26 **21.** 53.36
22. 286.74 **23.** 146.237 **24.** 578.336 **25.** 53 **26.** 970 **27.** 1,709
28. 8,400 **29.** 2.7 **30.** 670 **31.** 5.304 **32.** 0.3185 **33.** 0.06573
34. 0.1323 **35.** 276 **36.** 0.2 **37.** 0.0429 **38.** 0.5366 **39.** 0.0001
40. 0.2112 **41.** 38.2392 **42.** 1,815.516 **43.** 0.105 **44.** 0.16 **45.** 14.93952
46. 24.1351 **47.** 506.19024 **48.** 0.001984 **49.** 297,369.3 **50.** 7,592.7317

PRACTICE PROBLEM SET 12B: DIVISION OF DECIMALS

Do the indicated division.

1. $3\overline{)9.6}$
2. $2\overline{)10.8}$
3. $7\overline{)28.7}$
4. $10\overline{)7.4}$
5. $10\overline{)23.6}$
6. $12\overline{)61.2}$
7. $8\overline{)30.4}$
8. $11\overline{)46.2}$
9. $14\overline{)106.4}$
10. $15\overline{)139.5}$
11. $17\overline{)76.5}$
12. $0.5\overline{)6}$
13. $0.7\overline{)28}$
14. $0.3\overline{)3.9}$
15. $0.8\overline{)56.8}$
16. $0.6\overline{)0.96}$
17. $0.9\overline{)8.721}$
18. $0.4\overline{)27.88}$

19. $6.3\overline{)8.19}$ **20.** $5.6\overline{)15.12}$ **21.** $5.5\overline{)46.20}$

22. $3.9\overline{)8.034}$ **23.** $0.41\overline{)2.501}$ **24.** $0.8\overline{)77.6}$

25. 38.57 ÷ 9.5 = ______________

26. 478.24 ÷ 6.1 = ______________

27. 0.196 ÷ 1.4 = ______________

28. 17.898 ÷ 3.8 = ______________

29. 12.42 ÷ 0.92 = ______________

30. 40.5 ÷ 250 = ______________

31. 20.304 ÷ 3.76 = ______________

32. 0.242 ÷ 1.1 = ______________

33. 0.242 ÷ 0.11 = ______________

34. 0.020292 ÷ 0.0057 = ______________

35. 9.612 ÷ 2.7 = ______________

36. 29.007 ÷ 8.79 = ______________

37. 17.424 ÷ 0.132 = ______________

38. 15,625 ÷ 0.625 = ______________

39. 3,000 ÷ 0.375 = ______________

40. 0.00334 ÷ 0.167 = ______________

Answers to problems 41–50 are to be rounded off to the nearest hundredth.

41. 0.4 ÷ 52 = ______________

42. 3.7 ÷ 92 = ______________

43. 10.7 ÷ 65 = ______________

44. 82.6 ÷ 2.8 = ______________

45. 100.98 ÷ 16 = ______________

46. 55.04 ÷ 5.5 = ______________

47. 0.07 ÷ 3.4 = ______________

48. 6.08 ÷ 3.9 = ______________

49. 34.7 ÷ 0.012 = ______________

50. 19.90 ÷ 0.056 = ______________

ANSWERS

1. 3.2 **2.** 5.4 **3.** 4.1 **4.** 0.74 **5.** 2.36 **6.** 5.1 **7.** 3.8 **8.** 4.2 **9.** 7.6 **10.** 9.3 **11.** 4.5 **12.** 12 **13.** 40 **14.** 13 **15.** 71 **16.** 1.6 **17.** 9.69 **18.** 69.7 **19.** 1.3 **20.** 2.7 **21.** 8.4 **22.** 2.06 **23.** 6.1 **24.** 97 **25.** 4.06 **26.** 78.4 **27.** 0.14 **28.** 4.71 **29.** 13.5 **30.** 0.162 **31.** 5.4 **32.** 0.22 **33.** 2.2 **34.** 3.56 **35.** 3.56 **36.** 3.3 **37.** 132 **38.** 25,000 **39.** 8,000 **40.** 0.02 **41.** 0.01 **42.** 0.04 **43.** 0.16 **44.** 29.5 **45.** 6.31 **46.** 10.01 **47.** 0.02 **48.** 1.56 **49.** 2,891.67 **50.** 355.36

ARITHMETIC PRACTICE PROBLEMS: COLLECTION C

PRACTICE PROBLEM SET 1C: ADDITION OF WHOLE NUMBERS

Do the indicated addition.

1. 5 + 3	2. 8 + 6	3. 9 + 7	4. 8 + 8
5. 10 + 7	6. 12 + 7	7. 33 + 5	8. 54 + 3
9. 19 + 10	10. 25 + 14	11. 56 + 32	12. 73 + 24
13. 81 + 39	14. 56 + 28	15. 74 + 46	16. 98 + 17
17. 82 + 78	18. 66 + 56	19. 34 + 79	20. 42 + 59
21. 34 + 29	22. 73 + 88	23. 26 + 47	24. 85 + 67
25. 105 + 23	26. 276 + 123	27. 673 + 326	28. 345 + 552
29. 271 + 604	30. 127 + 842	31. 849 + 268	32. 962 + 375
33. 551 + 109	34. 686 + 475	35. 799 + 411	36. 1,076 + 687
37. 1,472 + 528	38. 2,368 + 276	39. 4,095 + 6,985	40. 7,867 + 7,287
41. 12,058 + 7,642	42. 46,584 + 3,676	43. 68,495 + 32,813	44. 39,720 + 68,581
45. 87,065 + 45,957	46. 338 47 + 609	47. 871 1,046 + 938	48. 4,309 7,654 + 824

49.
```
   9,246
     843
 + 2,276
```

50.
```
  42,036
   2,764
+ 39,980
```

ANSWERS

1. 8 **2.** 14 **3.** 16 **4.** 16 **5.** 17 **6.** 19 **7.** 38 **8.** 57 **9.** 29 **10.** 39 **11.** 88 **12.** 97 **13.** 120 **14.** 84 **15.** 120 **16.** 115 **17.** 160 **18.** 122 **19.** 113 **20.** 101 **21.** 63 **22.** 161 **23.** 73 **24.** 152 **25.** 128 **26.** 399 **27.** 999 **28.** 897 **29.** 875 **30.** 969 **31.** 1,117 **32.** 1,337 **33.** 660 **34.** 1,161 **35.** 1,210 **36.** 1,763 **37.** 2,000 **38.** 2,644 **39.** 11,080 **40.** 15,154 **41.** 19,700 **42.** 50,260 **43.** 101,308 **44.** 108,301 **45.** 133,022 **46.** 994 **47.** 2,855 **48.** 12,787 **49.** 12,365 **50.** 84,780

PRACTICE PROBLEM SET 2C: SUBTRACTION OF WHOLE NUMBERS

Do the indicated subtraction.

1. $9 - 5$ **2.** $8 - 6$ **3.** $8 - 4$ **4.** $7 - 2$

5. $6 - 5$ **6.** $7 - 7$ **7.** $18 - 7$ **8.** $39 - 8$

9. $26 - 5$ **10.** $37 - 4$ **11.** $76 - 32$ **12.** $99 - 76$

13. $52 - 41$ **14.** $68 - 38$ **15.** $84 - 50$ **16.** $92 - 51$

17. $74 - 54$ **18.** $38 - 17$ **19.** $89 - 49$ **20.** $44 - 21$

21. $168 - 53$ **22.** $375 - 242$ **23.** $402 - 101$ **24.** $679 - 542$

25. $325 - 221$ **26.** $15 - 9$ **27.** $12 - 5$ **28.** $28 - 19$

29. $40 - 21$ **30.** $52 - 36$ **31.** $192 - 56$ **32.** $346 - 237$

33. $560 - 352$ **34.** $482 - 367$ **35.** $720 - 415$ **36.** $800 - 92$

37. $100 - 47$ **38.** $752 - 684$ **39.** $812 - 254$ **40.** $206 - 129$

41. $1{,}000 - 356$ **42.** $8{,}040 - 3{,}212$ **43.** $5{,}432 - 2{,}678$ **44.** $5{,}000 - 3{,}221$

45. $6{,}252 - 4{,}553$ **46.** $9{,}632 - 7{,}488$ **47.** $6{,}000 - 2{,}804$ **48.** $4{,}767 - 3{,}298$

49. $3{,}666 - 2{,}777$ **50.** $10{,}000 - 7{,}643$

ANSWERS

1. 4 **2.** 2 **3.** 4 **4.** 5 **5.** 1 **6.** 0 **7.** 11 **8.** 31 **9.** 21 **10.** 33 **11.** 44 **12.** 23 **13.** 11 **14.** 30 **15.** 34 **16.** 41 **17.** 20 **18.** 21 **19.** 40 **20.** 23 **21.** 115 **22.** 133 **23.** 301 **24.** 137 **25.** 104 **26.** 6 **27.** 7 **28.** 9 **29.** 19 **30.** 16 **31.** 136 **32.** 109 **33.** 208 **34.** 115 **35.** 305 **36.** 708 **37.** 53 **38.** 68 **39.** 558 **40.** 77 **41.** 644 **42.** 4,828 **43.** 2,754 **44.** 1,779 **45.** 1,699 **46.** 2,144 **47.** 3,196 **48.** 1,469 **49.** 889 **50.** 2,357

PRACTICE PROBLEM SET 3C: MULTIPLICATION OF WHOLE NUMBERS

Do the indicated multiplication.

1. $\begin{array}{r} 7 \\ \times\ 3 \\ \hline \end{array}$
2. $\begin{array}{r} 8 \\ \times\ 6 \\ \hline \end{array}$
3. $\begin{array}{r} 9 \\ \times\ 4 \\ \hline \end{array}$
4. $\begin{array}{r} 5 \\ \times\ 2 \\ \hline \end{array}$
5. $\begin{array}{r} 6 \\ \times\ 6 \\ \hline \end{array}$
6. $\begin{array}{r} 4 \\ \times\ 8 \\ \hline \end{array}$
7. $\begin{array}{r} 8 \\ \times\ 9 \\ \hline \end{array}$
8. $\begin{array}{r} 10 \\ \times\ 5 \\ \hline \end{array}$
9. $\begin{array}{r} 12 \\ \times\ 7 \\ \hline \end{array}$
10. $\begin{array}{r} 11 \\ \times\ 4 \\ \hline \end{array}$
11. $\begin{array}{r} 5 \\ \times\ 7 \\ \hline \end{array}$
12. $\begin{array}{r} 9 \\ \times\ 9 \\ \hline \end{array}$
13. $\begin{array}{r} 10 \\ \times\ 8 \\ \hline \end{array}$
14. $\begin{array}{r} 11 \\ \times\ 3 \\ \hline \end{array}$
15. $\begin{array}{r} 12 \\ \times\ 8 \\ \hline \end{array}$
16. $\begin{array}{r} 12 \\ \times\ 11 \\ \hline \end{array}$
17. $\begin{array}{r} 30 \\ \times\ 5 \\ \hline \end{array}$
18. $\begin{array}{r} 22 \\ \times\ 3 \\ \hline \end{array}$
19. $\begin{array}{r} 34 \\ \times\ 2 \\ \hline \end{array}$
20. $\begin{array}{r} 21 \\ \times\ 8 \\ \hline \end{array}$
21. $\begin{array}{r} 53 \\ \times\ 3 \\ \hline \end{array}$
22. $\begin{array}{r} 65 \\ \times\ 5 \\ \hline \end{array}$
23. $\begin{array}{r} 72 \\ \times\ 8 \\ \hline \end{array}$
24. $\begin{array}{r} 60 \\ \times\ 5 \\ \hline \end{array}$
25. $\begin{array}{r} 76 \\ \times\ 8 \\ \hline \end{array}$
26. $\begin{array}{r} 29 \\ \times\ 9 \\ \hline \end{array}$
27. $\begin{array}{r} 53 \\ \times\ 7 \\ \hline \end{array}$
28. $\begin{array}{r} 64 \\ \times\ 6 \\ \hline \end{array}$

PRACTICE PROBLEM SET 6C: SUBTRACTION OF FRACTIONS

Do the indicated subtraction. State each answer as a proper fraction or as a mixed number. Reduce all answers to simplest form.

1. $\frac{3}{5} - \frac{2}{5} =$ ______
2. $\frac{6}{7} - \frac{3}{7} =$ ______
3. $\frac{5}{8} - \frac{2}{8} =$ ______
4. $\frac{11}{12} - \frac{5}{12} =$ ______
5. $\frac{7}{10} - \frac{2}{10} =$ ______
6. $\frac{6}{9} - \frac{4}{9} =$ ______
7. $\frac{9}{14} - \frac{3}{14} =$ ______
8. $\frac{13}{15} - \frac{4}{15} =$ ______
9. $\frac{11}{16} - \frac{7}{16} =$ ______
10. $\frac{19}{20} - \frac{7}{20} =$ ______
11. $\frac{5}{6} - \frac{1}{6} =$ ______
12. $\frac{23}{30} - \frac{7}{30} =$ ______
13. $\frac{1}{2} - \frac{3}{8} =$ ______
14. $\frac{7}{9} - \frac{1}{3} =$ ______
15. $\frac{11}{24} - \frac{1}{3} =$ ______
16. $\frac{9}{20} - \frac{1}{4} =$ ______
17. $\frac{24}{25} - \frac{3}{5} =$ ______
18. $\frac{17}{30} - \frac{1}{6} =$ ______
19. $\frac{2}{3} - \frac{1}{15} =$ ______
20. $\frac{5}{6} - \frac{23}{42} =$ ______
21. $\frac{5}{7} - \frac{4}{35} =$ ______
22. $\frac{13}{18} - \frac{5}{9} =$ ______
23. $\frac{11}{12} - \frac{11}{24} =$ ______
24. $\frac{7}{10} - \frac{13}{40} =$ ______
25. $\frac{1}{2} - \frac{1}{3} =$ ______
26. $\frac{3}{4} - \frac{2}{5} =$ ______
27. $\frac{4}{5} - \frac{1}{2} =$ ______
28. $\frac{5}{6} - \frac{3}{7} =$ ______
29. $\frac{2}{3} - \frac{1}{4} =$ ______
30. $\frac{6}{7} - \frac{3}{4} =$ ______
31. $\frac{7}{11} - \frac{2}{5} =$ ______
32. $\frac{3}{7} - \frac{1}{6} =$ ______
33. $\frac{5}{3} - \frac{1}{5} =$ ______
34. $\frac{7}{5} - \frac{1}{6} =$ ______
35. $\frac{5}{7} - \frac{5}{8} =$ ______
36. $\frac{6}{11} - \frac{4}{9} =$ ______
37. $1 - \frac{1}{2} =$ ______
38. $1 - \frac{5}{6} =$ ______
39. $2 - \frac{3}{4} =$ ______
40. $2 - \frac{5}{8} =$ ______
41. $\frac{3}{4} - \frac{1}{6} =$ ______
42. $\frac{5}{6} - \frac{7}{10} =$ ______
43. $\frac{5}{9} - \frac{1}{2} =$ ______
44. $\frac{3}{10} - \frac{1}{14} =$ ______
45. $\frac{4}{15} - \frac{1}{6} =$ ______
46. $\frac{8}{21} - \frac{3}{14} =$ ______
47. $\frac{7}{22} - \frac{1}{6} =$ ______
48. $\frac{7}{12} - \frac{3}{8} =$ ______
49. $\frac{13}{24} - \frac{3}{16} =$ ______
50. $\frac{11}{6} - \frac{5}{9} =$ ______

ANSWERS

1. $\frac{1}{5}$ **2.** $\frac{3}{7}$ **3.** $\frac{3}{8}$ **4.** $\frac{1}{2}$ **5.** $\frac{1}{2}$ **6.** $\frac{2}{9}$ **7.** $\frac{3}{7}$ **8.** $\frac{3}{5}$ **9.** $\frac{1}{4}$ **10.** $\frac{3}{5}$ **11.** $\frac{2}{3}$
12. $\frac{8}{15}$ **13.** $\frac{1}{8}$ **14.** $\frac{4}{9}$ **15.** $\frac{1}{8}$ **16.** $\frac{1}{5}$ **17.** $\frac{9}{25}$ **18.** $\frac{2}{5}$ **19.** $\frac{3}{5}$ **20.** $\frac{2}{7}$ **21.** $\frac{3}{5}$
22. $\frac{1}{6}$ **23.** $\frac{11}{24}$ **24.** $\frac{3}{8}$ **25.** $\frac{1}{6}$ **26.** $\frac{7}{20}$ **27.** $\frac{3}{10}$ **28.** $\frac{17}{42}$ **29.** $\frac{5}{12}$ **30.** $\frac{3}{28}$
31. $\frac{13}{55}$ **32.** $\frac{11}{42}$ **33.** $1\frac{7}{15}$ **34.** $1\frac{7}{30}$ **35.** $\frac{5}{56}$ **36.** $\frac{10}{99}$ **37.** $\frac{1}{2}$ **38.** $\frac{1}{6}$
39. $1\frac{1}{4}$ **40.** $1\frac{3}{8}$ **41.** $\frac{7}{12}$ **42.** $\frac{2}{15}$ **43.** $\frac{1}{18}$ **44.** $\frac{8}{35}$ **45.** $\frac{1}{10}$ **46.** $\frac{1}{6}$
47. $\frac{5}{33}$ **48.** $\frac{5}{24}$ **49.** $\frac{17}{48}$ **50.** $1\frac{5}{18}$

PRACTICE PROBLEM SET 7C: MULTIPLICATION OF FRACTIONS

Do the indicated multiplication. State each answer as a proper fraction or as a mixed number. Reduce all answers to simplest form.

1. $\frac{2}{3} \times \frac{1}{5} =$ ______ **2.** $\frac{1}{6} \times \frac{5}{2} =$ ______ **3.** $\frac{4}{7} \times \frac{2}{3} =$ ______

4. $\frac{5}{8} \times \frac{3}{4} =$ ______ **5.** $\frac{6}{7} \times \frac{3}{5} =$ ______ **6.** $\frac{7}{9} \times \frac{2}{3} =$ ______

7. $\frac{9}{4} \times \frac{3}{7} =$ ______ **8.** $\frac{11}{12} \times \frac{7}{8} =$ ______ **9.** $\frac{11}{14} \times \frac{3}{4} =$ ______

10. $\frac{8}{7} \times \frac{5}{9} =$ ______ **11.** $\frac{5}{11} \times \frac{3}{7} =$ ______ **12.** $\frac{6}{13} \times \frac{2}{5} =$ ______

13. $\frac{1}{2} \times \frac{4}{9} =$ ______ **14.** $\frac{2}{5} \times \frac{10}{7} =$ ______ **15.** $\frac{5}{8} \times \frac{4}{11} =$ ______

16. $\frac{7}{12} \times \frac{6}{5} =$ ______ **17.** $\frac{11}{20} \times \frac{5}{6} =$ ______ **18.** $\frac{16}{5} \times \frac{1}{8} =$ ______

19. $5 \times \frac{3}{20} =$ ______ **20.** $11 \times \frac{2}{33} =$ ______ **21.** $15 \times \frac{3}{5} =$ ______

22. $6 \times \frac{5}{12} =$ ______ **23.** $\frac{5}{2} \times \frac{4}{15} =$ ______ **24.** $\frac{10}{13} \times \frac{26}{25} =$ ______

25. $\frac{7}{8} \times \frac{4}{21} =$ ______ **26.** $\frac{11}{20} \times \frac{10}{55} =$ ______ **27.** $\frac{10}{7} \times \frac{21}{5} =$ ______

28. $\frac{12}{11} \times \frac{77}{18} =$ ______ **29.** $\frac{15}{8} \times \frac{16}{45} =$ ______ **30.** $9 \times \frac{7}{18} =$ ______

31. $\frac{8}{15} \times \frac{9}{16} =$ ______ **32.** $\frac{9}{22} \times \frac{33}{27} =$ ______ **33.** $\frac{8}{6} \times \frac{3}{12} =$ ______

34. $\frac{18}{7} \times \frac{21}{36} =$ ______ **35.** $\frac{12}{25} \times \frac{10}{21} =$ ______ **36.** $\frac{16}{35} \times \frac{45}{32} =$ ______

37. $\frac{12}{14} \times \frac{15}{9} =$ ______ **38.** $\frac{28}{21} \times \frac{9}{16} =$ ______ **39.** $\frac{9}{25} \times \frac{50}{27} =$ ______

40. $\frac{20}{32} \times \frac{16}{5} =$ ______ **41.** $\frac{33}{64} \times \frac{8}{22} =$ ______ **42.** $\frac{10}{9} \times \frac{45}{18} =$ ______

43. $\frac{25}{9} \times \frac{27}{75} =$ ______ **44.** $\frac{7}{18} \times \frac{24}{49} =$ ______ **45.** $\frac{7}{4} \times \frac{16}{21} =$ ______

46. $\frac{14}{36} \times \frac{6}{21} =$ ______ **47.** $\frac{10}{9} \times \frac{81}{45} =$ ______ **48.** $\frac{24}{36} \times \frac{15}{4} =$ ______

49. $\frac{18}{5} \times \frac{20}{27} =$ ______ **50.** $\frac{24}{56} \times \frac{14}{9} =$ ______

ANSWERS

1. $\frac{2}{15}$ **2.** $\frac{5}{12}$ **3.** $\frac{8}{21}$ **4.** $\frac{15}{32}$ **5.** $\frac{18}{35}$ **6.** $\frac{14}{27}$ **7.** $\frac{27}{28}$ **8.** $\frac{77}{96}$ **9.** $\frac{33}{56}$ **10.** $\frac{40}{63}$
11. $\frac{15}{77}$ **12.** $\frac{12}{65}$ **13.** $\frac{2}{9}$ **14.** $\frac{4}{7}$ **15.** $\frac{5}{22}$ **16.** $\frac{7}{10}$ **17.** $\frac{11}{24}$ **18.** $\frac{2}{5}$ **19.** $\frac{3}{4}$
20. $\frac{2}{3}$ **21.** 9 **22.** $2\frac{1}{2}$ **23.** $\frac{2}{3}$ **24.** $\frac{4}{5}$ **25.** $\frac{1}{6}$ **26.** $\frac{1}{10}$ **27.** 6 **28.** $4\frac{2}{3}$ **29.** $\frac{2}{3}$
30. $3\frac{1}{2}$ **31.** $\frac{3}{10}$ **32.** $\frac{1}{2}$ **33.** $\frac{1}{3}$ **34.** $1\frac{1}{2}$ **35.** $\frac{8}{35}$ **36.** $\frac{9}{14}$ **37.** $1\frac{3}{7}$ **38.** $\frac{3}{4}$
39. $\frac{2}{3}$ **40.** 2 **41.** $\frac{3}{16}$ **42.** $2\frac{7}{9}$ **43.** 1 **44.** $\frac{4}{21}$ **45.** $1\frac{1}{3}$ **46.** $\frac{1}{9}$ **47.** 2
48. $2\frac{1}{2}$ **49.** $2\frac{2}{3}$ **50.** $\frac{2}{3}$

PRACTICE PROBLEM SET 8C: DIVISION OF FRACTIONS

Do the indicated division. State each answer as a proper fraction or as a mixed number. Reduce all answers to simplest form.

1. $\frac{2}{3} \div \frac{5}{2} =$ ______
2. $\frac{1}{7} \div \frac{3}{4} =$ ______
3. $\frac{1}{6} \div \frac{4}{7} =$ ______
4. $\frac{1}{2} \div \frac{5}{7} =$ ______
5. $\frac{3}{8} \div \frac{1}{2} =$ ______
6. $\frac{5}{9} \div \frac{2}{3} =$ ______
7. $\frac{7}{8} \div \frac{9}{4} =$ ______
8. $\frac{5}{12} \div \frac{5}{6} =$ ______
9. $\frac{4}{7} \div \frac{8}{9} =$ ______
10. $\frac{5}{12} \div \frac{15}{6} =$ ______
11. $\frac{3}{10} \div \frac{6}{5} =$ ______
12. $\frac{9}{14} \div \frac{3}{7} =$ ______
13. $\frac{7}{9} \div \frac{35}{18} =$ ______
14. $\frac{10}{12} \div \frac{5}{6} =$ ______
15. $\frac{5}{21} \div \frac{15}{14} =$ ______
16. $\frac{35}{24} \div \frac{5}{12} =$ ______
17. $\frac{7}{18} \div \frac{21}{36} =$ ______
18. $\frac{5}{18} \div \frac{5}{6} =$ ______
19. $\frac{9}{35} \div \frac{27}{14} =$ ______
20. $\frac{7}{13} \div \frac{21}{26} =$ ______
21. $\frac{5}{12} \div \frac{25}{48} =$ ______
22. $\frac{10}{11} \div \frac{50}{33} =$ ______
23. $\frac{15}{22} \div \frac{40}{33} =$ ______
24. $\frac{9}{14} \div \frac{27}{42} =$ ______
25. $\frac{32}{25} \div \frac{16}{35} =$ ______
26. $\frac{11}{9} \div \frac{44}{63} =$ ______
27. $\frac{18}{5} \div \frac{36}{45} =$ ______
28. $\frac{8}{24} \div \frac{4}{27} =$ ______
29. $\frac{14}{15} \div \frac{49}{30} =$ ______
30. $\frac{9}{15} \div \frac{21}{25} =$ ______
31. $\frac{19}{8} \div \frac{38}{40} =$ ______
32. $\frac{12}{17} \div \frac{24}{68} =$ ______
33. $\frac{26}{15} \div \frac{39}{40} =$ ______
34. $\frac{30}{19} \div \frac{45}{38} =$ ______
35. $\frac{28}{45} \div \frac{21}{18} =$ ______
36. $\frac{32}{21} \div \frac{48}{63} =$ ______
37. $\frac{100}{33} \div \frac{25}{77} =$ ______
38. $\frac{84}{35} \div \frac{36}{63} =$ ______
39. $\frac{99}{50} \div \frac{44}{25} =$ ______
40. $\frac{75}{24} \div \frac{15}{16} =$ ______
41. $8 \div \frac{4}{5} =$ ______
42. $12 \div \frac{4}{3} =$ ______
43. $\frac{1}{2} \div 4 =$ ______
44. $\frac{2}{3} \div 6 =$ ______
45. $\frac{16}{21} \div \frac{80}{49} =$ ______
46. $\frac{63}{34} \div \frac{54}{51} =$ ______
47. $\frac{5}{2} \div 10 =$ ______
48. $\frac{8}{15} \div 6 =$ ______
49. $12 \div \frac{8}{9} =$ ______
50. $16 \div \frac{10}{3} =$ ______

ANSWERS

1. $\frac{4}{15}$ 2. $\frac{4}{21}$ 3. $\frac{7}{24}$ 4. $\frac{7}{10}$ 5. $\frac{3}{4}$ 6. $\frac{5}{6}$ 7. $\frac{7}{18}$ 8. $\frac{1}{2}$ 9. $\frac{9}{14}$ 10. $\frac{1}{6}$
11. $\frac{1}{4}$ 12. $1\frac{1}{2}$ 13. $\frac{2}{5}$ 14. 1 15. $\frac{2}{9}$ 16. $3\frac{1}{2}$ 17. $\frac{2}{3}$ 18. $\frac{1}{3}$ 19. $\frac{2}{15}$ 20. $\frac{2}{3}$
21. $\frac{4}{5}$ 22. $\frac{3}{5}$ 23. $\frac{9}{16}$ 24. 1 25. $2\frac{4}{5}$ 26. $1\frac{3}{4}$ 27. $4\frac{1}{2}$ 28. $2\frac{1}{4}$ 29. $\frac{4}{7}$
30. $\frac{5}{7}$ 31. $2\frac{1}{2}$ 32. 2 33. $1\frac{7}{9}$ 34. $1\frac{1}{3}$ 35. $\frac{8}{15}$ 36. 2 37. $9\frac{1}{3}$ 38. $4\frac{1}{5}$
39. $1\frac{1}{8}$ 40. $3\frac{1}{3}$ 41. 10 42. 9 43. $\frac{1}{8}$ 44. $\frac{1}{9}$ 45. $\frac{7}{15}$ 46. $1\frac{3}{4}$ 47. $\frac{1}{4}$
48. $\frac{4}{45}$ 49. $13\frac{1}{2}$ 50. $4\frac{4}{5}$

PRACTICE PROBLEM SET 9C: ADDITION OF DECIMALS

Do the indicated addition.

1. 9.5 + 8.4 = ______
2. 6.3 + 5.6 = ______
3. 12.5 + 9.2 = ______
4. 20.4 + 5.4 = ______
5. 16.5 + 14.7 = ______
6. 28.6 + 35.9 = ______
7. 43.8 + 21.5 = ______
8. 31.5 + 19.5 = ______
9. 10.36 + 15.42 = ______
10. 14.15 + 27.04 = ______
11. 37.61 + 48.35 = ______
12. 50.82 + 31.07 = ______
13. 45.75 + 26.46 = ______
14. 93.58 + 35.87 = ______
15. 24.55 + 39.45 = ______
16. 26.92 + 67.58 = ______
17. 34.63 + 78.39 = ______
18. 14.86 + 55.68 = ______
19. 15.09 + 51.95 = ______
20. 65.33 + 32.79 = ______
21. 12.8 + 25.36 = ______
22. 27.96 + 36.4 = ______
23. 75.8 + 19.55 = ______
24. 43.61 + 30.7 = ______
25. 9.8 + 25.27 + 14.6 = ______
26. 12.54 + 32.72 + 8.5 = ______
27. 74.63 + 18.2 + 31.59 = ______
28. 0.25 + 0.58 + 7.17 = ______
29. 0.75 + 15.48 + 28.63 = ______
30. 19.5 + 12.64 + 0.86 = ______
31. 5.675 + 10.432 = ______
32. 9.068 + 26.933 = ______
33. 14.665 + 32.549 = ______
34. 50.173 + 29.828 = ______
35. 15.682 + 4.235 + 19.176 = ______
36. 6.063 + 12.542 + 8.395 = ______
37. 24.354 + 40.293 + 35.317 = ______
38. 3.496 + 25.387 + 12.207 = ______
39. 35.205 + 0.859 + 7.152 = ______
40. 21.306 + 0.845 + 0.959 = ______
41. 9.56 + 8.059 + 10.886 = ______
42. 25.7 + 0.64 + 13.536 = ______
43. 42.06 + 85.92 + 22.025 = ______

44. 77.516 + 24.8 + 48.254 = ____________________

45. 35.64 + 19.9 + 30.607 = ____________________

46. 62.578 + 0.253 + 9.17 = ____________________

47. 0.87 + 14 + 5.84 = ____________________

48. 81 + 18.504 + 10.496 = ____________________

49. 60.58 + 6.0025 + 12.4275 = ____________________

50. 7.2356 + 16 + 51.98 = ____________________

ANSWERS

1. 17.9 **2.** 11.9 **3.** 21.7 **4.** 25.8 **5.** 31.2 **6.** 64.5 **7.** 65.3 **8.** 51
9. 25.78 **10.** 41.19 **11.** 85.96 **12.** 81.89 **13.** 72.21 **14.** 129.45 **15.** 64
16. 94.5 **17.** 113.02 **18.** 70.54 **19.** 67.04 **20.** 98.12 **21.** 38.16 **22.** 64.36
23. 95.35 **24.** 74.31 **25.** 49.67 **26.** 53.76 **27.** 124.42 **28.** 8 **29.** 44.86
30. 33 **31.** 16.107 **32.** 36.001 **33.** 47 214 **34.** 80.001 **35.** 39.093 **36.** 27
37. 99.964 **38.** 41.09 **39.** 43.216 **40.** 23.11 **41.** 28.505 **42.** 39.876
43. 150.005 **44.** 150.57 **45.** 86.147 **46.** 72.001 **47.** 20.71 **48.** 110,036
49. 79.01 **50.** 76.2156

PRACTICE PROBLEM SET 10C: SUBTRACTION OF DECIMALS

Do the indicated subtraction.

1. 10.9 − 6.4 = ________
2. 12.8 − 10.2 = ________
3. 14.7 − 9.2 = ________
4. 25.5 − 19.5 = ________
5. 50.5 − 22.1 = ________
6. 46.6 − 38.3 = ________
7. 14.2 − 9.5 = ________
8. 21.1 − 12.6 = ________
9. 30.3 − 15.8 = ________
10. 45.7 − 27.9 = ________
11. 62.5 − 35.6 = ________
12. 70.2 − 50.8 = ________
13. 16.45 − 9.23 = ________
14. 23.47 − 19.45 = ________
15. 31.65 − 20.41 = ________
16. 104.26 − 77.13 = ________
17. 12.83 − 6.57 = ________
18. 34.71 − 26.58 = ________
19. 40.57 − 21.49 = ________
20. 103.22 − 27.18 = ________
21. 15.25 − 10.56 = ________
22. 17.31 − 9.75 = ________
23. 30.05 − 17.66 = ________
24. 72.67 − 38.78 = ________
25. 84.31 − 55.62 = ________
26. 114.25 − 56.49 = ________
27. 20.01 − 13.23 = ________
28. 40.86 − 16.97 = ________
29. 50.62 − 18.8 = ________
30. 22.75 − 13.9 = ________
31. 62.08 − 34.5 = ________
32. 38.21 − 19.8 = ________
33. 28.2 − 15.46 = ________
34. 46.6 − 23.75 = ________
35. 55.5 − 41.82 = ________
36. 42.3 − 27.56 = ________
37. 71.7 − 52.91 = ________
38. 63.4 − 29.77 = ________
39. 12.351 − 8.241 = ________
40. 54.687 − 9.281 = ________
41. 36.204 − 18.153 = ________
42. 214.752 − 106.851 = ________
43. 18.003 − 12.237 = ________
44. 36.123 − 16.345 = ________
45. 0.000529 − 0.000376 = ________
46. 0.006283 − 0.005385 = ________
47. 0.000058 − 0.000039 = ________
48. 0.003002 − 0.002564 = ________

49. 9.000205 − 0.006347 = ______________

50. 4.01004 − 0.07335 = ______________

ANSWERS

1. 4.5 **2.** 2.6 **3.** 5.5 **4.** 6 **5.** 28.4 **6.** 8.3 **7.** 4.7 **8.** 8.5 **9.** 14.5 **10.** 17.8 **11.** 26.9 **12.** 19.4 **13.** 7.22 **14.** 4.02 **15.** 11.24 **16.** 27.13 **17.** 6.26 **18.** 8.13 **19.** 19.08 **20.** 76.04 **21.** 4.69 **22.** 7.56 **23.** 12.39 **24.** 33.89 **25.** 28.69 **26.** 57.76 **27.** 6.73 **28.** 23.89 **29.** 31.82 **30.** 8.85 **31.** 27.58 **32.** 18.41 **33.** 12.74 **34.** 22.85 **35.** 13.68 **36.** 14.74 **37.** 18.79 **38.** 33.63 **39.** 4.11 **40.** 45.406 **41.** 18.051 **42.** 107.901 **43.** 5.766 **44.** 19.778 **45.** 0.000153 **46.** 0.000898 **47.** 0.000019 **48.** 0.000438 **49.** 8.993858 **50.** 3.93669

PRACTICE PROBLEM SET 11C: MULTIPLICATION OF DECIMALS

Do the indicated multiplication.

1. $5.2 \times 6 =$ ____________
2. $3.5 \times 5 =$ ____________
3. $8.4 \times 7 =$ ____________
4. $6.3 \times 9 =$ ____________
5. $7.6 \times 4 =$ ____________
6. $10.7 \times 3 =$ ____________
7. $3.42 \times 8 =$ ____________
8. $12.54 \times 6 =$ ____________
9. $6.89 \times 9 =$ ____________
10. $12.56 \times 3 =$ ____________
11. $14.257 \times 4 =$ ____________
12. $25.647 \times 5 =$ ____________
13. $7.8 \times 0.6 =$ ____________
14. $9.2 \times 0.7 =$ ____________
15. $8.8 \times 4 =$ ____________
16. $51.3 \times 0.3 =$ ____________
17. $85.4 \times 0.8 =$ ____________
18. $102.5 \times 0.9 =$ ____________
19. $243.2 \times 0.7 =$ ____________
20. $586.4 \times 0.5 =$ ____________
21. $8.5 \times 4.3 =$ ____________
22. $9.42 \times 3.6 =$ ____________
23. $6.98 \times 2.3 =$ ____________
24. $7.03 \times 5.6 =$ ____________
25. $25.07 \times 4.5 =$ ____________
26. $38.2 \times 0.82 =$ ____________
27. $6.7 \times 0.66 =$ ____________
28. $40.9 \times 0.65 =$ ____________
29. $82.4 \times 0.37 =$ ____________
30. $26.6 \times 0.03 =$ ____________
31. $1.53 \times 0.08 =$ ____________
32. $5.07 \times 0.05 =$ ____________
33. $92.6 \times 0.02 =$ ____________
34. $4.04 \times 0.003 =$ ____________
35. $6.5 \times 0.012 =$ ____________
36. $2.17 \times 0.004 =$ ____________
37. $15.7 \times 0.005 =$ ____________
38. $75.8 \times 0.057 =$ ____________
39. $54.03 \times 6.07 =$ ____________
40. $36.6 \times 0.125 =$ ____________
41. $0.642 \times 0.38 =$ ____________
42. $0.705 \times 0.52 =$ ____________
43. $0.003 \times 0.065 =$ ____________
44. $0.058 \times 0.0047 =$ ____________
45. $0.0028 \times 1.005 =$ ____________
46. $5.002 \times 0.055 =$ ____________
47. $5.06 \times 100 =$ ____________
48. $3.008 \times 1{,}000 =$ ____________
49. $0.0087 \times 1{,}000 =$ ____________
50. $0.0064 \times 10 =$ ____________

ANSWERS

1. 31.2 **2.** 17.5 **3.** 58.8 **4.** 56.7 **5.** 30.4 **6.** 32.1 **7.** 27.36 **8.** 75.24 **9.** 62.01 **10.** 37.68 **11.** 57.028 **12.** 128.235 **13.** 4.68 **14.** 6.44 **15.** 35.2 **16.** 15.39 **17.** 68.32 **18.** 92.25 **19.** 170.24 **20.** 293.2 **21.** 36.55 **22.** 33.912 **23.** 16.054 **24.** 39.368 **25.** 112.815 **26.** 31.324 **27.** 4.422 **28.** 26.585 **29.** 30.488 **30.** 0.798 **31.** 0.1224 **32.** 0.2535 **33.** 1.852 **34.** 0.01212 **35.** 0.078 **36.** 0.00868 **37.** 0.0785 **38.** 4.3206 **39.** 327.9621 **40.** 4.575 **41.** 0.24396 **42.** 0.3666 **43.** 0.000195 **44.** 0.0002726 **45.** 0.002814 **46.** 0.27511 **47.** 506 **48.** 3,008 **49.** 8.7 **50.** 0.064

PRACTICE PROBLEM SET 12C: DIVISION OF DECIMALS

Do the indicated division.

1. $4\overline{)16.8}$ **2.** $3\overline{)40.5}$ **3.** $7\overline{)26.6}$

4. $8\overline{)97.6}$ **5.** $12\overline{)68.4}$ **6.** $15\overline{)109.5}$

7. $9\overline{)103.14}$ **8.** $6\overline{)107.04}$ **9.** $15\overline{)92.4}$

10. $18\overline{)74.7}$ **11.** $5.2\overline{)95.68}$ **12.** $6.3\overline{)130.41}$

13. $9.4\overline{)118.44}$ **14.** $7.5\overline{)182.25}$ **15.** $4.8\overline{)177.12}$

16. $8.7\overline{)394.11}$ **17.** $5.5\overline{)143}$ **18.** $4.8\overline{)312}$

19. $7.4\overline{)333}$ **20.** $6.5\overline{)468}$ **21.** $5.02\overline{)17.57}$

22. $4.36\overline{)37.496}$ **23.** $6.28\overline{)33.912}$ **24.** $7.82\overline{)52.394}$

25. $0.46\overline{)3.6984}$ **26.** $0.83\overline{)1.7098}$ **27.** $0.79\overline{)0.8532}$

28. $0.28\overline{)0.854}$ **29.** $5.02\overline{)81.826}$ **30.** $0.304\overline{)2.888}$

31. $1.52\overline{)9.576}$ **32.** $0.612\overline{)2.754}$ **33.** $40.3\overline{)22.568}$

34. $72.2\overline{)23.104}$ **35.** $6.41\overline{)6.2177}$ **36.** $2.55\overline{)1.683}$

37. $6.3\overline{)0.2835}$ **38.** $0.57\overline{)0.02394}$ **39.** $24.3\overline{)2.1627}$

40. $6.5\overline{)0.2314}$ **41.** $0.256\overline{)19.2}$ **42.** $0.481\overline{)1,178.45}$

43. $0.056\overline{)46.2}$ **44.** $3.27\overline{)3,073.8}$ **45.** $0.082\overline{)0.01681}$

46. $0.606\overline{)0.02727}$ **47.** $0.50\overline{)0.402}$ **48.** $0.0012\overline{)0.04032}$

49. $0.0804\overline{)0.06834}$ **50.** $0.3005\overline{)28.848}$

ANSWERS

1. 4.2 **2.** 13.5 **3.** 3.8 **4.** 12.2 **5.** 5.7 **6.** 7.3 **7.** 11.46 **8.** 17.84 **9.** 6.16 **10.** 4.15 **11.** 18.4 **12.** 20.7 **13.** 12.6 **14.** 24.3 **15.** 36.9 **16.** 45.3 **17.** 26 **18.** 65 **19.** 45 **20.** 72 **21.** 3.5 **22.** 8.6 **23.** 5.4 **24.** 6.7 **25.** 8.04 **26.** 2.06 **27.** 1.08 **28.** 3.05 **29.** 16.3 **30.** 9.5 **31.** 6.3 **32.** 4.5 **33.** 0.56 **34.** 0.32 **35.** 0.97 **36.** 0.66 **37.** 0.045 **38.** 0.042 **39.** 0.089 **40.** 0.0356 **41.** 75 **42.** 2,450 **43.** 825 **44.** 940 **45.** 0.205 **46.** 0.045 **47.** 0.804 **48.** 33.6 **49.** 0.85 **50.** 96

Appendix II

ARITHMETIC IN A NUTSHELL

CONCEPT	EXPLANATION	EXAMPLE

CHAPTER 1: WHOLE NUMBERS

CONCEPT	EXPLANATION	EXAMPLE
Reading Numbers	Starting from the right, group digits in sets of three. Last set may contain 1, 2, or 3 digits. Commas separate group of digits. Word "and" is not used. Numbers 21–99 (excluding multiples of 10) are written with hyphens.	50,232 fifty thousand, two hundred thirty-two
Rounding Off Numbers	Find digit in the place you are rounding. Look at digit to the right. (a) If less than 5, round down, replacing it and all digits to right with zeros. (b) If greater than 5, round up; add 1 to the digit you are rounding and replace all digits to right with zeros.	⑤37 to nearest hundred; 3 is less than 5; so 537 rounds down to 500 ⑥798 to nearest thousand; 7 is greater than 5; so 6,798 rounds up to 7,000
Addition of Whole Numbers	Write numbers in column with units below units, tens below tens, etc. Starting at right, add columns.	$\begin{array}{r} 725 \\ 3{,}042 \\ +\ 531 \\ \hline 4{,}298 \end{array}$ $\begin{array}{l} \left.\begin{array}{l} \\ \\ \\ \end{array}\right\} \text{addends} \\ \text{sum} \end{array}$
Subtraction of Whole Numbers	Write numbers in column with units below units, tens below tens, etc. Starting at the right, subtract bottom number from top number.	$\begin{array}{rl} 5{,}046 & \text{minuend} \\ -\ 2{,}329 & \text{subtrahend} \\ \hline 2{,}717 & \text{difference} \end{array}$
Multiplication of a Number by a Single-Digit Number	Multiply each digit of the number to be multiplied by the single-digit number.	$\begin{array}{rl} 312 & \text{multiplicand} \\ \times\ \ 6 & \text{multiplier} \\ \hline 1872 & \text{product} \end{array}$

CONCEPT	EXPLANATION	EXAMPLE
Multiplication of a Number by a Multiple-Digit Number	Multiply the number to be multiplied by the units digit, tens digit, etc. Add the partial products.	542 ×56 3252 } partial 2710 } products 30352
Multiplication of a Number by 10, 100, 1000, etc.	Add number of zeros to the number being multiplied.	$24 \times 10 = 240$ add one zero $24 \times 100 = 2400$ add two zeros
Division of a Number by a Single-Digit Number	7 cannot be divided into 2. 7 can be divided 4 times into 29. Place 4 above 9, multiply 7 by 4 and place partial product, 28, under 29. Subtract 28 from 29 (result should be less than divisor) and bring down next digit in dividend. Process is continued: 7 can be divided 2 times into 14.	↙quotient 42 divisor → 7)294 ← dividend 28 14 14 0
Division of a Number by a Multiple-Digit Number	25 cannot be divided into 4. 25 can be divided 1 time into 42. Place 1 above 2, multiply 25 by 1, and place partial product 25 under 42. Subtract 25 from 42 and bring down next digit in dividend. Continue process: 25 can be divided 7 times into 175.	17 25)425 25 175 175 0
Division of a Number by Zero	Cannot be done	$\frac{0}{0}$, $\frac{5}{0}$ } no solution
Exponent	Tells us how many times a number, called the base, is multiplied by itself. Exponential expression is called a *power*.	$4^3 = 4 \times 4 \times 4$ $= 64$
Order of Operations	1. Powers of numbers are evaluated in any order. 2. Multiplications and divisions are done in order from *left* to *right*. 3. Additions and subtractions are done in order from *left to right*.	$18 + 6 \div 3 - 2^2 + 5 \times 2 =$ $18 + 6 \div 3 - 4 + 5 \times 2 =$ $18 + 2 - 4 + 10 =$ $20 - 4 + 10 =$ $16 + 10 =$ 26

CHAPTER 2: FRACTIONS

Equal Fractions	$\frac{a}{b} = \frac{c}{d}$ when $a \cdot d = b \cdot c$	$\frac{1}{2} = \frac{3}{6}$ because $1 \cdot 6 = 2 \cdot 3$ $6 = 6$
Unequal Fractions	1. $\frac{a}{b} > \frac{c}{d}$ when $a \cdot d > b \cdot c$	1. $\frac{1}{2} > \frac{2}{5}$ because $1 \cdot 5 > 2 \cdot 2$ $5 > 4$

CONCEPT	EXPLANATION	EXAMPLE
	2. $\frac{a}{b} < \frac{c}{d}$ when $a \cdot d < b \cdot c$	2. $\frac{1}{3} < \frac{5}{7}$ because $1 \cdot 7 < 3 \cdot 5$; $7 < 15$
Types of Fractions	*Proper Fraction:* numerator less than denominator.	$\frac{4}{7}$
	Improper Fraction: numerator greater than or equal to denominator.	$\frac{9}{4}$
	Mixed Number: whole number added to a proper fraction.	$5\frac{1}{2}$
Converting a Mixed Number to an Improper Fraction	$\dfrac{\left(\textit{Whole Number} \times \textit{Denominator of Fraction}\right) + \textit{Numerator of Fraction}}{\textit{Denominator of Fraction}}$	$5\frac{1}{2} = \frac{(5 \times 2) + 1}{2} = \frac{10+1}{2} = \frac{11}{2}$
Converting an Improper Fraction to a Mixed Number	Divide denominator into numerator. Place remainder over denominator.	$\frac{11}{2} = 5\frac{1}{2}$
Greatest Common Divisor (GCD)	1. Write each number as a product of its prime factors.	GCD of 30 and 50 1. $30 = 2 \times 3 \times 5$ $50 = 2 \times 5 \times 5$
	2. List all prime factors *common* to the numbers.	2. 2 and 5
	3. Determine the *least* number of times each of the common factors is used in any one of the numbers.	3. 2, 5
	4. The GCD is the product of all the factors in step 3.	4. GCD $= 2 \times 5 = 10$
Reducing Fractions to Simplest or Lowest Terms	Divide both the numerator and denominator of the fraction by its greatest common divisor.	$\frac{25}{35} = \frac{5}{7}$ Both 25 and 35 were divided by 5
Equivalent Fractions	Are formed by either multiplying or dividing both the numerator and denominator of the fraction by the same nonzero number.	$\frac{21}{28} = \frac{3}{4}$ Both 21 and 28 were divided by 7 $\frac{4}{5} = \frac{16}{20}$ Both 4 and 5 were multiplied by 4
Least Common Denominator (LCD)	1. Write each denominator as a product of its prime factors.	LCD of 12, 18 1. $12 = 2 \times 2 \times 3$ $18 = 2 \times 3 \times 3$
	2. List each of the different prime factors.	2. 2, 3
	3. Determine the greatest number of times each of the prime factors is used in any of the denominators.	3. 2×2, 3×3

CONCEPT	EXPLANATION	EXAMPLE
	4. The LCD is the product of all the factors in step 3.	4. LCD = $2 \times 2 \times 3 \times 3 = 36$
Addition and Subtraction of Fractions	With common denominators: add (or subtract) the numerators; place sum (or difference) over common denominator; reduce to lowest terms.	$\frac{2}{5} + \frac{1}{5} = \frac{2+1}{5} = \frac{3}{5}$ $\frac{5}{9} - \frac{2}{9} = \frac{5-2}{9} = \frac{3}{9} = \frac{1}{3}$
	With unlike denominators: find the least common denominator; convert all fractions to fractions with LCD; add (or subtract) fractions.	$\frac{1}{4} + \frac{2}{3} = \frac{3}{12} + \frac{8}{12} = \frac{11}{12}$ $\frac{5}{6} - \frac{3}{8} = \frac{20}{24} - \frac{9}{24} = \frac{11}{24}$
Addition of Mixed Numbers	Add whole numbers. Add fractions, simplify, and reduce to lowest terms.	$5\frac{1}{2} = 5\frac{2}{4}$ $+\ 3\frac{1}{4} = 3\frac{1}{4}$ $8\frac{3}{4}$
Subtraction of Mixed Numbers	Subtract fractions; you may have to borrow from whole numbers. Subtract whole numbers. Simplify and reduce to lowest terms.	$3\frac{1}{5} = 3\frac{3}{15} = 2\frac{18}{15}$ $-\ 1\frac{2}{3} = 1\frac{10}{15} = 1\frac{10}{15}$ $1\frac{8}{15}$
Multiplication of Fractions	"Cancel"; multiply numerators; multiply denominators; reduce product to lowest terms if necessary.	$\frac{\not{5}^{1}}{\not{3}_{1}} \times \frac{\not{10}^{2}}{\not{9}_{3}} = \frac{1 \times 2}{1 \times 3} = \frac{2}{3}$
Multiplication with Mixed Numbers	Convert mixed numbers to improper fractions; and multiply fractions.	$3\frac{1}{3} \times 4\frac{1}{2}$ $\frac{\not{10}^{5}}{\not{3}_{1}} \times \frac{\not{9}^{3}}{\not{2}_{1}} = 15$
Division of Fractions	Invert divisor; multiply fractions.	$\frac{3}{5} \div \frac{27}{35} = \frac{\not{3}^{1}}{\not{5}_{1}} \times \frac{\not{35}^{7}}{\not{27}_{9}} = \frac{7}{9}$
Division with Mixed Numbers	Convert mixed numbers to improper fractions; invert divisor; multiply fractions.	$3\frac{3}{5} \div 2\frac{1}{4} =$ $\frac{18}{5} \div \frac{9}{2} = \frac{\not{18}^{2}}{5} \times \frac{2}{\not{9}_{1}} = \frac{4}{5}$
Ratio	Quotient of 2 numbers.	$\frac{5}{8}$ or 5:8
Proportion	Statement that 2 ratios are equal.	$\frac{3}{7} = \frac{9}{21}$ or 3:7 = 9:21
Terms of Proportion	*Extremes:* first and last terms.	3:7 = 9:21
	Means: second and third terms.	Extremes: 3 and 21 Means: 7 and 9

CONCEPT	EXPLANATION	EXAMPLE
Solving Proportions	Product of the means equals the product of the extremes.	$\frac{3}{x} = \frac{9}{21}$ $9 \cdot x = 3 \cdot 21$ $9x = 63$ $x = 7$

CHAPTER 3: DECIMALS

CONCEPT	EXPLANATION	EXAMPLE
Reading Decimal Numbers	Word "and" represents decimal point. "And" is not used when there aren't any digits to the left of the decimal point.	7.68 - Seven and sixty-eight hundredths. 0.603 - Six hundred three thousandths.
Types of Decimal Numbers	*Terminating:* digit zero repeats indefinitely.	0.35
	Nonterminating: digit, or group of digits, repeats indefinitely.	$0.2323\ldots = 0.\overline{23}$
Converting Fractions to Decimals	Divide numerator by denominator until remainder becomes 0 or repeats.	$\frac{3}{4} = 0.75$ $\frac{7}{11} = 0.6363\ldots = 0.\overline{63}$
Converting Decimals to Fractions	Place decimal number without the decimal point in numerator of fraction with a denominator of 1 followed by as many zeros as there are digits after the decimal point. Reduce to lowest terms.	$0.425 = \frac{425}{1{,}000} = \frac{17}{40}$
Addition of Decimal Numbers	Write numbers with decimal points in same column. Starting at right, add columns. Only like amounts can be added.	$\begin{array}{r} 325.24 \\ 17.603 \\ +\ 7.9 \\ \hline 350.743 \end{array}$
Subtraction of Decimal Numbers	Write numbers with decimal points in same column. Starting at right, subtract bottom number from top number	$\begin{array}{r} 407.850 \\ -\ 256.763 \\ \hline 151.087 \end{array}$
Multiplication of Decimal Numbers	Multiply the numbers. Count off, right to left, number of decimal places equal to the sum of decimal places in the numbers.	$\begin{array}{rl} 4.65 & \text{2 decimal places} \\ \times\ 3.5 & \text{1 decimal place} \\ \hline 2\ 325 & \\ 13\ 95 & \\ \hline 16.275 & 2 + 1 = 3 \text{ decimal places} \end{array}$
Multiplication of a Decimal Number by 10, 100, 1000, etc.	Move decimal point to the right in number being multiplied same number of places as zeros in multiplier.	$6.75 \times 100 = 675$ Decimal point moves 2 places to right.

CONCEPT	EXPLANATION	EXAMPLE
Multiplication of a Decimal Number by 0.1, 0.01, 0.001, etc.	Move decimal point to the left in number being multiplied same number of places as there are digits to the right of the decimal point in the multiplier.	$6.42 \times 0.001 = 0.00642$ Decimal point moves 3 places to the left
Division of a Decimal Number by a Whole Number	Place decimal point in quotient above decimal point in dividend.	$\begin{array}{r} 0.45 \\ 15\overline{)6.75} \\ \underline{6.0} \\ 75 \\ \underline{75} \\ 0 \end{array}$
Division of a Decimal Number by a Decimal Number	Move decimal point same number of places in divisor and dividend so the divisor becomes a whole number.	$\begin{array}{r} 0.056 \\ 3.4.\overline{)0.1.904} \\ \underline{1\ 70} \\ 204 \\ \underline{204} \\ 0 \end{array}$

CHAPTER 4: PERCENTAGE

CONCEPT	EXPLANATION	EXAMPLE
Converting Percents to Decimals	Drop percent sign and move decimal point 2 places to the left.	$45\% = 0.45$ $7.6\% = 0.076$ $300\% = 3$
Converting Percents to Fractions	Drop percent sign, place number over 100, and reduce to lowest terms.	$35\% = \frac{35}{100} = \frac{7}{20}$ $2\frac{1}{2}\% = \frac{2\frac{1}{2}}{100} = \frac{\frac{5}{2}}{100} = \frac{5}{200} = \frac{1}{40}$
Converting Decimals to Percents	Move decimal point 2 places to the right and add the percent sign.	$0.27 = 27\%$ $0.068 = 6.8\%$ $3.5 = 350\%$
Converting Fractions to Percents	Change fraction to a decimal, move decimal point 2 places to the right and add the percent sign.	$\frac{9}{20} = 0.45 = 45\%$ $\frac{1}{8} = 0.125 = 12.5\%$
Finding the Percentage	Percentage = Rate × Base $P = R \times B = RB$	Find P when $R = 15\%$ and $B = 300$: $P = 0.15 \times 300$ $= 45$
Finding the Base	$\text{Base} = \frac{\text{Percentage}}{\text{Rate}}$ $B = \frac{P}{R}$	Find B when $P = 45$ and $R = 15\%$: $B = \frac{45}{0.15}$ $= 300$
Finding the Rate	$\text{Rate} = \frac{\text{Percentage}}{\text{Base}}$ $R = \frac{P}{B}$	Find R when $P = 45$ and $B = 300$: $R = \frac{45}{300} = 0.15$ $= 15\%$

CONCEPT	EXPLANATION	EXAMPLE
Finding Percent of Increase	$\% \text{ of Increase} = \dfrac{\text{Amount of Increase}}{\text{Original Amount}}$	Class enrollment has increased from 30 to 36 students: Amount of Increase: $36 - 30 = 6$ $\% \text{ of increase} = \frac{6}{30} = \frac{1}{5}$ $= 20\%$
Finding Percent of Decrease	$\% \text{ of Decrease} = \dfrac{\text{Amount of Decrease}}{\text{Original Amount}}$	Salesman's weekly fuel consumption has decreased from 50 to 45 gallons: Amount of Decrease: $50 - 45 = 5$ $\% \text{ decrease} = \frac{5}{50} = \frac{1}{10}$ $= 10\%$
Simple Interest	Interest = Principal × Rate × Time $I = PRT$	Find interest on loan of $2,500 at 12% for 2 years: $I = \$2{,}500 \times 0.12 \times 2$ $= \$600$
Finding Interest Using Bankers' Rule	$\text{Time} = \dfrac{\text{Exact Number of Days}}{360 \text{ Days}}$	Find interest on loan of $5,000 at 14% from March 1 to May 30. $150 - 60 = 90$ days $I = \$5{,}000 \times \frac{14}{100} \times \frac{90}{360}$ $= \$175$
Finding Amount or Maturity Value	Amount = Principal + Interest $A = P + I$ or $A = P + PRT$	$6,000 invested at 14.5% for 6 years: $A = \$6{,}000 + \$6{,}000\ (0.145)\ (6)$ $= \$6{,}000 + \$5{,}220$ $= \$11{,}220$
Bank Discount (Interest on a Simple Discount Note)	Bank Discount = Maturity Value × Rate × Time $D = M \times R \times T = MRT$	$4,000 simple discount note at 15% for 120 days. Find D: $D = \$4{,}000 \times \frac{15}{100} \times \frac{120}{360}$ $= \$200$
Proceeds	Proceeds = Maturity Value − Bank Discount $S = M - D$ or $S = M - MRT$	$4,500 simple discount note at 16% for 90 days. Find S: $S = \$4{,}500 - \left(4{,}500 \times \frac{16}{100} \times \frac{90}{360}\right)$ $= \$4{,}500 - \$180$ $= \$4{,}320$

CONCEPT	EXPLANATION	EXAMPLE

CHAPTER 5: GEOMETRY

Perimeter (P)
Area (A)

Rectangle: $P = 2l + 2w$
$A = lw$

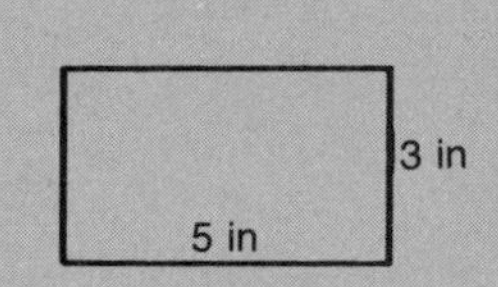

$P = 2(5 \text{ in}) + 2(3 \text{ in})$
$= 10 \text{ in} + 6 \text{ in}$
$= 16 \text{ in}$
$A = (5 \text{ in})(3 \text{ in})$
$= 15 \text{ sq in}$

Square: $P = 4s$
$A = s \times s = s^2$

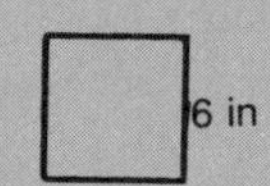

$P = 4(6 \text{ in}) = 24 \text{ in}$
$A = (6 \text{ in})^2 = 36 \text{ sq in}$

Triangle: $P = a + b + c$
$A = \frac{1}{2} bh$

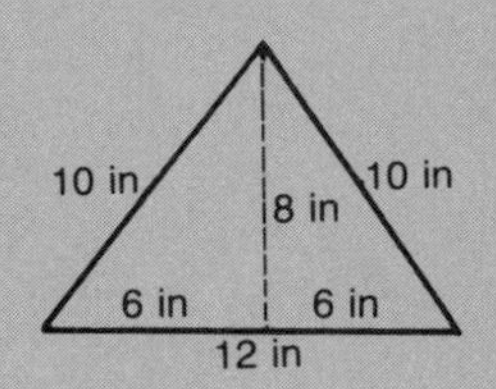

$P = 10 \text{ in} + 10 \text{ in} + 12 \text{ in}$
$= 32 \text{ in}$
$A = \frac{1}{2} \times 12 \text{ in} \times 8 \text{ in}$
$= 48 \text{ sq in}$

Circle: $C = \pi d$ or $2\pi r$
$A = \pi r^2$

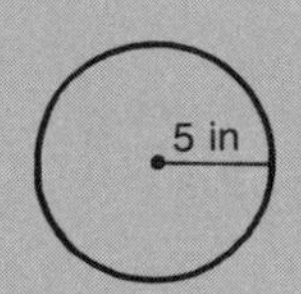

$C = 2(3.14)(5 \text{ in}) = 31.4 \text{ in}$
$A = (3.14)(5 \text{ in})^2 =$
$= (3.14)(25 \text{ sq in})$
$= 78.5 \text{ sq in}$

Parallelogram

$P = 2a + 2b$
$A = bh$

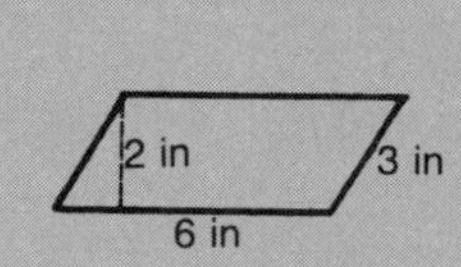

$P = 2(3 \text{ in}) + 2(6 \text{ in})$
$= 6 \text{ in} + 12 \text{ in}$
$= 18 \text{ in}$
$A = (2 \text{ in})(6 \text{ in})$
$= 12 \text{ sq in}$

Trapezoid

$A = \frac{1}{2}(b_1 + b_2)h$

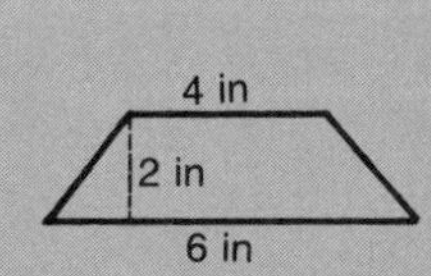

$A = \frac{1}{2}(4 \text{ in} + 6 \text{ in})(2 \text{ in})$
$= \frac{1}{2}(10 \text{ in})(2 \text{ in})$
$= 10 \text{ sq in}$

Volume

Rectangular Solid: $V = lwh$

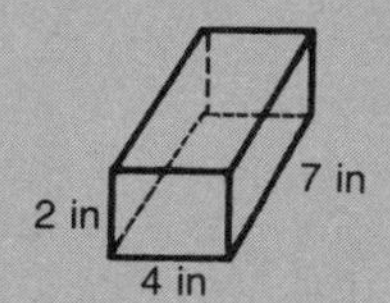

$V = 2 \text{ in} \times 4 \text{ in} \times 7 \text{ in}$
$= 56 \text{ cu in}$

Cube: $V = s \times s \times s = s^3$

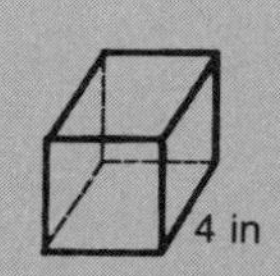

$V = 4 \text{ in} \times 4 \text{ in} \times 4 \text{ in}$
$= 64 \text{ cu in}$

CONCEPT	EXPLANATION	EXAMPLE
	Cylinder: $V = \pi r^2 h$	$V = 3.14 \times (3\text{ in})^2 \times 8\text{ in}$ $= 3.14 \times 9\text{ sq in} \times (8\text{ in})$ $= 226.08\text{ cu in}$
	Cone: $V = \frac{1}{3}\pi r^2 h$ 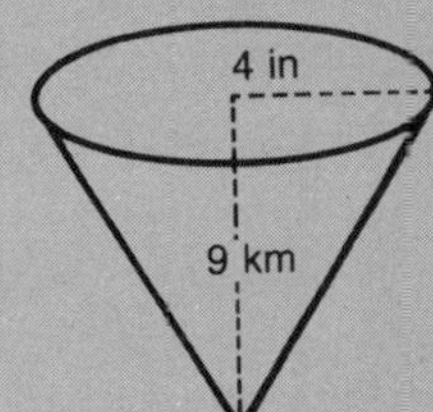	$V = \frac{1}{3} \times 3.14 \times (4\text{ in})^2 \times 9\text{ in}$ $= 3.14 \times 16\text{ sq in} \times 3\text{ in}$ $= 150.72\text{ cu in}$
	Sphere: $V = \frac{4}{3}\pi r^3$ 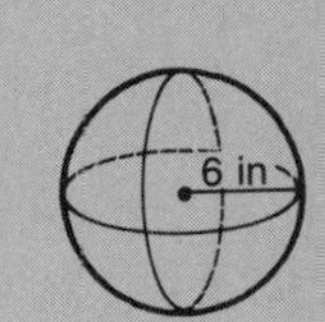	$V = \frac{4}{3}(3.14)(6\text{ in})^3$ $= \frac{4}{3} \times 3.14 \times 216\text{ cu in}$ $= 904.32\text{ cu in}$

CHAPTER 6: MEASUREMENT

CONCEPT	EXPLANATION	EXAMPLE
Converting from One Unit of Measure to Another in English System	Use conversion tables.	20 feet = ____ yards 1 yard = 3 feet Set up a proportion $\frac{20\text{ ft}}{x\text{ yd}} = \frac{3\text{ ft}}{1\text{ yd}}$ $3x = 20$ $x = 6\frac{2}{3}\text{ yd}$
		9 gals = ____ pt Step 1: 1 gal = 4 qt Set up a proportion: $\frac{9\text{ gal}}{x\text{ qt}} = \frac{1\text{ gal}}{4\text{ qt}}$ $x = 9 \times 4 = 36\text{ qt}$ Step 2: 1 qt = 2 pt Set up a proportion: $\frac{36\text{ qt}}{y\text{ pt}} = \frac{1\text{ qt}}{2\text{ pt}}$ $y = 36 \times 2 = 72\text{ pt}$
Simplifying Measurements	Convert smaller units of measure to larger units of measure.	31 in = 24 in + 7 in = 2 ft + 7 in = 2 ft 7 in
Addition of Measurement Numbers	1. Place the numbers to be added in columns with like units of measurement below like units of measurement. 2. Add the numbers in each column. 3. Simplify when possible.	5 gal 6 qt 1 pt 2 gal 3 qt 1 pt + 4 gal 1 qt 1 pt 11 gal 10 qt 3 pt 1 gal 2 qt 1 qt 1 pt 13 gal 3 qt 1 pt

CONCEPT	EXPLANATION	EXAMPLE
Subtraction of Measurement Numbers	1. Place the numbers to be subtracted in columns with like units of measurement below like units of measurement. 2. Subtract the numbers in each column, borrowing when necessary from the next nonzero unit of measure to the left. 3. Simplify when possible.	9 59 (60) 82 ~~10~~ h ~~0~~ min ~~22~~ s × 7 h 40 min 51 s 2 h 19 min 31 s
Multiplication of a Measurement Number by a Nonmeasurement Number	1. Multiply each part of the measurement number by the nonmeasurement number. 2. Simplify when possible.	6 ft 9 in × 5 30 ft 45 in 3 ft 9 in 33 ft 9 in
Multiplication of Measurement Numbers by Measurement Numbers	Sometimes can be done.	8 ft × 6 ft = 48 sq ft 10 hr × 3 ft would give 30 hr-ft, which has no meaning.
Division of a Measurement Number by a Nonmeasurement Number	1. Divide the largest unit in the measurement number by the nonmeasurement number.	1 yd 5)9 yd 2 ft 7 in 5 yd 4 yd 4 yd = 12 ft So 12 ft + 2 ft = 14 ft
	2. If it does not divide evenly, convert the remainder into the next smaller unit and add it to the units already there.	1 yd 2 ft 5)9 yd 14 ft 7 in 10 ft 4 ft 4 ft = 48 in So 48 in + 7 in = 55 in
	3. Divide the result of step 2 by the nonmeasurement number	1 yd 2 ft 11 in 5)9 yd 2 ft 55 in 55 in 0
	4. Repeat steps 2 and 3 until the division is complete	1 yd 2 ft 11 in Then: 5)9 yd 2 ft 7 in
Division of Measurement Numbers by Measurement Numbers	Sometimes can be done.	4 12 ft)48 ft Answer is a nonmeasurement number.

CONCEPT	EXPLANATION	EXAMPLE
Converting from One Unit of Measure to Another in Metric System	To change from a larger unit of measure to a smaller unit of measure move the decimal point an equal number of places to the right.	Convert 5 hectograms to decigrams. hg dag g dg 3 places So 5 hg = 5.000. dg = 5,000 dg
	To change from a smaller unit of measure to a larger unit of measure move the decimal an equal number of places to the left.	Convert 23 milliliters to decaliters da l dl cl ml 4 places 23 m = .0023. da = 0.0023 da
Converting English Units to Metric Units	Use English-to-metric conversion tables.	50 pounds = _____ grams 1 lb = 454 g Set up a proportion: $\frac{454 \text{ g}}{1 \text{ lb}} = \frac{x \text{ g}}{50 \text{ lbs}}$ $x = 454(50)$ $x = 22{,}700\text{g}$
Converting Metric Units to English Units	Use English-to-metric conversion tables.	8 liters = _____ pints 1 liter = 2.11 pints Set up a proportion: $\frac{2.11 \text{ pints}}{1 \text{ liter}} = \frac{x \text{ pints}}{8 \text{ liters}}$ $x = 2.11(8)$ $x = 16.88$ pints
Temperature Conversion	Celsius to Fahrenheit: $F = \frac{9}{5}C + 32$	25°C = _____ °F $F = \frac{9}{5}(25) + 32$ $F = 45 + 32$ $= 77°\text{F}$
	Fahrenheit to Celsius: $C = \frac{5}{9}(F - 32)$	86°F = _____ °C $C = \frac{5}{9}(86 - 32)$ $= \frac{5}{\cancel{9}_1}(\cancel{54}^6)$ $= 30°\text{C}$

CHAPTER 7: AN ALGEBRA WARMUP

Variable	A symbol denoting a quantity; may take on more than one value.	$x > -5$; x is an integer $x = -4, -3, -2, -1, 0, \ldots$
Absolute Value	$\lvert a \rvert = a$ if $a > 0$ $\lvert a \rvert = 0$ if $a = 0$ $\lvert a \rvert = -a$ if $a < 0$	$\lvert 5 \rvert = 5$ $\lvert 0 \rvert = 0$ $\lvert -5 \rvert = -(-5) = 5$

CONCEPT	EXPLANATION	EXAMPLE
Addition of Signed Numbers	1. *Numbers having like signs:* add their absolute values and prefix the common sign.	$(+7) + (+8) = +(\lvert+7\rvert + \lvert+8\rvert) = +(7 + 8) = 15$ $(-12) + (-4) = -(\lvert-12\rvert + \lvert-4\rvert) = -(12 + 4) = -16$
	2. *Numbers having unlike signs:* subtract the smaller absolute value from the larger absolute value and prefix the sign of the number having the larger absolute value.	$(-7) + (+12) = +(\lvert+12\rvert - \lvert-7\rvert) = +(12 - 7) = +5$ $(-8) + (+3) = -(\lvert-8\rvert - \lvert+3\rvert) = -(8 - 3) = -5$
Subtractions of Signed Numbers	1. Change the operation from subtraction to addition and change the sign of the subtrahend.	$10 - (-15) = 10 + (+15) = +25$
	2. Follow the rules for adding signed numbers.	
Multiplication of Signed Numbers	1. *Two numbers having like signs:* multiply their absolute values; the product is positive.	$(+8)(+9) = +(\lvert+8\rvert \cdot \lvert+9\rvert) = +(8 \cdot 9) = +72$ $(-4)(-10) = +(\lvert-4\rvert \cdot \lvert-10\rvert) = +(4 \cdot 10) = +40$
	2. *Two numbers having unlike signs:* multiply their absolute values; the product is negative.	$(-7)(+6) = -(\lvert-7\rvert \cdot \lvert+6\rvert) = -(7 \cdot 6) = -42$
Division of Signed Numbers	1. *Two numbers having like signs:* divide their absolute values; the quotient is positive.	$\frac{+10}{+5} = +\frac{\lvert+10\rvert}{\lvert+5\rvert} = +\frac{10}{5} = +2$ $\frac{-16}{-4} = +\frac{\lvert-16\rvert}{\lvert-4\rvert} = +\frac{16}{4} = +4$
	2. *Two numbers having unlike signs:* divide their absolute values; the quotient is negative.	$\frac{-18}{+6} = -\frac{\lvert-18\rvert}{\lvert+6\rvert} = -\frac{18}{6} = -3$
Exponent	Tells us how many times a number is multiplied by itself.	$5^4 = 5 \times 5 \times 5 \times 5$
Raising 10 to a Positive Power	The number of zeros following the digit 1 in the answer is the same as the exponent.	$10^{\boxed{3}} = 1\boxed{000}$ $10^{\boxed{5}} = 1\boxed{00,000}$
Multiplication of Exponential Numbers	$a^m \cdot a^n = a^{m+n}$	$5^3 \cdot 5^4 = 5^{3+4} = 5^7$
Definition of Zero Exponent	$a^0 = 1, a \neq 0$	$5^0 = 1$
Definition of Negative Exponents	$a^{-m} = \frac{1}{a^m}, a \neq 0$	$3^{-2} = \frac{1}{3^2} = \frac{1}{9}$

CONCEPT	EXPLANATION	EXAMPLE
Division of Exponential Numbers	$\frac{a^m}{a^n} = a^{m-n}$	$\frac{4^5}{4^3} = 4^{5-3} = 4^2 = 16$
Finding the Power of an Exponential Number	$(a^m)^n = a^{mn}$	$(2^4)^2 = 2^{4\cdot 2} = 2^8 = 256$
Finding the Power of a Product of Numbers	$(ab)^n = a^n b^n$	$(5a)^3 = 5^3 a^3 = 125a^3$
Raising a Fraction to a Power	$\left(\frac{a}{b}\right)^m = \frac{a^m}{b^m}$ $b \neq 0$	$\left(\frac{2}{3}\right)^2 = \frac{2^2}{3^2} = \frac{4}{9}$
Methods of Solving Equations	1. Adding the same number to each side of the equation.	$x - 6 = 5$ $x - 6 \boxed{+ 6} = 5 \boxed{+6}$ $x = 11$
	2. Subtracting the same number from each side of the equation.	$x + 4 = 12$ $x + 4 \boxed{- 4} = 12 \boxed{- 4}$ $x = 8$
	3. Multiplying each side of the equation by the same number.	$\frac{x}{5} = 2$ $\boxed{\overset{1}{\cancel{5}}} \cdot \frac{x}{\underset{1}{\cancel{5}}} = \boxed{5} \cdot 2$ $x = 10$
	4. Dividing each side of an equation by the same nonzero number.	$3x = 21$ $\frac{3x}{\boxed{3}} = \frac{21}{\boxed{3}}$ $x = 7$
	Using two or more of the above steps.	$3x - 6 = x - 18$ $3x - 6 \boxed{+ 6} = x - 18 \boxed{+ 6}$ $3x = x - 12$ $3x \boxed{- x} = x \boxed{- x} - 12$ $2x = -12$ $\frac{2x}{\boxed{2}} = \frac{-12}{\boxed{2}}$ $x = -6$

Appendix III

Answers to Pretests, Warmup Tests, and Cumulative Tests

ANSWERS

Chapter 1 Pretest

1. 18 < 22 **2.** 3 groups of a thousand **3.** 3 groups of ten **4.** four thousand sixty-two **5.** 40,220 **6.** 3,800 **7.** 38 **8.** 70,643 **9.** 48,306 **10.** 836 **11.** 18 **12.** 1,759 **13.** 19,209 **14.** 33,820 **15.** 752,400 **16.** 262,977 **17.** 963, $R = 5$ **18.** 0 **19.** no solution **20.** 110, $R = 2$ **21.** 2^3 **22.** 125 **23.** 0 **24.** 2

Chapter 1 Warmup Test

1. (a) > **(b)** < **2. (a)** fifteen thousand, twelve **(b)** two hundred forty-five thousand, three hundred ninety-eight **3. (a)** 65,346 **(b)** 3,040,978 **4. (a)** 4,300 **(b)** 30,000 **(c)** 210,000 **5. (a)** 1,941 **(b)** 215,079 **(c)** 649,803 **6. (a)** 204 **(b)** 37,215 **(c)** 24,488 **7. (a)** 118,465 **(b)** 9,963,666 **(c)** 10,680,000 **8. (a)** 48 **(b)** 208 **(c)** 73, $R = 127$ **9. (a)** 3^3 **(b)** 7^6 **10. (a)** 64 **(b)** 32 **11. (a)** 17 **(b)** 50 **12.** 163 miles **13.** $125 **14.** $1,010 **15.** 2,030,400 BTU

Chapter 2 Pretest

1. (a) improper **(b)** proper **(c)** proper **(d)** improper **2. (a)** $4\frac{1}{4}$ **(b)** $3\frac{1}{3}$ **(c)** $14\frac{2}{7}$ **3. (a)** $\frac{20}{3}$ **(b)** $\frac{60}{7}$ **(c)** $\frac{61}{4}$ **4. (a)** $\frac{2}{5}$ **(b)** $\frac{10}{7}$ **(c)** $\frac{6}{11}$ **5. (a)** 28 **(b)** 21 **(c)** 105 **6. (a)** $1\frac{5}{24}$ **(b)** $\frac{49}{60}$ **(c)** $\frac{25}{28}$ **(d)** $11\frac{2}{15}$ **7. (a)** $\frac{3}{10}$ **(b)** $\frac{3}{14}$ **(c)** $\frac{5}{21}$ **(d)** $1\frac{32}{35}$ **8. (a)** $\frac{2}{5}$ **(b)** $1\frac{1}{2}$ **(c)** $\frac{4}{7}$ **(d)** $5\frac{1}{4}$ **9. (a)** $\frac{2}{5}$ **(b)** $\frac{5}{7}$ **(c)** $\frac{5}{9}$ **(d)** $1\frac{1}{14}$ **10.** 6 gallons **11.** $21\frac{5}{8}$ **12.** 2:1 **13.** 98

Index